BIRDS

*their structure
and function*

BIRDS

their structure and function

SECOND EDITION

A. S. KING
BSc, PhD, MRCVS

*Professor of Veterinary Anatomy
Faculty of Veterinary Science, University of Liverpool*

AND

J. McLELLAND
BVMS, MVSc, PhD, MRCVS

*Reader in Veterinary Anatomy
Royal (Dick) School of Veterinary Studies
University of Edinburgh*

Baillière Tindall
London   Philadelphia   Toronto
Mexico City   Rio de Janeiro   Sydney   Tokyo   Hong Kong

Baillière Tindall   1 St Anne's Road
Eastbourne, East Sussex BN21 3UN, England

West Washington Square
Philadelphia, PA 19105, USA

1 Goldthorne Avenue
Toronto, Ontario M8Z 5T9, Canada

Apartado 26370—Cedro 512
Mexico 4, D.F., Mexico

Rua Evaristo da Veiga, 55–20° andar
Rio de Janeiro—RJ, Brazil

ABP Australia Ltd, 44 Waterloo Road
North Ryde, NSW 2064, Australia

Ichibancho Central Building, 22–1 Ichibancho
Chiyoda-ku, Tokyo 102, Japan

10/FL, Inter-Continental Plaza, 94 Granville Road
Tsim Sha Tsui East, Kowloon, Hong Kong

© 1984 Baillière Tindall

All rights reserved. No part of this publication may be reproduced, stored in a retrieval system or transmitted, in any form or by any means, electronic, mechanical, photocopying or otherwise, without the prior permission of Baillière Tindall, 1 St Anne's Road, Eastbourne, East Sussex BN21 3UN, England

First published 1975 as *Outlines of Avian Anatomy*
  German edition 1977 (Verlag Eugen Ulmer, Stuttgart)
Second edition 1984

Printed in Great Britain at The Pitman Press, Bath

British Library Cataloguing in Publication Data

King, A. S.
  Birds – their structure and function.
  – 2nd ed.
  1. Birds – Anatomy
  I. Title   II. McLelland, J.   III. King, A. S.
  Outlines of avian anatomy
  598.2'4   QL697

ISBN 0–7020–0872–9

Contents

	Preface	vii
1	Birds	1
2	External Anatomy	9
3	Integument	23
4	Skeletomuscular System	43
5	Coelomic Cavities	79
6	Digestive System	84
7	Respiratory System	110
8	Female Reproductive System	145
9	Male Reproductive System	166
10	Urinary System	175
11	Cloaca and Vent	187
12	Endocrine System	200
13	Cardiovascular System	214
14	Lymphatic System	229
15	Nervous System	237
16	Special Sense Organs	284
	Appendix	315
	Index	319

Preface

This book was first envisaged as a second edition of *Outlines of Avian Anatomy*. However, as the rewriting progressed our objectives widened, and this is reflected in the more general title that the book now carries. Originally, the main purpose of 'Outlines' was to meet the needs of university students and those who taught them, with particular attention to veterinary requirements; the interests of ornithologists took second place. The present volume is designed equally for the ornithologist on the one hand, and those engaged in university studies, research, or veterinary practice on the other. This has not resulted in any reduction in material utilized in university work, veterinary or otherwise. On the contrary such aspects have been up-dated and augmented by the results of recent research.

Much the greatest development in this new book has been the addition of many topics of interest to ornithologists. Among these is a detailed account of the external features essential for the identification and description of birds, including surface markings, variations in the beak and feet, and the structure and modifications of wings. We have also set out to supply the answers to questions which ornithologists often ask from their intense curiosity about birds. What is the anatomical basis of flight? How is it possible for a bird to perform the hard work of flying at high altitudes and over immense distances? What is the mechanism of voice production? Do birds see and hear better than we can? How good is their sense of smell? How do they navigate? Advanced scientific knowledge is not needed in order to follow the essential principles behind these problems. Some subjects are of course inevitably more complex, for example aspects of the nervous system, and here we have written for the more advanced reader.

A particular feature of the new book is the improved quality of the illustrations, all of which have been redrawn by a professional artist, Garry Martin. The references at the end of each chapter have been expanded to provide much more extensive sources than before. The anatomical nomenclature is now firmly based on the Nomina Anatomica Avium, 1979, the Latin terms being converted into their English equivalents. The scientific taxonomic nomenclature is that prepared by J. J. Morony, W. J. Bock and J. Farrand and published by the Department of Ornithology of the American Museum of Natural History, New York, in 1975. The English names of birds are those listed by E. S. Gruson in his *Check List of the Birds of the World*, Collins, London, 1976. The names of the common laboratory and domestic birds are as follows: duck, domestic forms of *Anas platyrhynchos*; goose, domestic forms of *Anser anser*; pigeon, domestic forms of *Columba livia*; turkey, domestic forms of *Meleagris gallopavo*; chicken or domestic fowl, domestic forms of *Gallus gallus*; quail, domestic forms of *Coturnix*.

We have had so much help from so many sources that it seems invidious to mention any particular individuals. We must, however, thank Mrs M. M. Thompson for typing the manuscripts and Mr M. Goldberg for photographing much of the artwork. Many biologists and publishing firms have generously allowed us to utilize their illustrations, and we are particularly indebted to Dr J. J. Baumel of Creighton University, Nebraska, Dr V. Komárek, Prague, and our Japanese colleagues, Professors M. Yasuda and T. Watanabe of Nagoya University. Scientific information has been drawn extensively from the literature, but we also acknowledge the particular stimulus of the very original Master's thesis by Dr Pat McCarthy, of the University of Sydney.

A. S. King
J. McLelland
February 1984

Chapter 1

BIRDS

The evolution of birds

The earliest known bird was *Archaeopteryx*. Five almost complete fossils of this ancient bird have been found in late Jurassic limestone about 150 million years old. As Fig 1–1 shows, these fossils coincide roughly with the middle of the 130 million year period of domination by the archosaurian reptiles, of

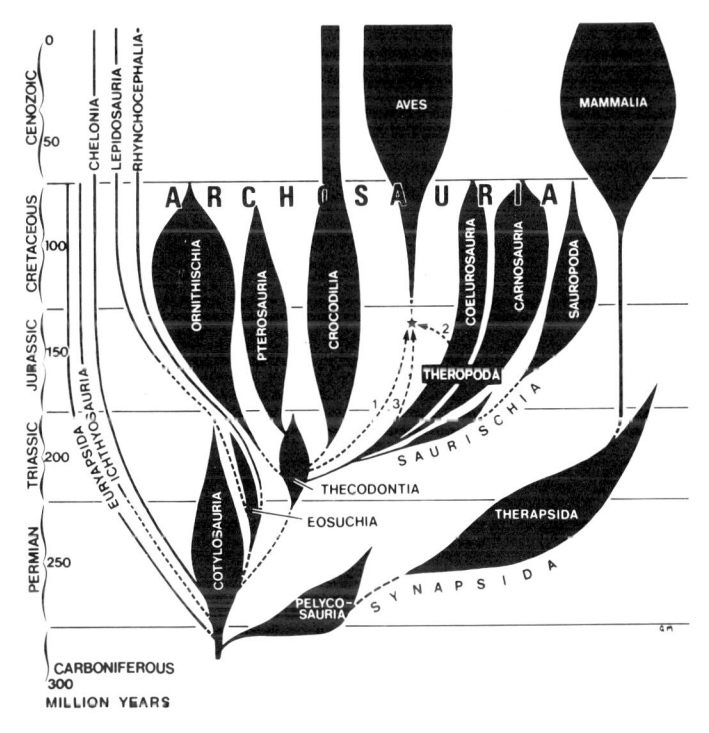

Fig 1–1 The phylogeny of reptiles, mammals and birds. 1, 2 and 3 indicate three possible reptilian sources of *Archaeopteryx**, line 1 from a primitive member of the thecodonts, and lines 2 and 3 from a coelurosaur. The five evolutionary lines on the left of the diagram are not drawn in proportion to the size of their populations. From King and King (1979), with kind permission of the publisher.

1

which the two orders Ornithischia and Saurischia form the group popularly known as the dinosaurs. Primitive mammals had probably been in existence since the end of the Triassic (Fig 1–1), but at that time were restricted to small, mainly insectivorous, and probably nocturnal and arboreal animals that were about the size of rats and mice.

The fossil imprints of *Archaeopteryx* clearly show feathers which were identical to those of modern flying birds, so there is no doubt whatever that it belonged to the class Aves. It is also generally believed that *Archaeopteryx*, which was about the size of the Common Magpie, was capable of some sort of flight as well as being arboreal. However, its fossil remains are a mosaic of reptilian and avian features. Among the reptilian characteristics were teeth, a long tail, a short sacrum with no more than six vertebrae, a hand (*manus*) with free metacarpal bones and claws on all three digits (Fig 1–2B), and a foot with free metatarsal bones (Fig 1–2C). Its avian features included feathers (primaries and secondaries on the wings, and contour feathers all over the body), paired clavicles joined to form a typical wish-bone (*furcula*), and a foot with the first digit opposed to the other three digits as in many modern perching birds (Fig 1–2C).

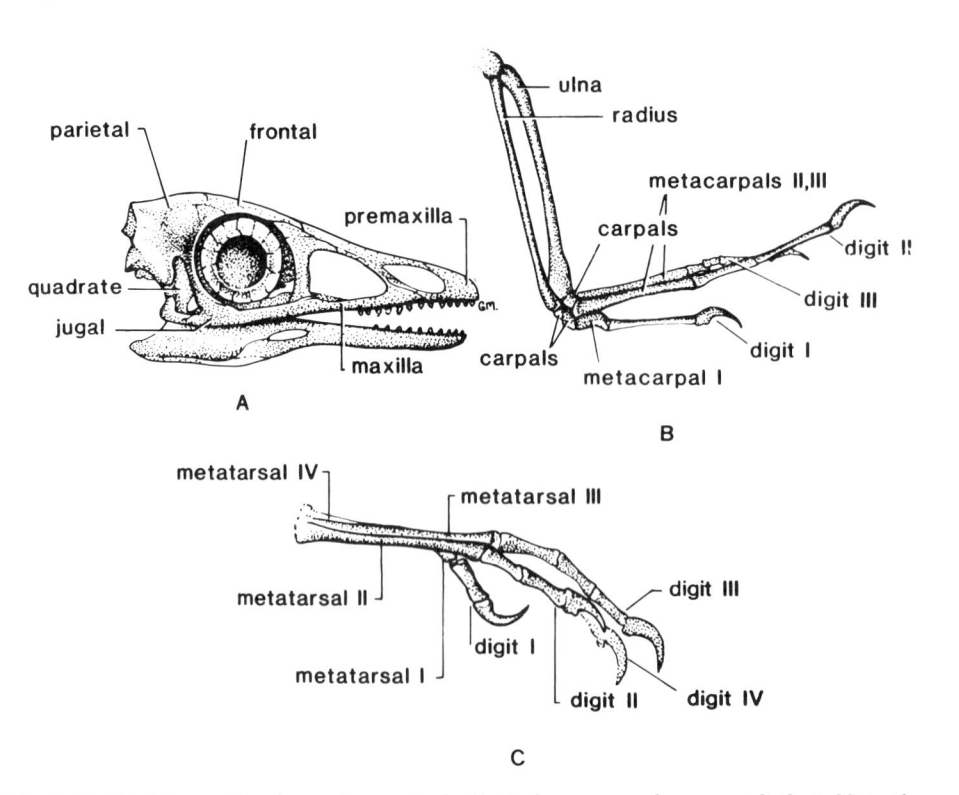

Fig 1–2 Skeleton of *Archaeopteryx*. **A**, skull; **B**, forearm and manus; **C**, foot. Note the reptilian features, which include teeth, free metacarpal bones and claws on all three digits of the manus, and free metatarsal bones of the foot. However, the first digit of the foot is opposed to the other three digits as in many modern perching birds. Redrawn from Heilmann (1926).

Although *Archaeopteryx* was undoubtedly descended from reptiles, the identity of its immediate ancestors is controversial. It is generally agreed that the progenitor was a slender, bipedal, fast running, predatory reptile, perhaps about one metre long (Fig 1–3). The three main possibilities for the origin of this progenitor are shown as lines 1, 2 and 3 in Fig 1–1. It may have been a primitive member of the thecodonts (line 1), the group which gave rise to the archosaurs. Alternatively, it may have been a coelurosaur (Fig 1–3), as indicated by lines 2 and 3 in Fig 1–1. It was certainly a close relative of the dinosaurs, so close that birds have been described as flying dinosaurs.

Fig 1–3 A possible ancestor of *Archaeopteryx*, the coelurosaur *Compsognathus longipes* which was about the same size as a large domestic cat. Redrawn from Heilmann (1926).

Unfortunately, the early fossil record of birds is so sketchy that *Archaeopteryx* cannot be regarded with certainty as the direct ancestor of modern birds. The next fossil bird was a wader (*Gallornis*), which appeared from 3 million to 25 million years after *Archaeopteryx*. It has been said that this period is too short for a wader to have evolved from a primitive tree-dwelling bird like *Archaeopteryx*; some modern birds had evolved by the early Cretaceous, and it is arguable that none of these could have arisen from *Archaeopteryx* in the time available. On the other hand, recent thinking about evolution suggests that it may proceed in leaps, and this is consistent with the view that *Archaeopteryx*, at least, lies very close to bird origins.

It is certain, however, that by the end of the Cretaceous various aquatic birds were well established, although terrestrial birds had still not entered the record. In the early Cenozoic there was a great radiation of both land and water birds. By the end of the Eocene (about 35 million years ago), 26 of the contemporary 32 avian orders were established. Finally, during the glaciations of the Pleistocene (the most recent one million years) many species dropped out and the contemporary avifauna became established.

Having arisen from thecodonts or coelurosaurs in the Triassic or Jurassic, birds have comparatively recent evolutionary relationships to reptiles. Therefore it is likely that the anatomy of birds will include reptilian characteristics. Since the Crocodilia also arose from thecodonts in the Triassic period (Fig 1–1) the evolutionary relationship between birds and crocodiles is relatively close, and therefore anatomical similarities between modern birds and crocodiles may be expected. On the other hand, birds have a more remote phylogenetic relationship to snakes (Lepidosauria) and turtles and tortoises (Chelonia), which diverged from the stem reptiles in the Permian period (Fig 1–1). The mammalian line (Therapsida in Fig 1–1) departed even earlier from the reptilian–avian line, so it is not surprising that contemporary descendants of the mammalian and avian evolutionary lines show only the most general anatomical similarities.

Another controversial matter is the evolution of flight. All the vertebrate classes have experimented with flight, and three of them, reptiles (pterosaurs), mammals (bats) and birds have mastered it. It seems very likely that feathers were an adaptation primarily for insulation and only secondarily for flight. Although all modern reptiles are ectothermic (i.e. they rely on external sources of heat to provide a body temperature that is adequate for their energetic requirements), it is probable that some degree of endothermy (the capacity to maintain an elevated body temperature by internal heat production) was widespread among the dinosaurs and was therefore also present in the avian progenitor. Endothermy must have preceded feathers, because the possession of an insulating covering of feathers by an ectothermic animal would deprive it of heat gain from solar radiation. Admirably designed as they were for insulation, the first feathers were also ideally preadapted for flight. All that was necessary to evolve an aerofoil was to enlarge the feathers of the wing and tail and attach them firmly to the skeleton.

Two main theories have been advanced to explain the evolution of flight, the cursorial theory and the arboreal theory. The cursorial theory proposes that the bipedal 'Proavis' ran along the ground extending and beating its feathered forelimbs until lift occurred; the beating of the 'wings' may have been an attempt to catch insects. The arboreal theory supposes the existence of a small bipedal feathered 'Proavis' with forelimbs that were adapted to predaceous grasping. In the search for food and perhaps to escape from larger dinosaurs, 'Proavis' may have jumped onto low branches or climbed the trunks of trees. Leaping between branches and parachuting to the ground followed, then gliding and finally flapping flight.

Flapping flight imposes many constraints on the anatomy of birds. Firstly, it severely restricts body size. Wing loading (p. 18) tends to increase as body weight increases. Larger birds therefore require relatively greater power in order to perform flapping flight. If bird A weighs twice as much as bird B it will require not twice as much power to fly at its minimum speed, but 2.25 times as much. Therefore, as birds get heavier, their reserves of power diminish until eventually a point is reached when the flight muscles are unable to provide enough power for flapping flight. The maximum body weight compatible with flapping flight has been estimated at approximately 12 kg, which is about the weight of the Mute Swan.

These restrictions on body weight can be moderated by drawing energy from external sources. The Pleistocene condor *Teratornis incredibilis*, which appears to have been the largest known flying bird, weighed about 20 kg. A giant Cretaceous pterosaur had an estimated wing span of about 15 m and must have weighed well over 20 kg. However, these huge flyers were almost certainly soarers, drawing energy from winds or thermals.

The anatomical requirements of flight include not only these limitations on total body weight, but also a general streamlining of the body, virtually total commitment of the forelimb to flight, a specialized pectoral girdle and wing bones, modification of the thoracic musculature for flight, and accentuation of the special sense organs, especially vision and balance, with corresponding enlargement and modification of the brain. Finally, the energetics of flight impose special demands on the respiratory and circulatory systems which are discussed immediately below.

So restrictive are the anatomical requirements of flight that the entire class of Aves (including also the flightless birds, all of which arose from flying ancestors), shows a greater uniformity of structure than many single orders of fishes, amphibians, reptiles and mammals. Paradoxically, however, the capacity for flight has led to an extraordinary diversity of anatomical detail. The ability to fly has enabled birds to penetrate a very wide variety of habitats. The result has been an extensive adaptive radiation, especially for locomotion and feeding. These detailed adaptations have resulted in 8948 living species of bird. Modern reptiles have about one-third less species, and mammals only about 4200 species. It is interesting to note, however, that among the mammals the only order capable of powered flight (Chiroptera, bats) has also experienced a very wide adaptive radiation leading to about 1000 species, which is a greater number than any other mammalian order except Rodentia.

The energetics of birds

It has been calculated that the average daily energetic cost of avoiding predators, procuring food and reproducing, is about 50 to 75 per cent higher in a 380 g bird than in a 380 g rodent. This extra cost in the bird is linked to the ability to fly, since flight is a highly energetic form of exercise. Thus a Budgerigar or raven flying at sea level in a wind tunnel uses one-and-a-half to three times as much oxygen as a mammal of similar size running in a wheel. During flight the birds increase their oxygen consumption by about 13 times above their standard metabolic rate. In turbulent air or when the bird is ascending, its oxygen consumption increases by 20 to 30 times for short periods. Human energetic performance gives a baseline for comparison; a good human athlete can achieve an increase of 15 to 20 times during maximum exercise, but only for a few minutes. Bats can increase their oxygen consumption during sustained flight by at least 20 times, and in general the energetics of bat flight appear to be comparable to those of birds, at least at sea level.

Some birds achieve remarkable energetic feats. Swallows migrating between Europe and Africa must fly for about 50 hours over the Sahara without resting. The American Golden Plover flies over 2000 miles non-stop between

Alaska and Hawaii. The Arctic Tern migrates from pole to pole, the round trip totalling some 20 000 miles. Although most birds such as the common passerine species fly at a leisurely 15 to 35 mph, swifts appear to reach 60 to 90 mph. The energetic performance of some non-flying species is also impressive. For example, the penguin 'flies' under water at about 20 mph and the Ostrich can reach 50 mph on the ground when pressed. However, the most notable feats of all are those performed by birds at high altitude. The migrating Bar-headed Goose rises from near sea level to traverse Himalayan peaks that are over 9000 m high, which in human energetic terms is rather like riding a bicycle over the summit of Mount Everest. Aircraft have collided with birds at truly prodigious altitudes. For instance, a vulture was reportedly struck at over 11 000 m. Migratory flights usually occur below 1500 m, but altitudes of 6000 m have been frequently observed.

It is at high altitudes that the energetic superiority of birds over mammals becomes really obvious. Far from being able to do heavy work (which is what flying is), an unacclimatized man is in a state of incipient hypoxic collapse after 10 minutes at 6100 m. In a hypobaric chamber (from which the air can be progressively pumped out, thus simulating ascent), mice become comatose at an 'altitude' of 6100 m, whereas ordinary House Sparrows can fly and even gain altitude; at 7600 m the sparrows can still fly but gaining altitude is now beyond them.

The remarkable energetic capacity of birds is based on a number of specialized adaptations of the respiratory and cardiovascular systems which will be summarized in Chapters 7 and 13. The respiratory adaptations combine to make the avian lung the most efficient gas-exchange system amongst all the vertebrates. This refinement of the mechanisms for the uptake of oxygen is matched by a cardiovascular system which excels in oxygen transport.

Classification of birds

The 8948 living species of birds are assigned to 27 orders. The number of species within an order varies greatly. By far the largest order is the Passeriformes which contains 5243 species (i.e. more than all the species in the remaining orders put together). The smallest orders are the Apterygiformes (three species), the Rheiformes (two species) and the Struthioniformes (one species).

The scientific name of an individual animal is a 'binomen', the first word being the name of the genus and the second the species (e.g. *Passer domesticus*, House Sparrow). The addition of a third term, as in *Meleagris gallopavo gallopavo*, denotes a subspecies; this constitutes the so-called trinomial system, 'subspecies' being an official term of the International Code of Zoological Nomenclature. Trinomial terms have often been used for the domestic species, notably the domestic fowl and turkey as above. However, the subspecies concept has been attacked by modern taxonomists, and it could be better to name each domestic species by its binomen followed by 'variant (or var.) *domesticus*'. It is certainly not correct to refer to the domestic fowl as *Gallus domesticus* since *domesticus* is not the name of the species.

According to custom, the following list of orders begins with those which are believed to be the most primitive and proceeds towards the most advanced, but the sequence is highly speculative.

Orders	Species
Struthioniformes	Ostrich
Rheiformes	rheas
Casuariiformes	cassowaries, emu
Apterygiformes	kiwis
Tinamiformes	tinamous
Sphenisciformes	penguins
Gaviiformes	loons
Podicipediformes	grebes
Procellariiformes	albatrosses, shearwaters, fulmars, petrels
Pelecaniformes	tropic birds, pelicans, boobies, Gannet, cormorants, shags, anhingas, frigate birds
Ciconiiformes	herons, egrets, bitterns, storks, ibises, spoonbills, flamingos
Anseriformes	screamers, ducks, geese, swans
Falconiformes	vultures, Osprey, kites, hawks, buzzards, eagles, harriers, Secretary Bird, falcons, kestrels
Galliformes	scrub hens, curassows, ptarmigans, partridges, quails, pheasants, domestic fowl, peacocks, grouse, turkeys, guineafowl, Hoatzin
Gruiformes	button-quails, cranes, rails, coots, waterhens, moorhens, bustards, Limpkin, trumpeters
Charadriiformes	jacanas, oystercatchers, avocets, stilts, phalaropes, thick-knees, pratincoles, coursers, plovers, dotterels, lapwings, snipe, curlews, godwits, woodcock, sandpipers, Dunlin, Sanderling, stints, knots, Ruff, gulls, terns, skuas, jaegers, razorbills, guillemots, skimmers, auks, auklets, puffins
Columbiformes	sandgrouse, pigeons, doves
Psittaciformes	lories, lorikeets, cockatoos, parrots, Kakapo, Budgerigar, lovebirds, parakeets, macaws
Cuculiformes	cuckoos, roadrunners, anis
Strigiformes	owls
Caprimulgiformes	Oilbird, frogmouths, potoos, goatsuckers, nightjars
Apodiformes	swifts, hummingbirds
Coliiformes	mousebirds
Trogoniformes	trogons
Coraciiformes	kingfishers, todies, motmots, bee-eaters, rollers, Hoopoe, hornbills
Piciformes	jacamars, puffbirds, barbets, honeyguides, toucans, woodpeckers, piculets
Passeriformes	woodcreepers, ovenbirds, antbirds, tapaculos, manakins, cotingas, lyrebirds, scrub-birds, larks, swallows, martins, wagtails, pipits, bulbuls, shrikes, waxwings, dippers, wrens, thrashers, mockingbirds, accentors, thrushes, babblers, warblers, flycatchers, tits, titmice, nuthatches, tree-creepers, flowerpeckers, sunbirds, white-eyes, honeyeaters, buntings, tanagers, honeycreepers, vireos, blackbirds, troupials, Brambling, canaries, finches, linnets, grosbeaks, waxbills, sparrows, weavers, starlings, mynas, drongos, orioles, butcher birds, bowerbirds, birds of paradise, crows, magpies, jays, ravens

In an earlier classification, now no longer formally recognized, birds were grouped together according to how extensively the keel of the sternum was developed. One group included the large flightless birds (i.e. the Ostrich, rheas, Emu, cassowaries and kiwis) and was given the name Ratitae (from Latin *ratis*, a raft) since its members possessed a sternum without a keel. All other birds with a keeled sternum were called Carinatae (from Latin *carina*, a keel). Although much controversy has surrounded the precise relationship of the large flightless birds, the available evidence now indicates that the ratites do in fact have a monophyletic origin and with the tinamous can be placed within the same group.

A list of the English common names of the birds referred to in the text together with their Latin scientific names is provided in the Appendix on page 315. The names of the domestic birds in the text refer to the following species: domestic duck, domestic form of Mallard; domestic goose, domestic form of Graylag Goose; domestic pigeon, domestic form of Rock Dove; domestic turkey, domestic form of Turkey; domestic fowl, domestic form of Red Jungle Fowl; domestic quail, domestic form of the genus *Coturnix*.

Further reading

Bock, W.J. (1965) The role of adaptive mechanisms in the origin of higher levels of organization. *Syst. Zool.*, 14, 272–287.

Bock, W.J. (1969) The origin and radiation of birds. *Ann. N.Y. Acad. Sci.*, 167, 147–155.

Brodkorb, P. (1971) Origin and evolution of birds. In *Avian Biology* (Ed.) Farner, D.S. & King, J.R. Vol. 1. New York and London: Academic Press.

Feduccia, A. (1980) *The Age of Birds*. Cambridge, MA: Harvard University Press.

Gruson, E.S. (1976) *Checklist of the Birds of the World*. London: Collins.

Heilmann, G. (1926) *The Origin of Birds*. London: Witherby.

King, A.S. & King, D.Z. (1979) Avian morphology: general principles. In *Form and Function in Birds* (Ed.) King, A.S. & McLelland, J. Vol. 1. London and New York: Academic Press.

Morony, J.J., Bock, W.J. & Farrand, J. (1975) *Reference List of the Birds of the World*. New York: American Museum of Natural History.

Ostrom, J.H. (1975) The origin of birds. *A. Rev. Earth Planet. Sci.*, 3, 55–77.

Pennycuick, C.J. (1972) *Animal Flight*. London: Arnold.

Tucker, V.A. (1968) Upon the wings of the wind. *New Scient.*, 38, 694–696.

Tucker, V.A. (1968) Respiratory physiology of House Sparrows in relation to high-altitude flight. *J. exp. Biol.*, 48, 67–87.

Tucker, V.A. (1969) Energetics of bird flight. *Scient. Am.*, 220, 70–78.

Chapter 2

EXTERNAL ANATOMY

EXTERNAL FEATURES OF A TYPICAL BIRD

For descriptive purposes the exterior of a bird can be divided into the following six main parts: the head, neck, trunk, tail, wing (thoracic limb) and pelvic limb. In ornithological descriptions these parts are subdivided into regions to which numerous names have been given. Some of the resulting terms are widely employed but others have fallen into disuse.

Head

The *bill*, or beak, is known anatomically as the *rostrum*. It includes the bones of the upper and lower jaws and their horny sheaths. The upper component of the bill is known as the maxillary rostrum and the lower component as the mandibular rostrum. The upper and lower horny sheaths can be called the *maxillary rhamphotheca* and the *mandibular rhamphotheca* (Fig 2–1); they are also sometimes known as the *rhinotheca* and *gnathotheca*. The midventral profile of the whole length of the mandibular rhamphotheca is designated as the *gonys* (Fig 2–1). Thus the gonys is formed entirely by horny (not soft) tissue, and bears the same relationship to the mandibular rhamphotheca as does the keel to a boat. The *interramal region* (or interramal space) is the zone of soft tissue which begins at the caudal end of the gonys and extends caudally between the mandibular rami as far as the caudal end of the mandible; the more caudal part of it is the gular region, forming the floor of the caudal part of the oropharynx. If the gonys is long (Fig 2–1) as in a puffin or House Sparrow, then the interramal region is correspondingly short. On the other hand, in birds such as pelicans, the gonys is very short and the interramal region then extends rostrally almost to the tip of the bill. The mid-dorsal profile of the maxillary rhamphotheca is the *culmen* (Fig 2–1), which corresponds to the ridge of the roof of a house (Latin *culmen*, a summit). The maxillary rhamphotheca has a cutting edge, or *tomium*, on the left and right side (the maxillary tomia), which matches the opposing mandibular tomium on each side (Fig 2–1). The entire opening of the mouth is called the *oral opening* (Fig 2–1) or *gape*. The term *rictus* is used in several different ways. As indicated in Fig 2–1 it refers to the caudal part of the oral opening, beginning at the caudal

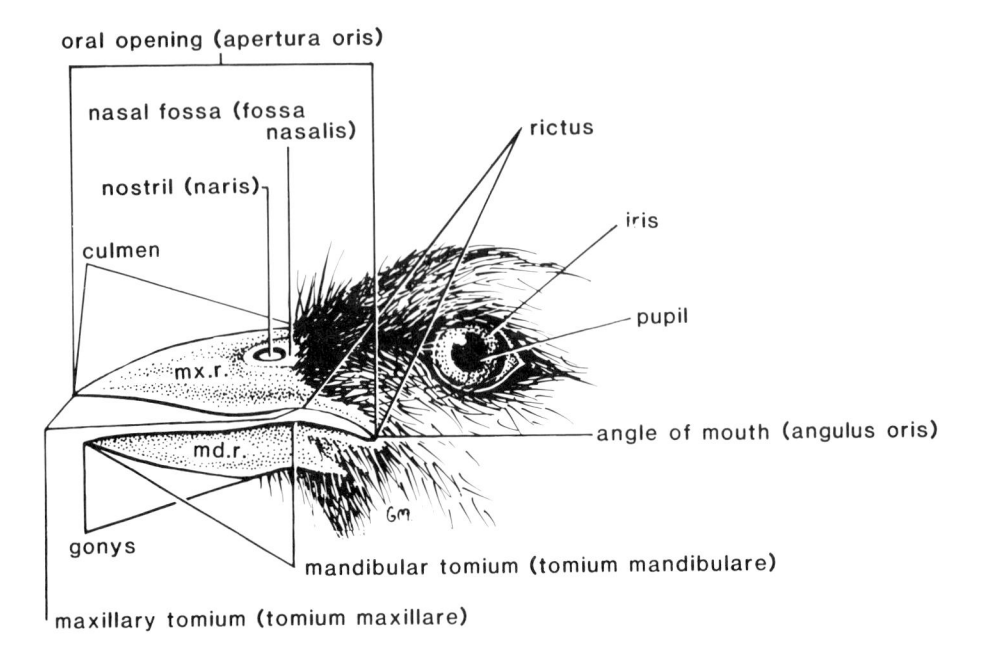

Fig 2-1 Diagram of the main components of the bill of a typical bird. The vernacular names of the components are followed by the anatomical Latin names in brackets where these differ. mx.r. = maxillary rhamphotheca; md.r. = mandibular rhamphotheca.

end of the tomia and ending caudally at the *angle of the mouth*. Alternatively, it means the entire oral opening (or gape). It can also refer to the area of skin immediately surrounding the angle of the mouth (as on p. 27). The maxillary rhamphotheca carries the paired *nostrils* (Fig 2–1). In most small passerine birds and some other species each nostril opens into a slight depression, the *nasal fossa* (Fig 2–1).

The remainder of the head can be divided into the regions shown in Fig 2–2, although the boundaries between them are not sharply defined. The top of the head is divided into the *forehead, crown* and the *back of the head*, the three regions together being sometimes known as the *pileum*. As Fig 2–3 shows, these three regions may carry a crown stripe (or median line), a lateral crown stripe (or head stripe) and a superciliary stripe (also known as the supercilium or superciliary line). The *orbital region* (Fig 2–2) is a narrow zone round the eye which includes the dorsal and ventral eyelids (palpebrae). The *ear region* (Fig 2–2) surrounds the external acoustic meatus (p. 302), and is recognizable by the particular texture of the *ear coverts* which overlie the meatus (Fig 14–2). In most birds the meatus is caudal and slightly ventral to the eye (Fig 2–2), but in woodcock and snipe it lies ventral to the eye. The opening is generally quite small, but in owls it is enormous (Fig 16–12) and in many strigiform species has a movable flap along its rostral edge, the operculum (p. 302). The *lore* (Latin *lorum*, a strap) is the narrow elongated area between the eye and the maxillary rhamphotheca. Much of it lies directly rostral to (in front of) the eye

EXTERNAL ANATOMY 11

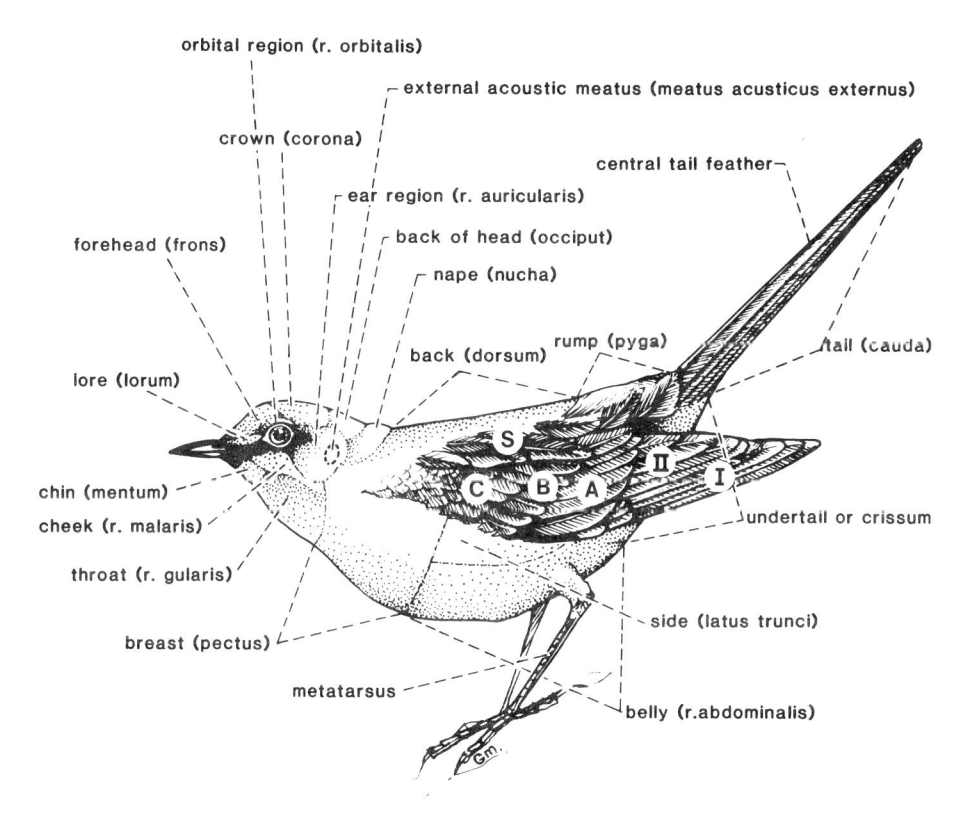

Fig 2–2 Diagram of the external features of a typical bird. The boundaries between the regions of the head are not sharply defined. The vernacular names are given first, with the anatomical Latin names in brackets. r. = regio; I = primaries; II = secondaries; A = major coverts; B = median coverts; C = minor coverts; S = scapulars.

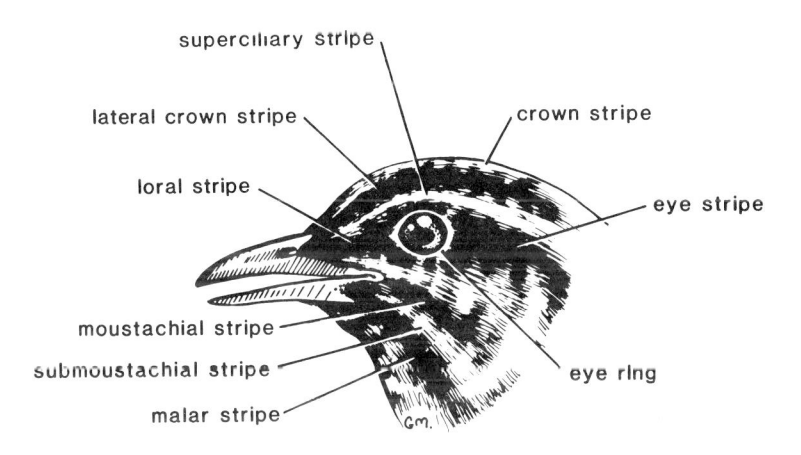

Fig 2–3 Some of the head markings that are commonly used in ornithological descriptions.

(Fig 2–2), and in many species shows a straplike line, the *loral stripe* or loral line (Fig 2–3), which runs from the maxillary rhamphotheca to the orbital region; caudal to the eye, the loral stripe continues as the *eye stripe* (Fig 2–3). The loral stripe plus the eye stripe are sometimes said to form the *lateral stripe*. Not only does the lore frequently carry a stripe, but it often has special feathers (such as loral bristles) or it can be naked in some species. The *chin* is the ill-defined rostral part of the under surface of the head, beginning at the caudal end of the interramal region and continuing caudally into the *throat* (Fig 2–2). The ventral part of the side of the head is the *cheek* (Fig 2–2), which is the region of soft tissue extending caudally from the mandibular rhamphotheca, and surrounding the angle of the mouth. The cheek region may carry as many as three stripes, as in Fig 2–3. The uppermost, the *moustachial stripe*, extends caudally from the angle of the mouth; below this stripe is the *malar stripe*, and between the two is the *submoustachial stripe*. Unfortunately, there is no agreement in ornithological manuals about the naming of the stripes on the heads of birds. Furthermore, 'line' is sometimes used instead of 'stripe', but the difference between a line and a stripe is not clear. In some instances, the word stripe or line is omitted, for example supercilium instead of superciliary line or stripe. In an attempt to eliminate confusion, at least among British ornithologists, the editors of 'British Birds' have suggested names, which differ from those used here in only two aspects as follows: (1) supercilium is proposed instead of superciliary line; (2) the loral stripe is regarded as part of the eye stripe, and the term lateral stripe is not used at all.

Adaptations of the bill for feeding are illustrated in Fig 2–4. Seed-cracking birds such as finches (Fig 2–4b) have a stout conical bill, and in crossbills (Fig 2–4c) the sharply-pointed upper and lower components of the bill cross over to hold the scales of fir cones apart while the tongue removes the seeds. Some parrots (Fig 2–4d) have transverse, rasplike, ridges inside the bill which can reduce the hardest fruit stones to fine particles. The bills of fruit-eating birds such as tanagers tend to be short and rather wide. In nectar- and pollen-feeders, of which the hummingbirds (Fig 2–4f) are a notable example, the bills are adapted to penetrate deeply into the throats of flowers. Leaf-eaters like the Hawaiian Goose crop vegetation with their lamellated bills. Insectivorous bills take various forms, including the short slender pointed bill of many warblers, the short wide bill of swifts and swallows, the short- and wide-gaped bill with rictal bristles giving a netlike function as in the nightjar (Fig 2–4h), and the chisel-like bill of some woodpeckers for penetrating wood (Fig 2–4e). Mud-probers have long bills which may be straight like that of the woodcock (Fig 2–4g), decurved as in the curlew, or recurved as in stilts. The long sturdy bill of the mollusc-opening oystercatchers (Fig 2–4j) is adapted for forcing open the shells of bivalves, as well as probing in sand or mud. Flamingos (Fig 2–4l) and many ducks (Fig 2–4k) filter out small organisms by straining water through a series of plates or lamellae. Of the fish-eaters, mergansers (Fig 2–4m) grasp their prey by means of saw-edged tomia, pelicans (Fig 2–4o) use their extensive interramal region as a pouchlike dip net, anhingas (Fig 2–4n) spear fish with a daggerlike beak and skimmers (Fig 2–4p) catch fish by ploughing

EXTERNAL ANATOMY

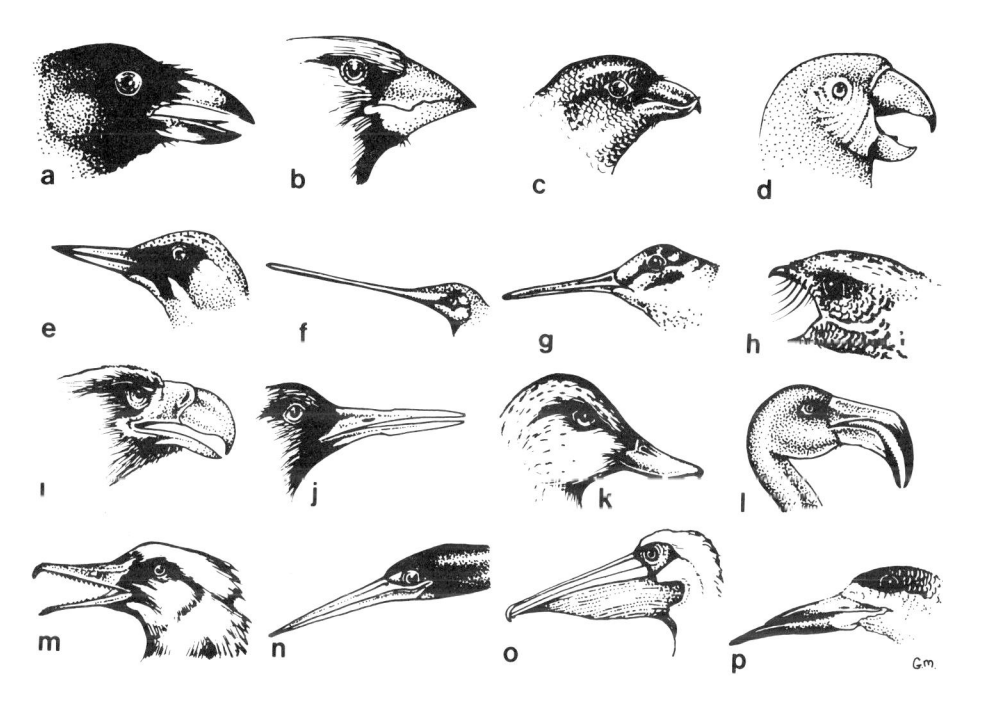

Fig 2–4 Examples of the avian bill showing adaptations for feeding. **a**, general purpose bill (raven); **b**, powerful seed-cracking bill (Hawfinch); **c**, bill for holding apart the scales of fir cones (Red Crossbill); **d**, bill for cracking fruit stones (parrot); **e**, chisel-like bill for penetrating wood (Green Woodpecker); **f**, bill for thrusting into flowers for pollen and nectar (Sword-billed Hummingbird); **g**, bill for probing mud (Eurasian Woodcock); **h**, short wide bill with rictal bristles for netting insects on the wing (European Nightjar); **i**, raptorial bill for tearing flesh (Steller's Sea Eagle); **j**, long sturdy bill for forcing open molluscs (Oystercatcher); **k**, filtering bill (Common Pochard); **l**, filtering bill (flamingo); **m**, saw-edged bill for catching fish (Red-breasted Merganser); **n**, daggerlike bill for spearing fish (Anhinga); **o**, pouchlike interramal region for catching fish by dip-net action (European White Pelican); **p**, knifelike projecting mandibular rostrum for ploughing water (Black Skimmer).

the water with a long knifelike mandibular rostrum. In nocturnal and diurnal birds of prey the hooked bill is powerful and sharp-pointed (Fig 2–4i).

Neck

The length of the neck varies with the number of (cervical) vertebrae, which ranges from 11 to 25 (p. 52). Its minimum length is long enough to enable the beak to reach the uropygial gland (oil gland) on the rump, and is usually proportional to the length of the legs. Thus long legs are generally matched by a long neck, to enable the bill to reach the ground. On the other hand, a long neck is not necessarily accompanied by long legs (consider their relative proportions in geese). Long necks occur in many waders, although some (e.g. plovers) have short necks. The length tends to be intermediate in swimming birds, but some (e.g. swans) have exceptionally long necks. The shortest necks occur in the great majority of the small passeriform species. Characteristically, the neck is carried in a double curve, in the form of an 'S'. Because the forelimb has been totally committed to flight in birds as a group, the bill assumes responsibility for grooming, nest building, and many other manipulative procedures which are generally the province of the mammalian forelimb. Consequently, the avian neck is extremely flexible and mobile, as well as being long enough to carry out these tasks. For descriptive purposes the neck can be divided into several regions. One of these, the *nape* (Fig 2–2), is applicable to birds in general and seems to be widely used by ornithologists. In birds with relatively long necks, such as waders, it is sometimes convenient to recognize the 'foreneck' (pars cranialis of the neck), 'side of the neck' (pars intermedia) and 'hindneck' (pars caudalis).

Trunk

The trunk is the whole of the body between the neck and the tail. It can be divided into thorax, abdomen and pelvis. The *thorax* is bounded externally by the rib cage, sternum and vertebral column. The *abdomen* and *pelvis* are not separated by any well-defined boundary, since in nearly all birds the bony pelvis (os coxae) is open ventrally. The dorsal part of the trunk is divided into the *back* and the *rump* (Fig 2–2). The region between the left and right scapulae is known as the *interscapular region*, and often carries distinctive streaks or colours. The whole back, together with the dorsal surface of the wings, may have a characteristic colour and is then known as the *mantle*. The lateral part of the trunk is called the *side* or flank (Fig 2–2). The ventral part consists of the *breast, belly*, and *undertail* (Fig 2–2). The term *crissum* refers to the general area round the vent, together with the undertail coverts; the term vent is often applied to this area, but should be restricted to the actual orifice (p. 192).

Tail

The flight feathers of the tail, or *rectrices* (Latin *rectrix, rectrices*, rudders), are always paired. They overlie each other as in Fig 2–5, the most dorsal one (the central tail feather, 1a in Fig 2–5; Fig 2–2) lying in the midline. In the great majority of birds, including the songbirds, there are twelve rectrices (six pairs),

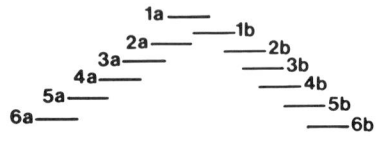

```
              1a ——
          2a ——     —— 1b
       3a ——          —— 2b
     4a ——              —— 3b
   5a ——                  —— 4b
 6a ——                      —— 5b
                              —— 6b
```

Fig 2–5 Diagrammatic transverse section of the twelve flight feathers (six pairs) of the tail (rectrices) of a typical bird. Rectrice 1a lies in the midline, and can be called the central tail feather. It is the most dorsal of the flight feathers of the tail, partly overlying feathers 1b and 2a, each of which in turn partly overlies its successor. The tail is spread by lateral movement, 6b sliding laterally under 5b, 6a under 5a, and so on. From Coues (1890).

but in other species the number ranges from 6 to 32. Numerous *tail-coverts* overlie and underlie the rectrices. These are nearly always small feathers, but in the peacock they become greatly enlarged to form the eyed feathers of the train.

Wing

The feathers of the wing can be divided into *flight feathers* or *remiges* (Latin *remex, remiges*, rowers) and *coverts* or *tectrices* (Latin *tectrix, tectrices*, coverers).

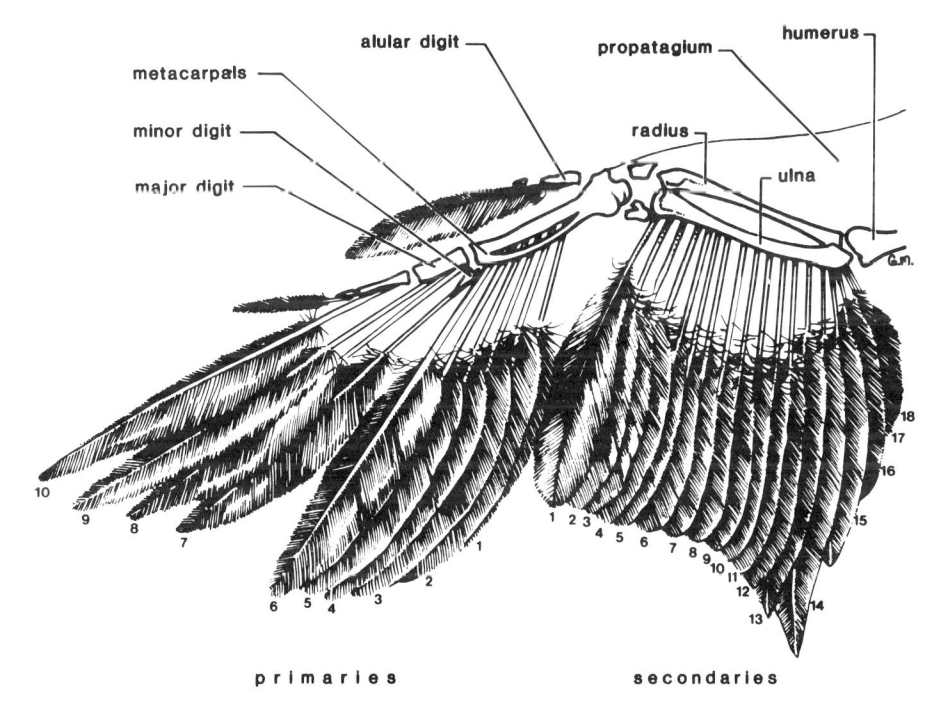

Fig 2–6 Ventral view diagram of the bony attachments of the primary and secondary flight feathers of the right wing. In this specimen, primaries 1–6 are metacarpal primaries, primaries 8–10 are digital primaries attaching to the major digit, and primary 7 is a digital primary attaching to the minor digit. The rudimentary primary attached to the terminal phalanx of the major digit is known as a remicle; this is present in many species but is absent in others. The alular digit is shown with only one flight feather, but 2–7 are present depending on the species. There is no carpal remex in this wing. From Wray (1887).

There are two groups of flight feathers, namely the primary flight feathers and the secondary flight feathers. The primaries are attached to the manus (the hand) and are themselves divided into metacarpal primaries (primaries 1–6 in Fig 2–6) which attach to the metacarpal bones, and digital primaries which attach to the phalanges of the major digit (primaries 8–10 in Fig 2–6) and to the single phalanx of the minor digit (primary 7 in Fig 2–6); the secondaries attach to the ulnar (Fig 2–6). In some species feathers which are attached to the humerus become enlarged and function as flight feathers, but these are modified coverts and should not be called 'tertiaries' or 'tertials'. All flight feathers overlap each other in the same manner (Figs 2–6, 2–7 and 2–8). Thus, when the dorsal surface of the wing is examined, as in Fig 2–7, the trailing edge of primary 10 is seen to be covered by the leading edge of primary 9, and so on.

The *primaries* should always be numbered in a proximal (medial) to distal (lateral) direction as in Fig 2–6. When primaries have disappeared, as has occurred in some groups of birds, they have been lost from the *distal* end of the series. Therefore homologous primaries have the same number only if their numbering begins at the *proximal* end of the series. Species that are capable of flight have from 12 to 9 primaries. Of the flightless birds the rheas and Ostrich have more than this number, for decoration, while other flightless birds have less. In many species the outermost (most distal) primary is so reduced in size that it is hard to see, and is then called a *remicle* (Fig 2–6). Passerine birds have 10 primaries, but in some families the tenth is such a short remicle that the wing is said to have 9 'functional' primaries. All members of a family usually have the same number of functional primaries. The number of metacarpal primaries is rather constant, six being the general rule (Fig 2–6), but a few species have seven.

The *secondaries* are much more variable in number than the primaries, ranging from 6 to 32 depending on the length of the forearm. They attach to the ulna at quill knobs (p. 59). In many species there is a relatively wide space between the fourth and fifth secondaries, with an extra major covert above; this was originally believed to be due to the loss of a secondary and gave rise to

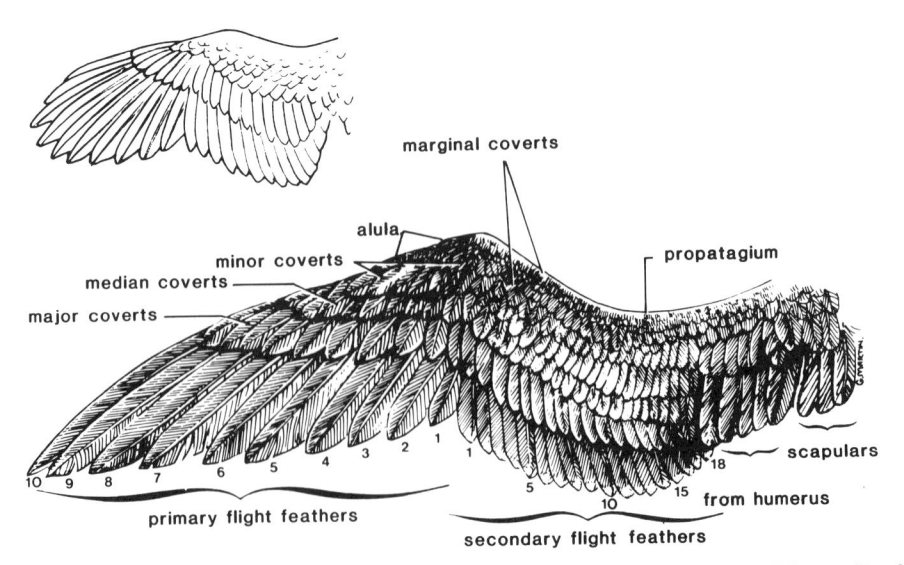

Fig 2–7 Drawing of the dorsal surface of the left wing of a Common Murre. In the region of the secondaries some of the median coverts overlap each other in the opposite manner to the flight feathers (contrary overlap). This wing has a fairly high aspect ratio. The inset shows a galliform wing (Willow Ptarmigan), with a low aspect ratio.

the concept of *diastataxy*, meaning arranged with a gap. This feature is not shown in Figs 2-6, 2-7 and 2-8.

Between the primaries and secondaries there is an extra *carpal flight feather* (carpal remex) in many species, complete with a major upper covert. This is not present in the wing shown in Fig 2-6.

The alular digit bears from two to seven quills forming the *alula* or *bastard wing*. Like the primaries these feathers should be numbered proximally to distally.

The shaft of each flight feather is covered, on both the dorsal and the ventral surface, by a series of *wing coverts*. The coverts covering the primaries are known as *primary coverts*, and those covering the secondaries are called *secondary coverts*. On the dorsal surface (Fig 2-7) there is a single row of *major coverts*, a single row of *median coverts*, several rows of *minor coverts* and an indeterminate number of *marginal coverts* on the leading edge of the propatagium. All the major coverts overlap each other in the same manner as the flight feathers (Figs 2-7 and 2-8), and are therefore said to have a *conforming overlap*. Several of the rows, or parts of the rows, of the smaller coverts overlap in the reverse manner to the flight feathers, i.e. the trailing edge overlaps the leading edge of the adjacent, more proximal, feather (as in the first row of median coverts covering the secondaries in Fig 2-7); this arrangement is designated as a *contrary overlap*.

On the ventral surface (the underneath) of the wing the coverts are arranged essentially like those of the dorsal surface (Fig 2-8), but the rows are less regular.

The triangular fold of (feathered) skin on the leading edge of the wing between the shoulder and carpal joints is known as the *propatagium* (Figs 2-6 and 2-7). In its cranial free edge is the elastic propatagial ligament. The *metapatagium* is the fold of skin on the trailing edge of the wing between the trunk and the brachium. The *postpatagium* is the fold of skin on the caudal margin of the forearm and manus. The *alular patagium* is the fold of skin which unites the alular and major digits. In the ringing or banding of birds, especially young Anatidae, wing clips are sometimes passed through the propatagium.

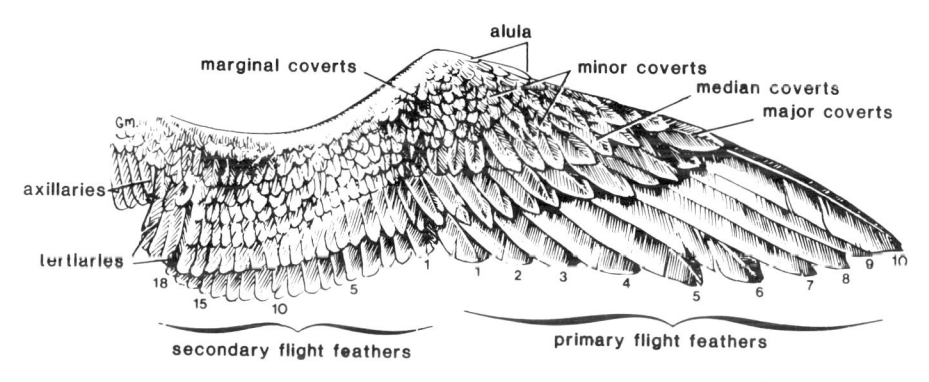

Fig 2-8 Drawing of the ventral surface of the left wing in Fig 2-7.

There are four general types of wing, with numerous intermediate forms. Many passeriform species and some galliforms and pigeons manoeuvre with extreme precision through narrow spaces in vegetation, and these birds have an *elliptical wing*. The shape is short and broad with a low aspect ratio as in the inset in Fig 2–7 (aspect ratio = wing span divided by average width). The outline of the elliptical wing tends to resemble that of the mark II Spitfire of the Second World War. Wing loading is moderate or low (wing loading = body weight divided by the surface area of the wing). There is a large alula, and additional wing slots are formed by separation of the primaries to prevent stalling at low speeds. The wing beat is fairly fast and the amplitude of each beat is moderately great. Manoeuvrability is good. The *broad soaring wing* occurs in eagles, pelicans and New World vultures, enabling these birds to soar at low speeds. This type of wing is fairly long and broad with a medium aspect ratio and moderate wing loading. The alula and wing slots are conspicuous. The *long soaring wing* is restricted to oceanic species such as albatrosses, the Gannet and gulls. The shape is long, slender and pointed (Fig 2–7), with a high aspect ratio and high wing loading. The alula is sometimes large but there are no wing slots. Such wings allow gliding at high speed, but also satisfy the competing needs of flapping flight. However, their relative fragility and clumsiness demand a habitat that is free from obstacles. The *high speed wing* of swifts, falcons and hummingbirds, and to a lesser extent ducks and terns, is relatively small with a moderately high aspect ratio and high wing loading. The wing tip is tapered and may be swept back. There are no wing slots, except in falcons, and these birds can close them in fast flight. The wing beat is rapid and the amplitude of each stroke is small.

Pelvic limb

The pelvic limb articulates with the trunk at the hip joint. The first segment of the limb is the *thigh*, and contains the femur. The first (most proximal) joint within the limb itself is the *knee joint*, between the femur and tibia (Fig. 4–11). The next segment is the *leg* proper, or anatomically the *crus*, and this is based on the tibiotarsus and fibula (Fig 4–11). In scientific ornithology this segment is sometimes designated as the 'lower leg', a term which is not entirely accurate anatomically but is reasonably precise. Unfortunately, in bird identification manuals the part of this region that shows below the feathers of the belly is sometimes called the 'thigh', but this term is so inaccurate anatomically that it really ought to be avoided. The *intertarsal joint* (*ankle joint* or *hock joint*) comes next (Fig 4–11). Identification manuals sometimes call this the 'knee', but again it would be better not to use such an anatomically inaccurate term. The intertarsal joint is succeeded, in strict anatomical terms, by the foot or pes. The first part of the *pes* is formed by the *tarsometatarsus* (Fig 4–11), which corresponds to the instep of the human foot. In birds, this region is commonly called either the 'tarsus' or the 'metatarsus'. Neither of these terms is entirely correct anatomically, since the skeletal basis of this region is constituted by the fusion of the distal row of tarsal bones with the metatarsal bones thus forming the composite tarsometatarsal bone; however, the term 'metatarsus' is reasonable, since the metatarsal component forms the great majority of the tarsometatarsus (Fig 4–11). The same region is sometimes popularly known as the 'shank', but this term is not clearly defined and is occasionally applied either to the whole limb or to all of the visible part of the limb. The next joint is between the distal end of the metatarsal bone and the toes (Fig 4–11). Unfortunately, some identification manuals call this the 'ankle', which is another very inaccurate term since it is the *intertarsal joint* that corresponds to the human ankle. The *toes* or *digits* (Fig 4–11), form the final component of the anatomical pes, but through long-established usage in scientific ornithology the term 'foot' is generally restricted to the toes. In birds in general the (true) thigh, the (true) knee joint and the proximal (upper) part of the (true) leg are entirely concealed by the feathers of the belly and flank as in Fig 2–2. Thus, only the distal (lower) part of the (true) leg, the intertarsal (hock or ankle) joint, the tarsometatarsus and the toes are visible. In descriptive ornithology these *visible* components together are commonly referred to as the 'leg', but this term is not satisfactory since it is also used for the whole limb and is anatomically inaccurate anyway.

Foot

The 'foot' of a typical bird has four toes (e.g. Fig 2–9e), and no bird has more than this number. Species from several orders, including rheas, cassowaries and the Emu, many waders, diving-petrels, auks, some woodpeckers and one passerine species, have only three functional toes; in most of these it is the first digit (hallux, homologous to the big toe of man) which is either vestigial (Fig 2–9a) or absent (Fig 2–9b), but in some kingfishers it is the second toe that is lost. Thus true or functional *tridactylism* occurs in running birds, wading birds (some of which, like certain plovers, are fast running), birds that use mainly their wings to swim under water and climbing birds. One species alone, the Ostrich, has only two toes, digits I and II being lost (Fig 2–9d).

The arrangement of the toes when all four are present depends on function. Most birds have three toes pointing forwards and one (digit I) backwards (the *anisodactyl foot*), as in songbirds (Fig 2–9e) and in *Archaeopteryx* (Fig 1–2C). Among the species with four toes the main variation is to have two (digits II and III) directed forwards and the other two (I and IV) backwards (Fig 2–9f) as in woodpeckers, toucans, cuckoos and parrots (the *zygodactyl* or yoke-toed foot). Owls, touracos and the Osprey have a basically zygodactyl foot, but nevertheless can move the outer toe (digit IV) readily from the backwards to the forwards position. In one group of birds, the trogons (which adopt an arboreal habitat), it is digits I and II that are turned backwards (sometimes known as the *heterodactyl foot*). Essentially, the zygodactyl foot is adapted for climbing and grasping. In most swifts (Fig 2–9g) all four toes point forwards as an adaptation for climbing vertically, and in mousebirds the first digit can be turned forwards to join the other three forward-pointing toes (both of these forms constituting the *pamprodactyl foot*). Lastly, digits III and IV may be partly united as in some kingfishers (Fig 2–9c), forming the *syndactyl foot*.

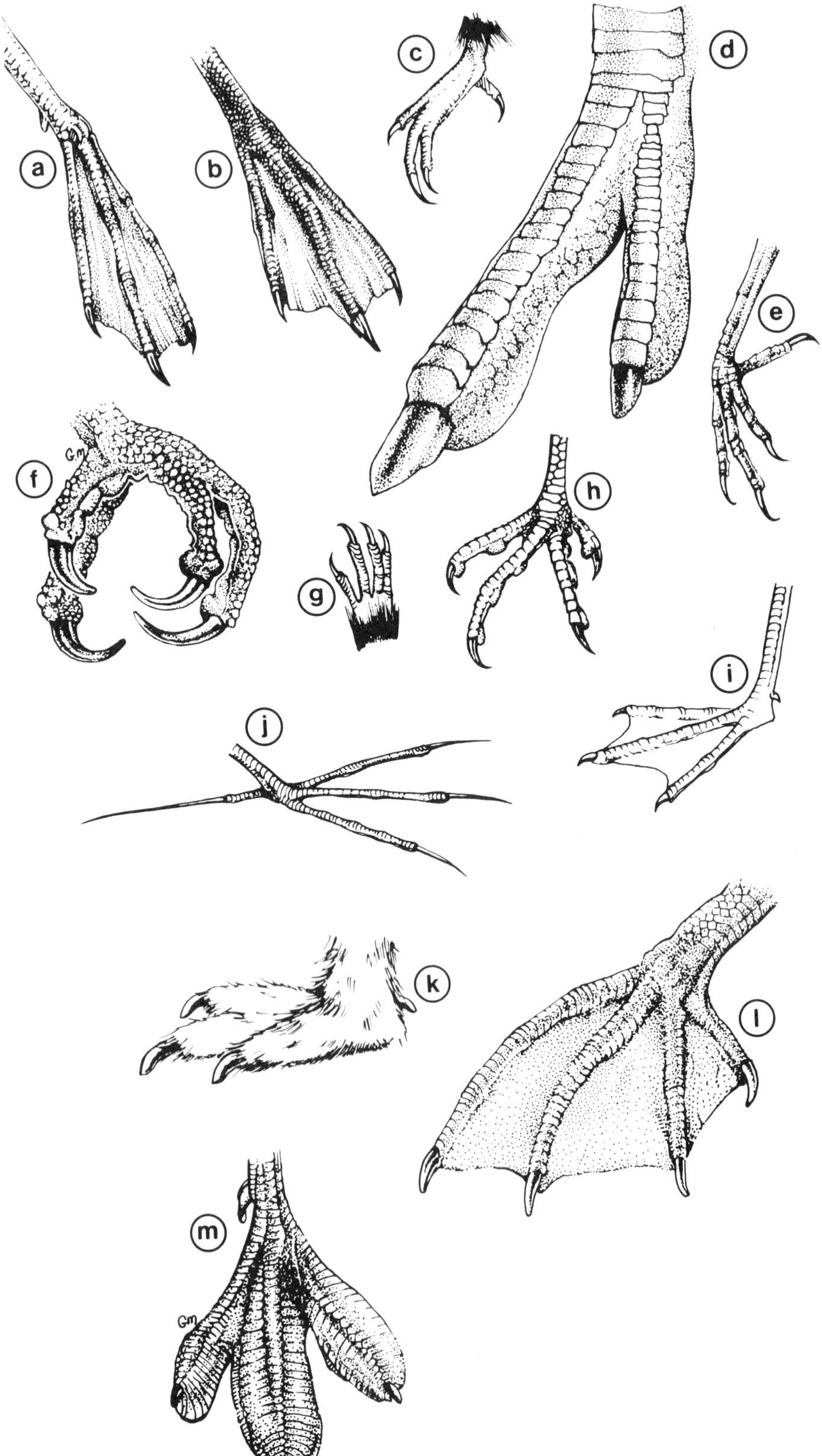
a
b
c
d
e
f
g
h
i
j
k
l
m

Functionally, there are three main types of foot, although there are also many intermediate forms.

1. *The grasping foot.* This is adapted for holding either perch or prey. In the anisodactyl *perching foot* of passerines (Fig 2–9e) all the toes are freely mobile, and the well-developed backward toe is fully apposable giving a firm grip. An even better hold is obtained by the zygodactyl foot with its *two* pairs of apposable toes (Fig 2–9f), a device which parrots use like a hand when feeding. The *raptorial foot* (anisodactyl) has widely spread toes and needle-sharp curved claws with formidable holding powers (Fig 2–9h).

2. *The walking and wading foot.* The capacity to grip is largely sacrificed in such feet, the backward pointing first digit losing contact with the ground and being reduced in size or lost altogether. The toes are partly or completely webbed in some wading birds (flamingos, storks and avocets), so enabling the bird to walk over soft surfaces and also to swim well (Fig 2–9a, –9b and –9i). The toes are greatly elongated in some rails and especially in the 'lily-trotter' jacanas (Fig 2–9i), thus spreading the weight over a wide surface and enabling the bird to walk on floating vegetation. The ptarmigans that walk on snow have raised the first digit off the ground, but have greatly increased the surface area of the foot by forming 'snowshoes' in winter consisting of a dense mat of stiff feathers on their toes (Fig 2–9k) even on the under surface.

3. *The swimming foot.* In the fully adapted swimming foot all four toes are webbed as in cormorants (Fig 2–9l), this being known as the *totipalmate foot*. In gulls and ducks, flamingos, storks, and avocets (Fig 2–9a, –9b and –9i) the three toes are webbed to varying extents, but the first digit is free and usually underdeveloped; this variation is designated as the *palmate*

Fig 2 9 Examples of the avian foot. **a**, vestigial first digit (the short projection visible on the upper left side of the foot), but the other three toes are webbed forming the palmate foot for swimming (left foot of Little Gull); **b**, another palmate foot for swimming, but the first digit is lost altogether (left foot of Razorbill); **c**, digits III and IV are partly united forming the syndactyl foot (left foot of a kingfisher); **d**, digits I and II are lost, leaving only two toes (right foot of Ostrich); **e**, a typical avian foot for perching, with three toes pointing forwards and one (digit I) backwards, the anisodactyl foot (Greek *anisos*, unequal; *dactyl*, toes), (left foot of Blackbird); **f**, the main variant from the typical four-toed foot, digits II and III pointing forwards and I and IV backwards, the zygodactyl foot (Greek, yoke-toed) for grasping (left foot of Blue-and-yellow Macaw); **g**, all four toes point forwards for vertical climbing, the pamprodactyl foot (Greek, *pam*, all; *pro*, front; *dactyl*, toes) (right foot of a swift); **h**, another anisodactyl foot, but with widely-spread needle-sharp curved raptorial claws for grasping (right foot of Common Kestrel); **i**, another palmate foot for walking over soft surfaces (right foot of Avocet); **j**, an anisodactyl foot with very long toes for walking on floating vegetation (right foot of a jacana); **k**, an anisodactyl foot with a vestigial first digit, but with surface area greatly increased by feathered toes for walking on snow (right foot of Rock Ptarmigan); **l**, all four toes pointing forwards and webbed together, forming a totipalmate foot that is totally adapted for swimming (right foot of Great Cormorant); **m**, an anisodactyl foot, but all four digits possess broad lobes forming a lobate foot for swimming (left foot of Little Grebe).

foot. Grebes, coots and phalaropes have broad lobes along the sides of all four toes (Fig 2–9m), and this type is called the *lobate foot*.

Further reading

Coues, E. (1890) *Handbook of Field and General Ornithology*. London: Macmillan.

Harrison, J.G. (1964) Leg. In *A New Dictionary of Birds* (Ed.) Thomson, A.L. London: Nelson.

Lucas, A.M. (1979) Anatomica topographica externa. In *Nomina Anatomica Avium* (Ed.) Baumel, J.J., King, A.S., Lucas, A.M., Breazile, J.E. & Evans, H.E. London and New York: Academic Press.

Mountfort, G. (1964) Bill. In *A New Dictionary of Birds* (Ed.) Thomson, A.L. London: Nelson.

Savile, D.B.O. (1957) Adaptive evolution in the avian wing. *Evolution, Lancaster, Pa.* II. 212–224.

Van Tyne, J. & Berger, A.J. (1976) *Fundamentals of Ornithology*. 2nd Edn. New York: Wiley.

Wray, R.S. (1887) On some points in the morphology of the wings of birds. *Proc. zool. Soc. Lond.*, 343–357.

Chapter 3

INTEGUMENT

SKIN

The skin is generally much thinner and more delicate than in mammals. It is attached to the muscles in relatively few places, but has extensive attachments to the skeleton, for example to the bones of the manus (hand) and pes (foot). It consists of the epithelium of the epidermis, and the connective tissue of the dermis and subcutaneous layer.

The *epidermis* consists of a deep layer of living cells and a superficial layer of cornified dead cells. The living layer or germinal layer comprises three strata: the basal, intermediate and transitional layers. The basal layer is adjacent to the dermis and constantly produces cells to replace those which are lost at the surface. Next comes the intermediate layer, consisting of enlarged polygonal cells which are characterized by desmosomes and are homologous to the prickle cell layer of mammalian skin. This layer merges with the transitional cell layer in which keratinization is almost complete. Superficial to these living layers lies the cornified layer (stratum corneum) consisting of horny dead cells which contain mainly keratin and keratin-bound substances. In feathered areas the whole epidermis is only about ten cells thick, these being more or less equally divided between the living layer and the dead horny layer. In the beak and foot pads the epidermis is greatly thickened to resist mechanical stresses.

Compared to that of mammals the *dermis* is thin. It is made up of superficial and deep layers. The superficial layer varies in thickness according to age and the part of the body. Within the deep layer an outer compact component and an inner loose compartment can be distinguished. The loose component contains the smooth muscles of the feathers and of the non-feathered regions of skin (apteria), these muscles being interconnected by elastic tendons. Within the dermis lie the feather follicles. Movement of the feathers (apart from the filoplumes) is brought about by the feather muscles, which attach to the walls of the follicles (Fig 3–2). The apterial muscles exert tension on the skin lying between the tracts of feathers. The dermis contains many blood vessels and in some sites, such as the wattles, these are especially numerous.

The *subcutaneous layer* is formed mainly by loose connective tissue. It contains fat, both as a layer and as discrete fat bodies, the latter being attached by fascia to the underlying muscles; in the domestic fowl and some passerines fat bodies have been found in definite locations. Subcutaneous fat is especially abundant in aquatic birds as well as in passerine birds during migration. The subcutaneous layer also contains the striated subcutaneous muscles; these are attached to the skeleton or the skeletal musculature and regulate the tension of the skin.

In chickens and turkeys a synovial bursa, the *sternal bursa*, develops in the skin ventral to the cranial part of the keel when the birds are about four weeks old. Enlarged bursae are known as 'breast blisters' and may become secondarily infected. Such blisters have to be trimmed from the carcase in the processing plant.

Horny structures

Horny bill. The horny bill (beak or rhamphotheca) is a hard keratinized epidermal structure covering the rostral parts of the upper and lower jaws, and functionally replacing the lips and teeth of mammals (Fig 2–1). Its basic anatomy, and the variations in its external appearance with the mode of feeding (Fig 2–4), are described in Chapter 2. Despite the wide range of shapes of bill in the adult bird, the bill in the chick is remarkably uniform between species. Although the bill is usually formed by a single sheath, in some groups (e.g. albatrosses, fulmars, pelicans and the Gannet) it is divided into a number of distinct sections. The base of the culmen in some birds is enlarged to form ornamental excrescences. Examples of these are the frontal shield of coots and the massive casque of hornbills. Moulting of the bill occurs in puffins in which the decorative outer sheath is shed after the breeding season.

Histologically the horny bill resembles skin, since it consists of dermis (which is closely attached to the periosteum of the jaw bones) and epidermis. However, the epidermis is modified. Firstly, its stratum corneum is very thick. Secondly, the cells of its stratum corneum contain free calcium phosphate and orientated crystals of hydroxyapatite, in addition to the abundant keratin and keratin-bound phospholipids and calcium which characterize the cells of the stratum corneum of ordinary skin. This combination of characteristics gives the horny bill its typical hardness.

Although the entire bill of birds is usually thick and hard, most waders (charadriiforms) have a relatively soft leathery bill. In ducks and geese the bill is hard only at the tip, the upper and lower hard portions forming the maxillary and mandibular nails respectively. The nail and its associated dermis form a complex sensory organ. The horn of the nail is perforated by canals which open through minute pores on the occlusal surface of the nail. Each of these canals is filled by a dermal core, which corresponds to a highly specialized dermal papilla since it contains a profuse array of mechanoreceptor nerve endings. These dermal papillae project slightly through the pores on the occlusal surface of the nail, their projecting tips being capped by a thin covering of keratinized epithelium. Since the tips of the dermal papillae reach the occlusal surface of the nail, they are directly accessible to tactile stimuli during feeding. In ducks and geese the whole complex is known as the *bill tip organ* (one in the upper and one in the lower bill) and plays an important role in the sensory discrimination between food and other particles. The principal type of mechanoreceptor in the bill tip organ is the *Herbst corpuscle*. It has a complex onionlike structure with a central axonal ending surrounded by numerous slender concentric cellular lamellae; it is highly sensitive to vibration and capable of following very fast frequencies (up to 1000 Hz). The number and site of Herbst corpuscles in the avian bill is related to the way the bill is used as a tactile exploratory organ during feeding. The flexible beak of waders (Charadriiformes) is probably packed with Herbst corpuscles. In grain-eating birds, especially fringillids, which cut open seeds, the tip and lateral ridges of the bill are well supplied. The tongue of woodpeckers also contains these corpuscles. They are in fact, the most widely distributed mechanoreceptors in the avian body, occurring wherever there is mechanical disturbance, i.e. at the base of feather follicles, along the bones of the wing and pelvic limb, in muscles, tendons and joint capsules, near large blood vessels, and even in association with the alimentary canal and especially the cloaca. Other sensory corpuscles are distributed in the dermis that is associated with the soft horn of the anseriform bill, especially near the sides and tip. These include *Grandry corpuscles*, which consist of one or two granular cells associated with numerous axonal endings, closely resembling mammalian Merkel cells. These are rapidly adapting, velocity-sensitive mechanoreceptors. They are also present in the dermis of the bill of grain-eating passerine species. In birds generally the nerve endings of the upper bill are innervated by the maxillary and ophthalmic divisions of the trigeminal nerve and those of the lower bill by the mandibular division (see Chapter 15). The swollen, highly sensitive, sometimes feathered skin at the base of the upper bill in owls, parrots and pigeons is known as the *cere* (Fig 6–3).

On the rostral part of the upper bill the newly-hatched chick has a small pointed keratinized process, the *egg tooth*. It is used at hatching to break the shell and is shed soon afterwards. The actual pipping is achieved by the tooth through a strong upward and backward thrust of the head. The hole is then enlarged by the beak. The process seems to be associated with a rise in the CO_2 or a fall in the O_2 content of the gas in the air space of the egg. A minute egg callosity also occurs in the lower beak of the domestic chick and some wild birds. An egg tooth is absent in megapodes which seemingly escape from the egg simply by kicking their way out. Keratinized egg teeth that are similar to those in birds also occur in some reptiles, but in others (snakes and lizards) the structure is a true tooth, containing cement and dentine. In the nestlings of parasitic honeyguides, both the upper and lower bill carry a pair of sharp hooks. The honeyguides use these hooks to kill the nestlings of the host species.

Despite its hardness, the horny tissue of the bill is normally lost by wear and is continuously being replaced. In caged birds the bill is sometimes not subjected to the usual wear and consequently may have to be clipped. *Beak trimming* has been used in the poultry industry to prevent feather-pecking and cannibalism, but the profuse sensory innervation of the beak should be taken into account before advocating this procedure. When carried out correctly, the upper beak is trimmed not more than one-third of the distance from the tip to the entrance to the nostrils. In adult poultry, beak trimming results in a temporary fall in the intake of food and a reduction in body weight that lasts at least six weeks. The effects are greater if half the beak is removed rather than one-third. Feeding efficiency (the number of pecks per gram of pellets ingested) is also reduced by beak trimming.

Anomalies of the bill are not uncommon, various forms arising from hereditary defects in the domestic fowl, pigeons and wild anseriforms. The growth of the upper and lower beaks may be interrupted, with twisting and misalignment.

Claws. A horny claw encloses the terminal phalanx of each digit of the foot (Fig 2–9). It is made up of a strongly keratinized dorsal plate forming the dorsal ridge and lateral walls of the claw, and a softer ventral plate forming the sole of the claw. The dorsal plate grows faster than the ventral plate thus producing curved claws for digging, scratching, perching, climbing, fighting and grasping prey. In caged birds, the claws, like the beak, may have to be clipped. In species from several different orders (e.g. owls, bitterns and the Gannet) one claw on each foot is modified into a comblike structure to form a toilet claw.

In the great majority of birds the digits of the wing are devoid of claws. However, the alular digit usually has a claw in nestlings of some gaviiform, ciconiiform, anseriform, falconiform, galliform, gruiform, charadriiform and strigiform species but later it is absorbed and for this reason is regarded as a primitive remnant. The young Hoatzin carries large claws that are movable by special muscles on both its alular and major digits and uses them like a climbing reptile to grasp branches while clambering about in the nest. Claws are present in adult ratites. They occur on the major digit in cassowaries, kiwis and the Emu, on the alular digit and sometimes also on the other two digits in rheas, and on all three digits in the Ostrich. Three large digital claws were also present in *Archaeopteryx* (Fig 1–2B).

Spurs. A spur occurs on the caudomedial surface of the tarsometatarsal region of most Phasianidae including the domestic fowl and turkey, being well developed and pointed in the male but small in the female (an exception is the female bantam which has a large spur). In the domestic fowl the cock can be aged by the metatarsal spur, which elongates by about 1 cm each year, reaching a maximum of about 6 cm; a bony base develops by about six months, and subsequently fuses with the tarsometatarsus.

Some Phasianidae have more than one spur on each leg, examples being Indian game fowl which can have two, and peacock-pheasants which have up to four. Entrapment of the spur in the floor of the cage has been reported as a problem in battery pullets. In the male the spurs, as well as the claws, may cause flank wounds in the female during mating.

Spurs also occur in the carpometacarpal region of a number of species, including the Spur-winged Goose, the Spur-winged Lapwing, jacanas and sheathbills.

Scales. In most species scales cover the lower part of the hindlimbs instead of feathers (Fig 2–9), this non-feathered skin of the foot being termed the *podotheca*. The scales are raised areas of highly keratinized epidermis separated by folds of less keratinized skin. They tend to become looser in older birds and can be split or become dislodged with wear and tear, to be subsequently regrown. In the condition 'scaly leg' which affects domestic fowl, the scales are lifted by mites which tunnel beneath them. In some species feathers emerge between the scales.

The pattern of the scales has been used in taxonomy. When the podotheca is undivided or has only a small number of scales near the toes, the foot, as in most passerines, is described as being holothecal or 'booted'. Overlapping of the scales (imbricated scales) is seen in kingfishers and woodpeckers. In some birds, including Pallas's Sandgrouse, ptarmigan and owls, the foot is largely covered by feathers (Fig 2–9k).

Comb, ricti, wattles, ear lobes, frontal process and caruncles

These ornamental outgrowths of skin are characterized by a thickened and exceptionally vascular dermis with many arteriovenous anastomoses.

The *comb* is the bright red vertical projection from the forehead and crown in the domestic fowl (Fig 14–2) and is highly variable in form. Dubbing or removal of the comb is sometimes performed in day-old chicks in the general hope of improving egg production later, avoiding bruising when the comb is large or preventing frostbite in birds on free range in cold climates. The appearance of the comb may be used to assess the health of the bird. For example, in severe haemorrhage it is usually blanched, while in jaundice it has a characteristic yellow appearance. Among wild birds a comb is present in cassowaries and in the male Andean Condor.

The *ricti* are the triangular folds of skin at the angles of the mouth (Fig 2–1) and in some species are brightly coloured. In the chicken the rictus has two parts, the maxillary rictus which is roughened externally and the mandibular rictus which is relatively smooth.

The *wattles* are naked folds of skin that in the chicken hang down from the ventral surfaces of the mandible (Fig 14–2). Similar structures also occur in the rictal region of some birds such as wattlebirds and the Wattled Starling. In the Bantam Silkie and the domestic turkey there is a median wattle or dewlap. In cold climates the wattles of chicks are cropped to avoid frostbite.

The *ear lobes* are folds of skin in the chicken which hang down ventral to the external ear openings and may be red or white in colour. In Silkies the ear lobes are purple because of the presence of melanin pigment.

The *frontal process* or *snood* is the distensible fleshy process arising in the turkey from the dorsal surface of the head between the eyes and nostrils, its turgidity and colour varying with the emotional state of the bird.

The *caruncles* are small protuberances of the skin on the head and upper neck of the turkey, and are basically similar to the frontal process although they are much less distensible.

Uropygial gland and other cutaneous glands

The principal cutaneous gland of birds, the *uropygial gland* (preen gland or oil gland), is present in most birds and is relatively large in some aquatic species, but is absent in other species including the Ostrich, Emu, cassowaries, bustards, frogmouths and many pigeons, woodpeckers and parrots. It is a bilobed gland lying dorsally near the tip of the tail. In the domestic fowl the gland is drained by a pair of ducts, one duct for each lobe, each duct opening through a single narrow, median, nipplelike papilla (Fig 3–5). Other species have up to eighteen orifices. The papilla is usually bare, except for a tuft of down feathers at the tip in most species which is known as the uropygial wick. The secretion of the gland is a lipoid sebaceous material, holocrine in type, and consists of a combination of sudanophilic secretory granules and fragments of cells.

The secretion of the gland is spread over the feathers during preening (the procedure in which the individual feathers are drawn through the mandible) and helps to keep the feathers, beak and scales supple and waterproof. It is also important in preventing the growth of microorganisms, and skin infections are relatively rare in birds. The secretion is odorous in the female and nestling Hoopoe, and in the Musk Duck and petrels.

Glands in the external ear secrete a waxy material containing masses of desquamated cells. Other glands at and around the vent secrete mucus.

The cells of the epidermis contain similar lipoid sebaceous secretions which are presumably released all over the skin. Although true cutaneous glands are absent, except for the uropygial gland and those of the external ear and vent, possibly the skin as a whole should be regarded as a gland.

Birds have no sweat glands, and therefore for dissipation of heat they rely extensively on evaporative cooling from the respiratory tract (see p. 113) to supplement heat transfer (radiation, conduction and convection) from the body surface, chiefly from the unfeathered parts of the body. Nevertheless substantial cutaneous evaporative water loss has been recorded in several avian species. This occurs especially in the early stages of hyperthermia and may then exceed respiratory water loss. It has been suggested that these unexpectedly high rates of cutaneous water loss may be possible in birds because the skin is so thin and vascular, thus enabling a rapid diffusion of fluid from the cutaneous blood capillaries.

Incubation patches

In most birds the dermis of the breast becomes modified during the brooding period to form the *incubation (brood) patch* or *patches*. In these regions the dermis becomes thickened and very vascular, and the feathers are lost. These modifications promote the transfer of heat from parent to eggs. A single median patch occurs in grebes, pigeons and many passerines. In auks and skuas there are two lateral patches. Gulls possess a single median patch as well as two lateral patches. Incubation patches occur in both sexes in some species (e.g. grebes, pigeons, plovers and cranes), in the male only (e.g. in some charadriiforms, gruiforms and tinamiforms) or in the female only (e.g. in most galliforms, and in owls, falconiforms, apodiforms and some passeriforms). Among those birds which do not develop incubation patches are penguins which hold the eggs on top of their feet and enclose them in a fold of abdominal skin, and Gannets which incubate the eggs by covering them with the highly vascular webs of the feet. Ducks and geese also have no true incubation patch but warm the eggs by covering them in the nest with down feathers plucked from their breast.

Foot pads

Foot pads are thickenings of the skin on the ground surface of the foot and are specialized to withstand compression. In the domestic fowl there is a metatarsal pad at the metatarsophalangeal articulations, and several digital pads on each of the main digits. In young woodpeckers, toucans and barbets, a pad is also present on the back of the tarsus and is moulted when the bird leaves the nest. Bacterial infection of the metatarsal pad leads to foot abscesses or 'bumblefoot'.

FEATHERS

There are seven main types of feather, namely contour, semiplume, down, powder down, hypopenna, filoplume and bristle feathers. The most conspicuous are the contour feathers which cover the surface of the body and are therefore externally visible.

Contour feathers

Contour feathers arise from feather follicles and can be divided into flight feathers and body feathers. The flight feathers include the remiges of the wing and the rectrices of the tail, the remiges being subdivided into 'primaries' and 'secondaries' (see p. 16). Body feathers are the contour feathers which cover most of the body. The contour feathers which cover the bases of the remiges and rectrices dorsally or ventrally are called coverts. The ear coverts (Fig 14–2) are the rows of small contour feathers which screen the external ear opening and appear to improve the ability to hear (see p. 302).

Structure of a mature contour feather. The attached end of the feather, or calamus, is a short tube that is embedded in the follicle (Fig 3–1A). In cross-section it is oval in shape. The embedded tip of the calamus has a circular opening, the proximal or inferior umbilicus (Fig 3–1C). In the embedded feather, the dermis of the follicle projects slightly into the proximal umbilicus and becomes continuous inside the calamus with a very small mound of pulp, the dermal papilla (see below); this small mound is covered by a layer of living epidermal cells which will contribute to the next feather when moulting occurs. Beyond (distal to) this minute mound of pulp the calamus of the mature

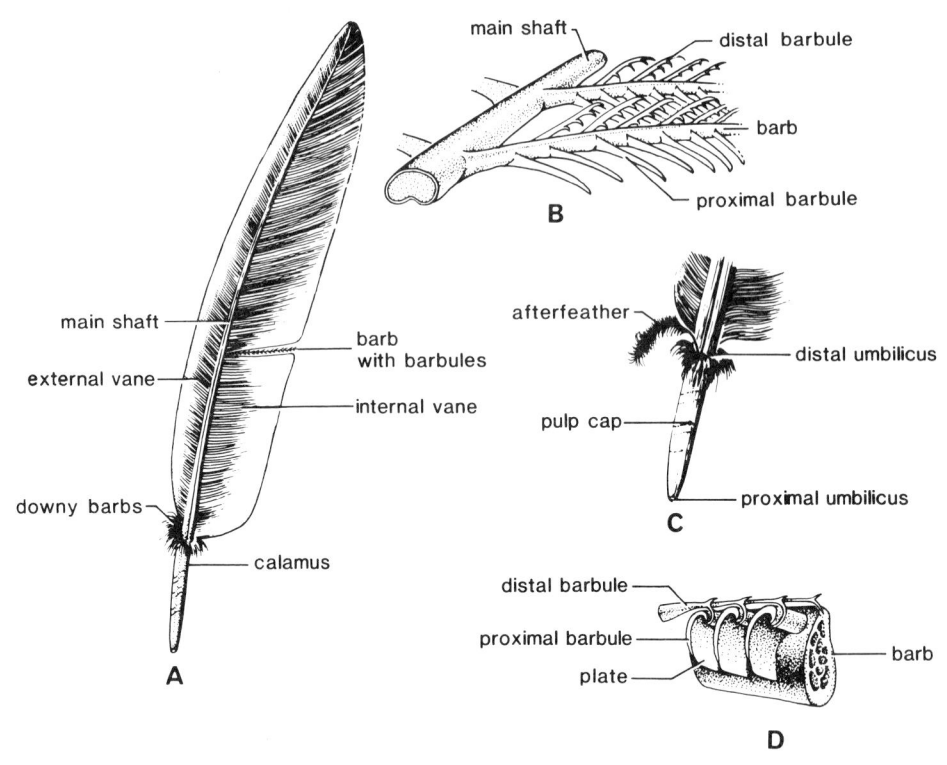

Fig 3–1 **A**, a flight feather; **B**, diagram of the interlocking barbules of the barbs of a flight feather; **C**, the proximal part of a contour feather with an afterfeather at the distal umbilicus; **D**, a view of the hooked distal barbules locking onto the curved dorsal edges of the proximal barbules, as seen with the scanning electron microscope. In some species the proximal barbules form curved plates, such as is shown in D. These plates are sufficiently flexible to be somewhat rotated, their thin lower edges being swivelled to the left in the diagram. When this happens the curved plates make contact with each other, thus creating a virtually continuous surface downstroke of the wing. During the upstroke of the wing the lower edges of the plates are pushed apart by the air pressure, thus allowing the wing to be elevated at a minimal energetic cost.

feather is mainly hollow. However, in the growing feather the calamus is filled with pulp, which is a loose reticulum of mesoderm, and there is also an axial artery and vein (Fig 3–3). As the feather matures, the vessels degenerate and the pulp dies and is progressively resorbed towards the proximal umbilicus. This process of resorption would be expected to produce a completely hollow calamus, but in fact a series of thin epidermal partitions remain which divide the cavity of the calamus into a sequence of isolated chambers. These partitions are dome-shaped and called pulp caps (Fig 3–2). By holding the feather against the light it is possible to see the caps through the wall of the calamus.

The calamus is continued peripherally by the main shaft of the feather, or rachis (Fig 3–1A). The rachis bears two series of slender but stiff filaments, the barbs, at about 45 degrees to the main shaft (Fig 3–1B). Each barb also carries two series of even finer filaments, again at 45 degrees; these are the barbules (Fig 3–1B). Therefore the barbules of adjacent barbs cross each other at 90 degrees. The barbules which slope towards the free tip of the feather (the distal barbules) carry tiny hooks which loosely engage the other (proximal) barbules (Fig 3–1B and 3–1D). When their hooks are engaged the barbules and barbs on each side of the main shaft form the two load-bearing, almost airtight, vanes of the feather (Fig 3–1A). If the hooks become disengaged, they can easily be rehooked like a zip-fastener, by simply passing the vane through the bird's beak, thus forming a much less vulnerable surface than the wing membrane of a bat or pterodactyl. Furthermore, the flight feathers are light but yet stiff enough to require support at only one end (the calamus), whereas the bat's membrane stretches between the tips of the five digits of the manus. It is important that the mass of the wing should be kept to a minimum, since the lower the mass the less the inertia during the wing stroke. It is particularly important to reduce the mass of the *tip* of the wing, since the amplitude and therefore the speed of movement of the various parts of the wing during the wing stroke increase with the distance from the shoulder joint. Bats overcome this problem by tapering the phalanges towards the wing tip, but in birds the wing tip contains no bones and consists solely of the very light feather vanes. At the base of the contour feather the barbs are not hooked together but form downy barbs (Fig 3–1A); this part of the feather is known as the fluffy or plumaceous part. The rest of the feather in which the barbs are united is firm and constitutes the pennaceous part (Fig 3–1A).

The undersurface of the rachis has a groove along its whole length (see the transverse section of the rachis in Fig 3–1B). This groove ends at the junction of the rachis with the calamus. At the junction there is an opening, the distal or superior umbilicus (Fig 3–1C). This would lead into the hollow inside of the calamus, except for the presence of a pulp cap at this point. A small extra feather is often attached to the rim of the distal umbilicus; this is the afterfeather (Fig 3–1C). The distal umbilicus and afterfeather are logical consequences of the way in which the barbs are formed in the growing feather (see section **Growth of a feather**).

The domestic fowl and other species perform *dust bathing*. Apparently, this removes the frayed tips of the contour feathers. As a result, the plumage of domestic fowl is neater if dust baths are available, compared to battery birds where the procedure is not possible. Some wild species change the colour of the plumage during the breeding season by eliminating the tips of contour feathers by means of dust baths.

The vanes of the primary flight feathers are asymmetric in their shape in that the external vane is always narrower than the inner (Fig 3–1A). On the dorsal surface of the wing the internal vane is covered by the external vane of an adjacent feather (p. 16). This asymmetry in the vanes is also seen in the primary flight feathers of *Archaeopteryx* indicating that this ancient bird had an aerodynamically designed wing and was capable of at least gliding flight. In a number of primary remiges and rectrices of many birds, one or both vanes abruptly narrow towards the tip, the narrowing being known as the *notch*. This is especially well developed in the primaries of vultures, eagles, buzzards and hawks. Because of the notches, the feather tips in the spread wing are separated by *slots* (rather like the slots on some aeroplane wings) so that the induced drag of the wing during flight is lessened. In some species (e.g. pigeons) part of the vanes of the flight feathers of the wing and tail have specialized barbules which hold overlapping feathers in place next to one another.

Structure of a mature follicle. The follicle is a cylindrical pit in the skin and closely fits the calamus just as a hair follicle tightly encloses the hair of a mammal (Fig 3–2). Like the rest of the skin, the wall of the follicle consists of epidermis and dermis. At the proximal umbilicus, the dermis forms a mound, the dermal papilla, which carries a very small hump of pulp within the tip of

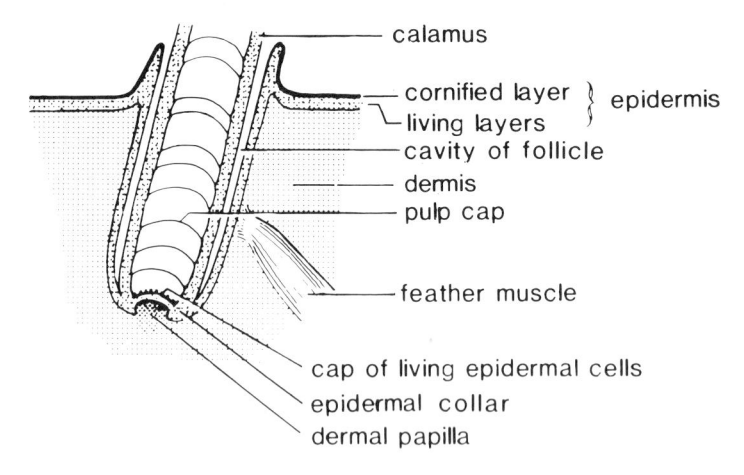

Fig 3–2 A mature follicle.

the calamus. The follicle is lined by epidermis formed of the layer of living cells and the layer of cornified dead cells. At the dermal papilla the epidermis of the follicle becomes continuous with the wall of the calamus around the circumference of the proximal umbilicus; at this junction, there has to be a transition between the living epidermal cells of the follicle and the dead epidermal cells of the calamus. At the dermal papilla the epidermis of the follicle is also continuous with the thin cap of epidermis which covers the minute mound of pulp within the inferior umbilicus.

The main anchorage of the calamus is by means of the epidermis rather than the dermis. However, if the feather is pulled out, not only will the epidermis be torn at the base of the follicle in the region of the epidermal collar (see section **Growth of a feather**), but there will also be some damage to the dermal papilla, and since the dermis is vascular this will cause bleeding into the empty follicle. Also, pieces of the dermis immediately ooze into the follicle together with this blood. If the plucked feather is young, the epidermal lining of the follicle may be completely pulled out of the follicle before it breaks at its junction with the calamus.

Between the outer surface of the calamus and the epidermis of the follicle there is a narrow space, the cavity of the follicle.

Inserted on the dermis of the follicle are the elastic tendons of several substantial bundles of smooth muscle, the feather muscles (p. 23). The latter have the action of raising or lowering the feather.

Growth of a feather. The first sign of a developing feather (sometimes known as a pinfeather) is a disclike thickening of the epidermis under which is a condensation of dermis. The feather then develops as a pointed projection from

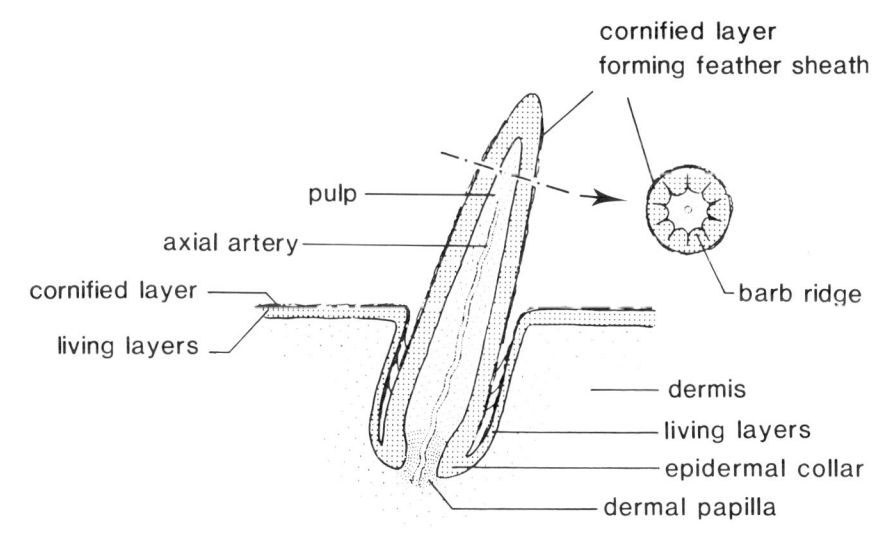

Fig 3–3 An early stage in the development of a feather, in longitudinal section. The feather consists of an axial core of dermis forming the pulp, enclosed by an epidermal covering derived from the epidermal collar. The base has sunk into the skin, forming the follicle. The transverse section shows incipient barb ridges and the axial artery.

the surface of the skin. The projection consists of a core of dermis and a covering of epidermis. As the projection grows its base sinks into the skin, thus forming the follicle (Fig 3–3). The elongating feather still consists of an axial core of dermis and an outer covering of epidermis. The epidermal part is formed by proliferation of the epidermal zone at the base of the follicle, this zone being known as the epidermal collar. The cornified layer of this proliferating epidermal tissue becomes the feather sheath, initially enclosing the developing feather. The central core of dermis constitutes a basic difference between a feather and a hair, since a hair consists of a column of epidermal cells only. As mentioned above, a single artery (axial artery) and vein run through the axis of the dermal core. At this stage, the embryonic feather is simple, and basically resembles a mammalian horn both in its shape and in possessing an outer ectodermal covering and inner mesodermal core.

From here onwards the special featherlike qualities begin to appear. The epidermal cells group themselves into two series of spiral barb ridges (Fig 3–3),

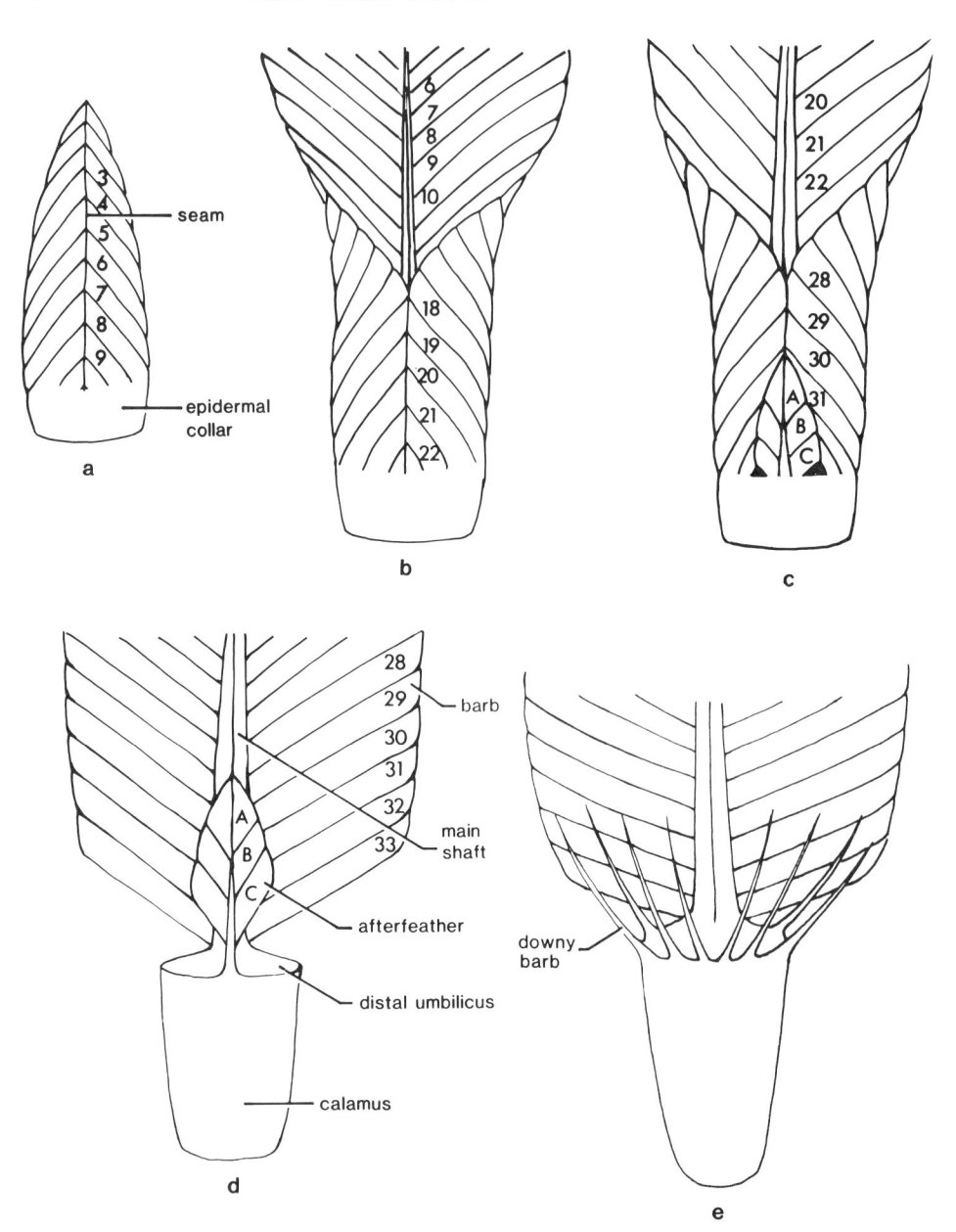

Fig 3–4 Intermediate and final stages in the development of a contour feather. **a, b, c** and **d,** body feather; **e,** flight feather. All the diagrams show the ventral surface of the feather. The feather sheath is omitted.

a The first barbs (1–7) have formed but are still enclosed in the feather sheath. Barb ridges 8, 9 and 10 are arising on the ventral side and growing spirally towards the dorsal side of the feather. The tips of the barbs will presently separate along the ventral 'seam'.

b The first barbs (e.g. 6–10) have separated along the ventral 'seam', escaped from the feather sheath and moved into their definitive positions. Additional barb ridges (e.g. 20, 21 and 22) are developing in a typical spiral form.

foreshadowing the two series of barbs. The tips of these epidermal ridges end along a longitudinal line on the ventral aspect of the feather; this line resembles a seam (Fig 3–4a) in that the tips of the barbs eventually part company along this seam and then move apart. On the dorsal side of the developing feather, i.e. opposite to the 'seam', a thicker longitudinal ridge of epidermal cells foreshadows the rachis. At this stage, the feather sheath progressively ruptures (Fig 3–4b and –4c), starting at the free tip of the feather. This releases the barbs, the tips of which have separated along their 'seam'. Where the barbs split apart the pulp is exposed and is eventually rubbed away. The seam ends at the junction of rachis and calamus, and from here the feather continues as an unsplit, basically hollow, cylinder; at the junction of the split and the unsplit regions, a hole must form, that is the distal umbilicus. In a developing body feather, the afterfeather is a diminutive feather which forms, at the ventral rim of the distal umbilicus, by the splitting apart of a similar, though miniature, series of epidermal ridges on the ventral aspect of the developing feather (Fig 3–4d). In a developing flight feather a circlet of downy barbs arises in place of the afterfeather (Fig 3–4e), but otherwise its development resembles that of a body feather. In the condition 'clubbed down' in the domestic fowl chick the feathers fail to uncurl; this may be caused by riboflavin deficiency.

Replacement of a feather. Most birds replace all their feathers at least once a year by moulting (see below). The discarded feather is pushed out by proliferation of the epidermal collar at the base of the follicle. This mechanism of extrusion essentially resembles the way in which a temporary milk tooth of a mammal is pushed out by the budding permanent tooth beneath it, except that the cycle of moulting is repeated throughout the life of the bird.

If a feather is plucked, the dead epidermal cells of the embedded tip of the calamus are torn away from the living cells of the epidermal collar at the base of the follicle; some of these living cells will presumably remain on the end of the plucked feather and be lost. The mesodermal tissues within the dermal papilla will also be damaged, and will bleed into the empty follicle as already stated. Nevertheless, the lost feather will be replaced immediately or at the next moult. The process of replacement begins with proliferation of epidermal cells which may still remain on the dermal papilla despite the damage caused

---

c Further splitting of the feather sheath has released more barbs. Barbs up to 22 have moved into their definitive positions. Barbs 24, 25, 26 and 27 have been partially released. Additional barb ridges (e.g. 31, 32 and 33) are forming. At the level of A, the ventral 'seam' has divided to form the afterfeather (A, B and C). The calamus is developing below the afterfeather.

d The feather is almost fully developed, apart from further lengthening of the calamus. All the barbs (up to 33) have been released from the feather sheath and have taken up their definitive positions. The barbs (A, B and C) of the afterfeather are joined to the shaft of the afterfeather, which is attached to the ventral side of the distal umbilicus, and are about to be released from the feather sheath.

e In the flight feather, the circlet of downy barbs has been formed in place of the afterfeather, but otherwise the development resembles that of the body feather.

by plucking, and by proliferation of the remaining cells of the epidermal collar. The subsequent growth and differentiation of the replacement feather follows the same procedure as the growth of a new feather during moulting.

Pterylosis. In most birds the contour feathers are distributed over the body in well-defined tracts called *pterylae* (Fig 3–5). The number of these varies with the species, there being nearly seventy in the domestic fowl. The main groups of pterylae include those of the head, spine, lateral trunk, ventral body wall, tail, wing and pelvic limb. The bare spaces between the tracts are *apteria*, the names of which basically correspond to those of the tracts (Fig 3–5). Attempts to use taxonomically the differences between species in the pattern of pterylae have been disappointing.

Other types of feather

Semiplumes are characterized by a wholly fluffy vane and a rachis which is always larger than the longest barb (Fig 3–6B). Most semiplumes lie along the margins of the feather tracts whilst some are distributed singly within the tracts. They act as thermal insulation and in aquatic birds increase buoyancy.

Down feathers include the natal downs of the newly-hatched chick, as well as the definitive downs of the adult. They are entirely fluffy (Fig 3–6C) and the rachis is either shorter than the longest barb or absent altogether (except in penguins, adult down feathers differ structurally from natal down feathers). Down feathers may be evenly distributed over the entire body (penguins and ducks), sparsely or unevenly distributed (gulls and owls), restricted to pterylae (tinamous), restricted to apteria (most galliform species) or sparsely distributed in apteria or even entirely absent (ratites, pigeons, hummingbirds and passerines).

Powder down feathers occur in herons, toucans, pigeons, parrots and bowerbirds, and shed a fine white waxy powder that consists of minute granules of keratin which are about 1 mm in diameter. The powder down appears to form a waterproof dressing for the contour feathers. Most powder feathers have the structure of down feathers but some of them are semiplumes or contour feathers, and indeed all contour feathers are reported to produce minute amounts of powder. True powder down feathers are generally concentrated in patches, as in herons, but in some species (e.g. parrots) they are scattered. Powder down feathers are absent in the domestic fowl.

Hypopennae or afterfeathers are generally small feathers projecting from the distal umbilicus of plumaceous and pennaceous feathers (Fig 3–1C) and are highly variable in form. In the Emu and cassowaries the hypopenna is as long as the main feather. Hypopennae are generally not associated with the rectrices or the larger remiges.

Very close to the follicle of each contour feather is a delicate *filoplume*. This has a long fine shaft with a tuft of short barbs or barbules at the free end (Fig 3–6A). Their follicles have many free nerve endings and several encapsulated nerve endings (e.g. Herbst corpuscles) lie nearby. Possibly, the filoplumes produce a proprioceptive sensory input which is needed to keep the contour

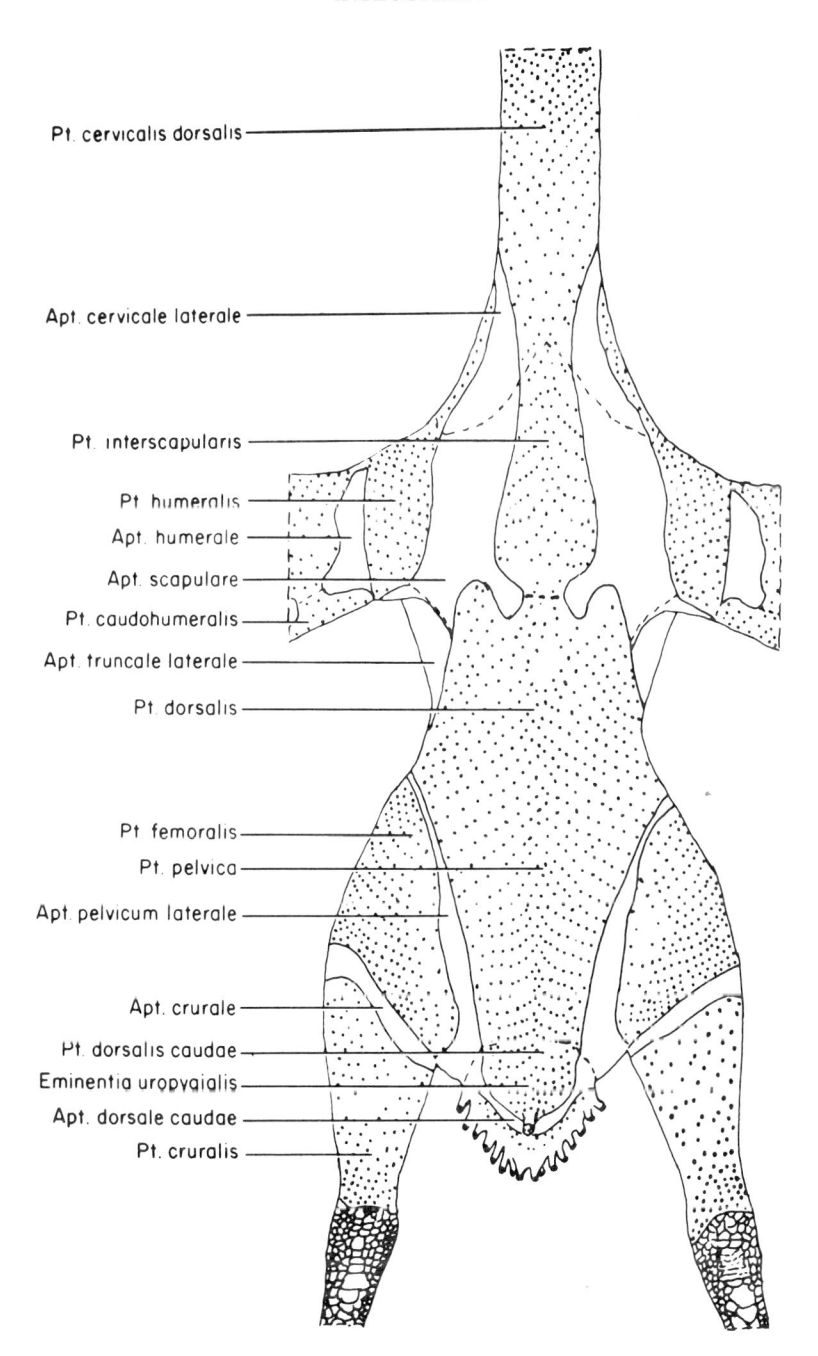

Fig 3–5 Dorsal view of the pterylae and apteria of the domestic fowl. The ducts of the uropygial gland open through the nipplelike papilla on the uropygial eminence. Pt. = pteryla; Apt. = apterium. From Lucas and Stettenheim (1972).

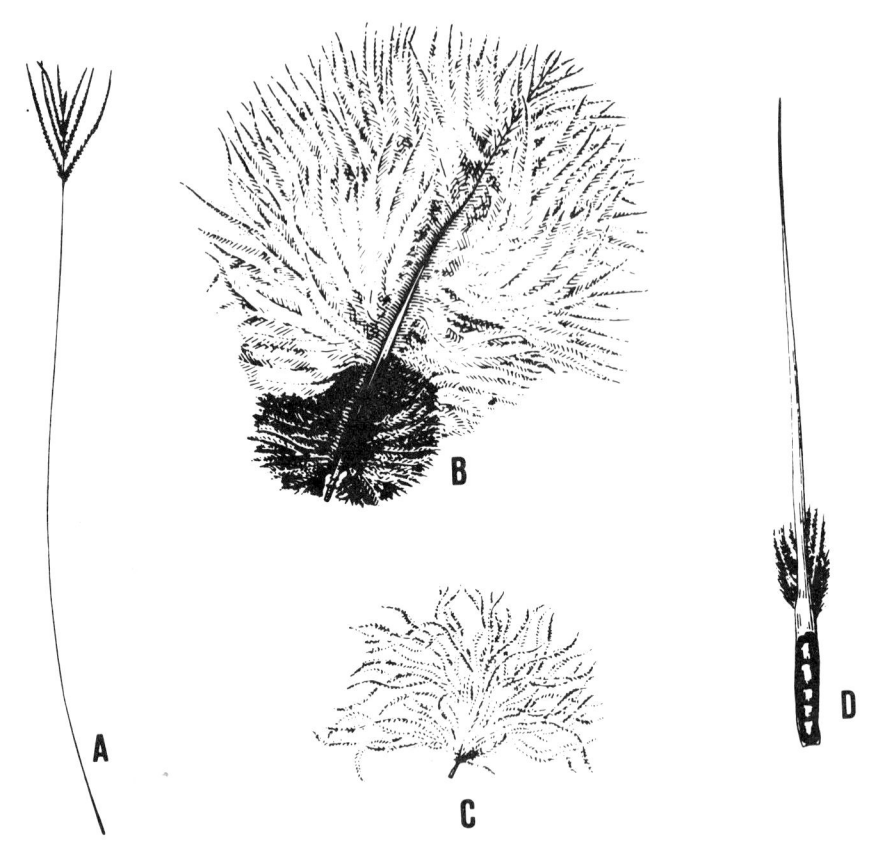

Fig 3–6 Feather types. **A**, filoplume; **B**, semiplume; **C**, down; **D**, bristle. From Van Tyne and Berger (1976), with kind permission of the publisher.

feather in optimum positions. Another type of filoplume has no barbs and extends like a hair beyond the contour feather of the nape and breast of some species like the American Robin and bulbuls. These may be decorative. In the poultry processing plant filoplumes remain after the carcase has passed through the plucking machine and have to be removed by singeing. Filoplumes are absent in the Ostrich, Emu and cassowaries.

Bristles have a stiff rachis and a few barbs at the proximal end, or no barbs at all (Fig 3–6D). Semibristles have barbs along most of the length of the rachis. Usually bristles are situated about the mouth (Fig 2–4h), nostrils and eyes (eyelashes, like those of mammals, occur in a number of species including the Ostrich, hornbills and many cuckoos). Since the follicles of bristles are surrounded by numerous encapsulated sensory corpuscles it is assumed that they have a tactile function like a cat's whiskers.

Feather colour

The colour of feathers is due to the reflection of some components of the incident white light. According to the way in which the white light is reflected feather colours are either pigmentary or structural. Some colours are a combination of both these forms.

Pigmentary colours or biochromes are due to the presence of pigments in the feathers mainly melanins, carotenoids and porphyrins. *Melanin* pigments occur in granules in the skin and feathers, and are responsible for dull-yellow, red-brown, brown and black colours (e.g. the black of Blackbirds and crows and the yellow of the down of domestic chicks). *Carotenoids* occur in the feathers in a diffused form dissolved in fat globules. Birds obtain most of the carotenoid pigments either directly or indirectly from plants. They are responsible for some yellow, orange and red colours (e.g. the yellow of canaries and orioles, the red of cardinals and the pink of flamingos). In the absence of dietary carotenoids the plumage becomes drab and some species in captivity may have to be fed special diets. *Porphyrins* are nitrogenous pigments synthesized by birds and occur in a diffused form. They are responsible for the green and red of touracos.

Structural colours or schemochromes are due to the physical nature of the feather surface. They may result either from interference producing iridescent colours which change with the angle of view, or from scattering of light producing non-iridescent colours. *Iridescence* can be caused by a variety of mechanisms. In hummingbirds and starlings, for example, the feather barbules are twisted so that the flat surfaces are directed towards the viewer. The iridescence arises through the interference of light reflected from the surface of granules of melanin in the barbules. In the peacock the iridescent barbules are covered by three very fine layers of keratin which reflect colours like a soap bubble. *Non-iridescent* structural colours result from the Tyndall scattering of shorter waves in white light by very small particles. Most blue colours in birds are due to this scattering (e.g. the blue of the Budgerigar) the particles responsible for the scattering in this case being airfilled cavities within the keratin of the barbs. White colour in birds is caused by the reflection and refraction of all wavelengths of white light striking the feather.

Pigmentary/structural colours are due to the combined action of pigment and structure. The best example is the colour green which is mainly a combination of yellow carotenoid pigment and physical scattering.

The colour of feathers must be regarded as a major attribute and is utilized in a variety of ways including display, concealment, protection from injurous light and temperature, and prevention of excessive wear (feathers containing melanin, for example, are much more resistant to wear than white feathers).

Moulting of feathers

Moulting is the shedding and renewal of plumage and serves to replace worn out feathers as well as, in some species, to provide the brightly coloured plumage for courtship and mating. After hatching, birds pass through a series of moults which culminate in the acquisition of the adult plumage. Thereafter all the feathers are replaced at least once a year in all birds (generally after the breeding season). In some species the feathers are replaced twice a year and in a few species three times a year. In the domestic fowl, after the end of the first year there is usually one complete moult each autumn. In the following account of the sequence of moults the term 'basic' is applied to plumage which is the only one in the annual cycle or to the winter (non-breeding) plumage; 'alternate' is applied to the second, spring (breeding) plumage; and 'supplementary' to a third plumage. The sequence of plumages is as follows:

>Natal down plumage
>(Postnatal moult)
>Juvenile plumage
>(Prealternate I moult)
>Alternate I plumage
>(Prebasic II moult)
>Basic II plumage
>(Prealternate II moult)
>Alternate II plumage

At hatching the chicks of some species are covered entirely by natal down plumage (e.g. ducks and galliform birds; ptilopaedic species), whilst others are almost naked (e.g. passerines; psilopaedic species), and only develop a downy coat a few days later. Some birds (e.g. woodpeckers and kingfishers) never develop a natal down plumage. The down plumage is moulted twice in penguins, divers and many owls. In megapodes the down plumage has prejuvenile remiges so that the chicks are able to fly at the time of hatching. Similar prejuvenile remiges develop slightly later in galliform species, so that these birds too are precocious with regard to flight. The juvenile plumage consists of contour feathers of the adult type, but has a looser texture than adult plumage. It is short-lived. The basic I plumage of the first winter is duller and has immature feathers compared to the adult plumage. In a number of species the juvenile remiges and retrices are retained in the basic I plumage. The alternate I plumage of the second spring and the basic II plumage of the second winter are very similar to the adult plumage. Most species take two years to reach the adult plumage, but some take longer, e.g. Herring Gulls (four years) and albatrosses (seven to eight years).

The pattern of moulting within an individual follows an orderly progression. In a complete moult the sequence is usually inner primaries, outer primaries, secondaries and tail feathers, and body feathers. The moulting of the different groups is overlapping. The moulting of the inner primaries (matching pairs are shed at the same time so that flight is still possible) moves *outwards* with the result that when the outer primaries begin to be shed the inners are fully developed. In contrast, moulting of the secondaries is basically in an *inward* direction. At this time the inner pair of tail feathers is usually shed, the tail moult then proceeding outwards. Exceptions include treecreepers which retain the relatively long central pair of tail feathers until all the other tail feathers have been replaced. In birds in which a moult occurs before the breeding season, only the body feathers are generally replaced. Differences in the pattern of moulting are sometimes seen between males and females of the same species.

The sequence of moulting in ducks (most of which have two plumages a year) is unusual in that the prealternate moults in many males and some females take place seven to eight months earlier than in other birds due to the fact that courtship occurs in autumn and winter. Because of this, the period in which the basic plumage is worn is usually extremely short. Since the basic plumage is a dull female type of plumage it is popularly known as the 'eclipse plumage', and strongly contrasts in males with the distinctive alternate plumage. During the prebasic moult the remiges are all shed within hours so that the bird is flightless for a few weeks and has to remain hidden or out at sea. Nearly all the north west European Shelduck, which are flightless for 25 to 31 days, migrate at the end of the breeding season to the sandbanks in the Heligoland Bight where they can survive flightlessness in comparative safety. Sometimes loss of remiges also occurs in flamingos, grebes, divers and darters. Other species which have an eclipse plumage include bee-eaters, cuckoo-shrikes, weavers and sunbirds.

Further reading

Auber, L. (1957) The structures producing 'non-iridescent' blue colour in bird-feathers. *Proc. zool. Soc. Lond.*, 129, 455–486.

Auber, L. (1957) The distribution of structural colours and unusual pigments in the class Aves. *Ibis*, 99, 463–476.

Bailey, R.E. (1952) The incubation patch of passerine birds. *Condor*, 54, 121–136.

Berkhoudt, H. (1976) The epidermal structure of the bill tip organ in ducks. *Neth. J. Zool.*, 26, 561–566.

Brooks, W.S. & Garrett, J.E. (1970) The mechanism of pipping in birds. *Auk*, 87, 458–466.

Brush, A.H. (1981) Carotenoids in wild and captive birds. In *Carotenoids as Colorants and Vitamin A Precursors* (Ed.) Bauernfiend, J.C. New York and London: Academic Press.

Cane, A.K. & Spearman, R.I.C. (1967) A histochemical study of keratinization in the domestic fowl (*Gallus gallus*). *J. Zool.*, 153, 337–352.

Chandler, A.C. (1916) A study of the structure of feathers, with reference to their taxonomic significance. *Univ. Calif. Publs Zool.*, 13, 243–446.

Clark, G.A., Jr (1961) Occurrence and timing of egg tooth in birds. *Wilson Bull.*, 73, 268–278.

Cohen, J. (1966) Feathers and pattern. In *Advances in Morphogenesis* (Ed.) Abercrombie, M. & Bracket, J. New York and London: Academic Press.

Elder, W.H. (1969) The oil gland of birds. *Wilson Bull.*, 66, 6–31.

Feduccia, A. & Tordoff, H.B. (1979) Feathers of *Archaeopteryx*: asymmetric vanes indicate aerodynamic function. *Science*, 203, 1021–1022.

Fisher, H.I. (1940) The occurrence of vestigial claws on the wings of birds. *Am. Midl. Nat.*, 23, 234–243.

Greenewalt, C.H., Brandt, W. & Friel, D.D. (1960) Iridescent colors of hummingbird feathers. *J. opt. Soc. Am.*, 50, 1005–1016.

Harrison, J.H. (1964) Moult. In *A New Dictionary of Birds* (Ed.) Thomson, A.L. London: Nelson.

Hodges, R.D. (1974) *The Histology of the Fowl*. London and New York: Academic Press.

Humphrey, P.S. & Clark, G.A., Jr (1964) The anatomy of waterfowl. In *The Waterfowl of the World* (Ed.) Delacour, J. London. Country Life Limited.

Jacob, J. & Ziswiler, V. (1982) The uropygial gland. In *Avian Biology* (Ed.) Farner, D.S. & King, J.R. Vol. VI. New York and London: Academic Press.

Kingsbury, J.W., Allen, V.G. & Rotheram, B.A. (1953) The histological structure of the beak in the chick. *Anat. Rec.*, 116, 95–115.

Lucas, A.M. (1968) Lipid secretion in the avian epidermis. *Anat. Rec.*, 160, 386–387.

Lucas, A.M.(1975) Aves common integument. In *Sisson and Grossman's The Anatomy of the Domestic Animals* (Ed.) Getty, R. Vol. 2. 5th Edn. Philadelphia: Saunders.

Lucas, A.M. (1979) Integumentum commune. In *Nomina Anatomica Avium* (Ed.) Baumel, J.J., King, A.S., Lucas, A.M., Breazile, J.E. & Evans, H.E. London and New York: Academic Press.

Lucas, A.M. (1980) Lipid secretion by avian epidermis. In *The Skin of Vertebrates*. Symposium of the Linnean Society, No. 9 (Ed.) Spearman, R.I.C. & Riley, P.A. London and New York: Academic Press.

Lucas, A.M. & Stettenheim, P.R. (1972) *Avian Anatomy. Integument.* Agricultural Handbook 362. Washington, D.C.: US Dept of Agriculture.

Matoltsy, A.G. (1969) Keratinization of the avian epidermis. An ultrastructural study of the newborn chick skin. *J. Ultrastruct. Res.* 29, 438–458.

Menon, G.K., Agarwal, S.K. & Lucas, A.M. (1981) Evidence for the holocrine nature of lipid secretion by avian epidermal cells: a histological and fine structural study of rictus, toe web and the uropygial gland. *J. Morph.*, 167, 185–199.

Spearman, R.I.C. (1971) Integumentary system. In *Physiology and Biochemistry of the Domestic Fowl* (Ed.) Bell, D.J. & Freeman, B.M. Vol. 2. London and New York: Academic Press.

Spearman, R.I.C. (1983) The integumentary system. In *Physiology and Biochemistry of the Domestic Fowl* (Ed.) Freeman, B.M. Vol. 4. London and New York: Academic Press.

Spearman, R.I.C. & Hardy, J.A. (1984) Integument. In *Form and Function in Birds* (Ed.) King, A.S. & McLelland, J. Vol. 3. London and New York: Academic Press.

Stettenheim, P.R. (1972) The integument. In *Avian Biology* (Ed.) Farner, D.S. & King, J.R. Vol. 2. New York and London: Academic Press.

Van Tyne, J. & Berger, A.J. (1976) *Fundamentals of Ornithology.* 2nd Edn. New York: Wiley.

Voitkevich, A.A. (1966) *The Feathers and Plumage of Birds.* London: Sidgewick and Jackson.

Chapter 4

SKELETOMUSCULAR SYSTEM

HEAD

Skull

The avian skull has several reptilian features, including a single occipital condyle, movable quadrate and pterygoid bones, the quadrate bone articulating with an articular bone which is part of the lower jaw, and a lower jaw consisting usually of six small bones instead of one large one. The overall shape of the skull is modified, however, by the enormous eyes and by the relatively large brain. Except in ratites, the sutures between many of the individual skull bones have become obliterated soon after hatching, perhaps because of the extensive invasion of the skull bones by air spaces (pneumatization). The result is a light compact box with a huge bony orbit and a vaulted cranial cavity. Attached to the rostral aspect is a facial component which is characteristically avian in its beak shape and hingelike mobility. In its essential features the skull remains remarkably uniform throughout birds generally.

The upper jaw. The upper jaw (Figs 4–1A and 4–2) is formed mainly by the premaxillary and nasal bones; the maxilla is usually small. In most birds the upper jaw is a rigid triangular block which may be elevated or depressed, hingeing at the flexible junction of the upper jaw with the braincase (Fig 4–2D). This type of jaw movement is known as *prokinesis* (see below). In the prokinetic upper jaw the bony nasal openings are relatively small; the outline of each opening is oval (the holorhinal nostril) and characteristically ends rostral to the junction of the upper jaw and the braincase (Fig. 4–2D). A different type of upper jaw is found in pigeons, hummingbirds, cranes, bustards, coots and most waders (shorebirds); the bones are arranged in the form of a framework, consisting of a median dorsal bar, a pair of ventral bars, a pair of nasal bars and the bony tip (Figs 4–2A and –2B). Movement in this type of upper jaw occurs *rostral* to the junction of the jaw and braincase and is known as *rhynchokinesis* (see below). In the rhynchokinetic upper jaw the bony nasal opening is relatively large and slitlike (the schizorhinal nostril), and characteristically ends caudal to the junction of the jaw and braincase (Figs

4–2A and –2B). Between the prokinetic and rhynchokinetic upper jaws intermediate forms occur. The bony nasal openings in some pelicaniform and sphenisciform species are very reduced. In hummingbirds the ventral bar of the rhynchokinetic jaw is doubled and the basal bar is slightly twisted. The nasal bar in the Ostrich and other ratites is incomplete.

In birds generally, movement of the upper jaw is made possible by the presence of flexible *elastic zones* in the facial bones. In these zones several structural alterations are generally present which permit bending of the bone. Firstly, in the region of an elastic zone pneumatization (p. 46) is absent. Secondly, in the region of a zone the bone is generally extremely flattened.

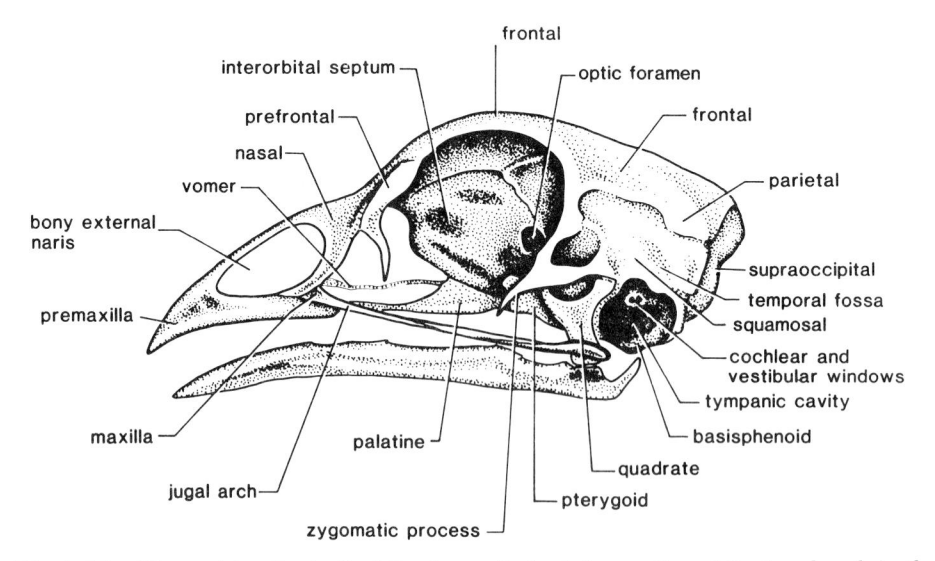

Fig 4–1A The prokinetic skull of the domestic fowl. Inspection of the jugal arch in the drawing shows that it consists of three bones (unlabelled) which are fused together. The most caudal of these is the quadratojugal bone, which articulates with the quadrate bone. The middle one is the jugal bone. The most rostral bone is formed by the jugal process of the maxillary bone.

However, in large species a flattened form of bone would be either too thick to be flexible or too thin to resist the forces acting on it. The zones in these species, therefore, are composed of many thin bony laminae that are separated from each other by connective tissue to form a multilayered structure that is similar to that of plywood. Such multilayering is present in some of the facial bones of pelicans and the Ostrich.

In prokinetic birds a short elastic zone, the craniofacial elastic zone, occurs at the junction between the braincase and the nasal and premaxillary bones. In large parrots, which have extreme mobility of the upper jaw, this elastic zone is converted into a synovial articulation (craniofacial hinge in Fig 4–2D). Ventral to the craniofacial elastic zone is a gap in the nasal-interorbital septum. In some large anseriforms and in pelicans and caprimulgids, a ball and socket synovial joint is present on either side of the craniofacial elastic zone.

In the rhynchokinetic upper jaw an elastic zone, the premaxillonasal elastic zone, is present in the dorsal bar (Fig 4–2B). Sometimes, as in larids and alcids, it lies in the caudal part of the bar so that almost the entire length of the upper jaw can be moved (*proximal rhynchokinesis*). In the Ostrich and in oystercatchers the zone is situated near the middle of the dorsal bar so that the rostral half of the upper jaw is movable (*intermediate rhynchokinesis*). In woodcocks, curlews, godwits and snipes, the zone is located at the rostral end of the dorsal bar with the result that only the tip of the upper jaw is mobile (*distal rhynchokinesis*). In avocets the elastic zone is a relatively long indistinct region (*elongated rhynchokinesis*). Two elastic zones occur in the dorsal bar of plovers and cranes (*double rhynchokinesis*). In the rhynchokinetic jaw additional short elastic zones are present in the ventral and nasal bars (Fig 4–2B). Unlike the nasal-interorbital septum of prokinetic species, the septum of rhynchokinetic birds extends rostral to the base of the upper jaw and is fused to the dorsal bar so that bending cannot occur at the junction of the jaw and the braincase. However, where bending does occur in the jaw a gap exists between the interorbital and nasal septa.

The palate. The skeleton of the palate (Figs 4–1A and 4–2) is formed by the palatal processes of the premaxillae and maxillae, the palatine bones, the vomers, the jugal arches, and the pterygoids and quadrate bones. It does not form a complete shelf between the nasal and oral cavities as in mammals. It is best developed in the so-called 'desmognathous' birds such as swans. The internal nares lie between the *palatines* and the *vomers*, but in kiwis in which the palate is relatively well developed, the palatines and vomers are situated rostral to the internal nares. In galliform species the palatines and vomers are fused. The palatine bar is the rod-shaped bone extending from the main part of the palatine bone to the upper jaw with which it is fused. The *pterygoid* caudally forms a hinge joint with the quadrate bone (Fig 4–2A). Rostrally, it articulates with the palatine (Fig 4–2A). The pterygoid constitutes a link in the so-called '*palatoquadrate bridge*'. Gliding joints occur between the palatoquadrate bridge and the braincase, and involve either the pterygoid or palatine bones and sometimes both as in Fig 4–2A. The *jugal arch* consists of three bones fused together, namely the quadratojugal and jugal bones and the jugal process of the maxillary bone (Fig 4–1A). At its caudal end it forms a ball and socket joint with the quadrate bone (Fig 4–2A). The *quadrate bone* is highly mobile and articulates with the lower jaw, the braincase, the jugal arch and the pterygoid bone (Figs 4–2A and –2C). The joint formed with the braincase is a double articulation. The right and left quadrate bones play a central role in all movements of the upper jaw. Thus rotation of the quadrate bones rostrally pushes on the pterygoid and palatine bones and the jugal arches so that the upper jaw swings upward at the elastic zone (Fig 4–2). Likewise, rotation of the quadrate bones caudally causes the upper jaw to swing downwards. In the prokinetic upper jaw elastic zones are present at the rostral ends of the palatine bar and the jugal arch. In parrots the zone in the palatine bar is converted into a syndesmosis (a fibrous joint). Similarly, the zone in the jugal arch is represented by a syndesmosis in parrots and cardueline finches. The palate has been used as an important factor in classifying birds.

The braincase. The roof of the cranial cavity is formed largely by the frontal and parietal bones (Figs 4–1A and 4–2). In hornbills and cassowaries the frontal bones (along with the nasals) may contribute to the horn-covered casques. The main bones of the floor of the cranial cavity are the basioccipital, basisphenoid and parasphenoid. The principal bones of each lateral wall are the squamosal, orbitosphenoid and prootic bones. The caudal end of the cranium includes the supraoccipital (in part) with associated epiotic and opisthotic bones, and the exoccipital and basioccipital bones. The braincase articulates with the upper jaw, the palatovomeral complex, the pterygoids and the quadrate bones (Fig 4–2A), and in some species it forms a joint directly with the mandible (joint 3 in Fig 4–2A). Since the upper jaw can be moved independently of the cranium of birds, unlike in other terrestrial vertebrates, the cranium may remain stationary during movements of the jaws.

The skull bones of the majority of birds are invaded by a system of *pneumatic diverticula* arising from the tympanic and nasal cavities probably as an adaptation for reducing the weight of the skull for flight. Pneumatization begins about the ninth day of incubation and is continued in some species for about two years after hatching. The pneumatized bones consist essentially of thin outer and inner laminae enclosing a honeycomb of air spaces supported by delicate spicules. Four different designs of air spaces have been described. Pneumatization appears to be maximally developed in large owls and in nightjars. It is very reduced or even absent in species which subject the skull to excessive forces, e.g. birds which dive from the surface of the water (many ducks), birds which plunge into the water from the air (kingfishers and terns), and birds which use the bill for hammering (woodpeckers).

Bony orbit and temporal fossa. The extremely large orbits are incomplete except in parrots. In birds generally the enormous size of the eyeball has compressed the bone at the depth of the bony orbit to a thin midline plate (Fig 4–1A), the bony *interorbital septum* (mesethmoid). In kiwis, which have very small eyes but an extremely well-developed olfactory sense (p. 311), there is no interorbital septum and the orbits are separated from each other by the enlarged cartilaginous nasal capsules rather as in mammals. The various orbital nerves emerge from the caudal edge of the interorbital septum. The dorsal region of the orbit is formed mainly by the frontal bone. In many birds (e.g. falconiforms, galliforms, piciforms, strigiforms and passeriforms) there is also a distinct prefrontal bone, the prefrontal and frontal bones being unfused (Fig 4–1A). The prefrontal bone in nightjars forms a syndesmosis with the upper jaw and can move against the braincase. In the Whale-headed Stork it is totally fused with the upper jaw. Most birds have a large temporal fossa (Fig 4–1A) connecting with the bony orbit. Its main bone is the squamosal.

The lower jaw. The lower jaw (Figs 4–1A and 4–2) is derived embryologically from the fusion of five small membrane bones (dental, supra-angular, angular, splenial and prearticular) and the articular bone which is derived from among the cartilages of the pharyngeal (visceral) arches. In mammals the lower jaw consists of a single dentary bone. The two rami of the mandible in birds are fused in the rostral midline at the symphysis. The articular bone articulates with the quadrate bone (Fig 4–2A) which is also derived from the pharyngeal arch system. In mammals the articular and quadrate bones have been withdrawn from the articulation of the jaw and converted into the auditory ossicles, the malleus and incus respectively. In birds the quadrate bone forms the primary link between the lower jaw and the cranium, and as already indicated above, is the basis of the mechanism of kinesis which is described more fully below. At the caudal end of the mandible is a short process, the medial mandibular process, which in some birds (e.g. the Black Skimmer and the Adelie Penguin) articulates with the ventrocaudal part of the braincase to form a secondary jaw articulation (joint 3 in Fig 4–2A). This joint seems to act as an extra brace for the mandible when the jaws are opened. In melaphagid honeyeaters the ramus of the mandible has a dorsal process which forms a synovial joint with the ectethmoid plate. This joint acts as a brace when the mandible is raised, and permits the mandible to be supported against the braincase independent of the quadratomandibular articulation. Within each ramus of the mandible of many birds, including certain sphenisciforms, procellariiforms, pelecaniforms, ciconiiforms, charadriiforms and caprimulgiforms, there are two zones of bending (elastic zones) which allow the interramal space to be widened. In potoos and nightjars the caudal elastic zone is developed as a mobile syndesmotic joint; in the lateral spreading movement of the mandible which occurs when these birds are feeding the rami are bent outwards at the caudal zones and inwards at the rostral zones.

Ligaments of the jaw apparatus. In addition to the usual articular ligaments, the jaw apparatus of birds possesses a second type of ligament, the *linkage ligament*. Linkage ligaments characteristically lie separate from the articulations and generally span more than one joint. Although the precise function of the linkage ligaments of the jaw apparatus has not been established two of the ligaments at least are involved in linking the upper and lower jaws together mechanically; this results in the forms of cranial kinesis (see below) known as coupled kinesis in which the movements of the upper and lower jaws are dependent on one another.

The jaw muscles. In most birds seven pairs of muscles act on the upper and lower jaws. Two muscles open the jaws, the protractor pterygoidei et quadrati by raising the upper jaw, and the depressor mandibulae by lowering the mandible (Figs 4–6 and 6–3). Two muscles simultaneously close both the upper and lower jaws, pterygoideus and pseudotemporalis profundus. Three muscles raise the mandible (adductor mandibulae externus, pseudotemporalis superficialis and adductor mandibulae caudalis) (Fig 4–6). The pterygoid muscle in parrots is also important in moving the lower jaw forward and in

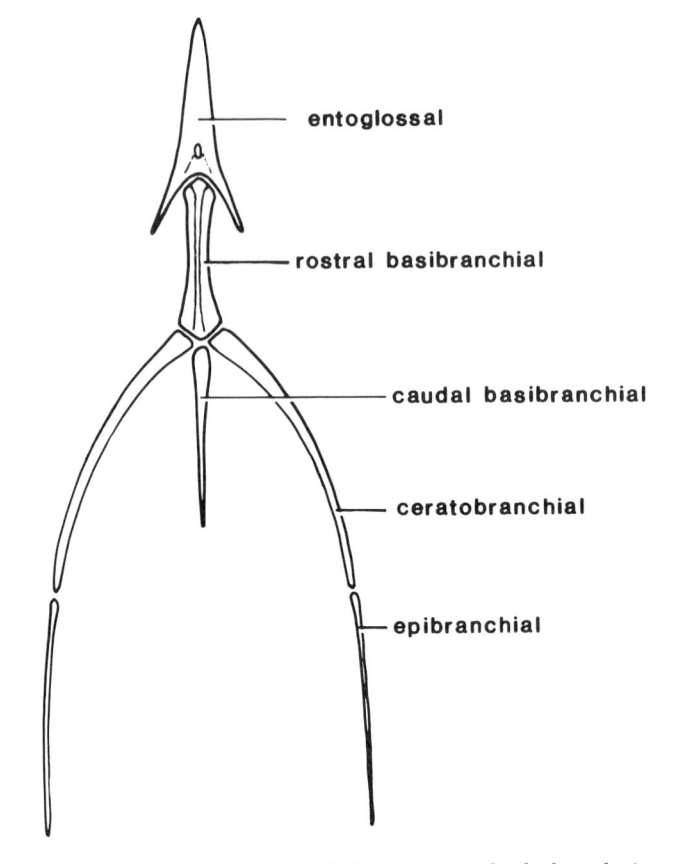

Fig 4–1B The hyoid bone of the domestic fowl, dorsal view.

lowering the upper jaw during chewing. In birds which have elastic zones in the mandibular rami the pterygoid muscles cause the rami to move laterally so that interramal distance is increased. The adductor mandibulae externus is exceptionally well developed in skimmers and acts to prevent depression of the mandible when objects are struck during skimming. In parrots an additional muscle is present, the ethmomandibularis, which closes and protracts the lower jaw. An extra role for the pterygoid muscle occurs in woodpeckers in which it is very strong and helps to absorb the shock produced by pounding the bill on a tree.

Cranial kinesis. Cranial kinesis is the mobility of the upper jaw in relation to the braincase and is possessed to a greater or lesser extent by all birds.

Prokinesis. This movement involves elevation or depression of the upper jaw as a unit. The power for the movement is derived from muscles acting on the palatine, pterygoid and quadrate bones. The rostral and caudal forces exerted by these muscles are transmitted via the *palatoquadrate bridge* (p. 45) and the jugal arch to the ventrocaudal end of the upper jaw which is elevated or depressed at the craniofacial elastic zone.

Rhynchokinesis. In this form of cranial kinesis the movement of the upper jaw occurs further rostrally so that only the rostral part of the upper jaw moves. Basically, rhynchokinesis involves the same movements of the quadrates, pterygoids and palatine bones as in prokinesis. However, in rhynchokinetic species the muscle forces are transmitted to the rostral part of the upper jaw via the ventral bar (Fig 4–2A and –2B); this is because the ventral bar is kinematically separated from the dorsal bar by the presence of the elongated slitlike schizorhinal bony nasal opening (Fig 4–2B), which ends caudal to the base of the upper jaw thus permitting the nasal bars (which have elastic zones) to move with the ventral bars.

Hyobranchial apparatus

The skeleton of the tongue consists of a rostral median rod, the entoglossal bone and a more caudal median rod, the rostral basibranchial bone (Fig 4–1B). The latter articulates on each side with a long horn formed by the ceratobranchial and epibranchial bones; caudally, it articulates in the midline with a short median rod, the caudal basibranchial bone. In parrots the entoglossal bone is usually paired. The two entoglossal bones in the Budgerigar bifurcate rostrally. In birds generally, the rostral and caudal basibranchial bones are fused in the adult. In the Wryneck and woodpeckers, which have extremely long protrusible tongues enabling them to probe considerable distances for food, the hyobranchial horns are remarkably elongated and wind round the back of the skull. The tips of the horns may end near the base of the bill, as in the Orange-backed Woodpecker. However, in some species (e.g. the Green Woodpecker) the two horns enter the right nasal cavity via the right nostril, and in the Wryneck the horns terminate in the left nasal cavity. Sometimes the horns end by encircling the right orbit, as in the Malaysian Gray-breasted Woodpecker. As a result of the elongation of the hyobranchial apparatus, the tongue in the Green Woodpecker can be protruded four times the length of the upper beak.

Other muscles of the head and neck

Four straight muscles (rectus dorsalis, ventralis, medialis and lateralis) and two oblique muscles (obliquus dorsalis and ventralis) move the *eyeball* much as in mammals (Figs 15–9 and 16–9). Movements of the *third eyelid* across the surface of the eye are controlled by two muscles (quadratus and pyramidalis membranae nictitantis) (Figs 16–9 and 16–10). The *eyelids* are moved by the levator of the upper lid (levator palpebrae dorsalis) and the depressor of the lower lid (depressor palpebrae ventralis). Tension on the orbital fascia is regulated by the tensor periorbitae. Muscles of the *hyobranchial apparatus* (intermandibularis, serpiohyoideus, stylohydoideus, branchiomandibularis, interceratobranchialis, ceratoglossus, hypoglossus, mesoglossus and

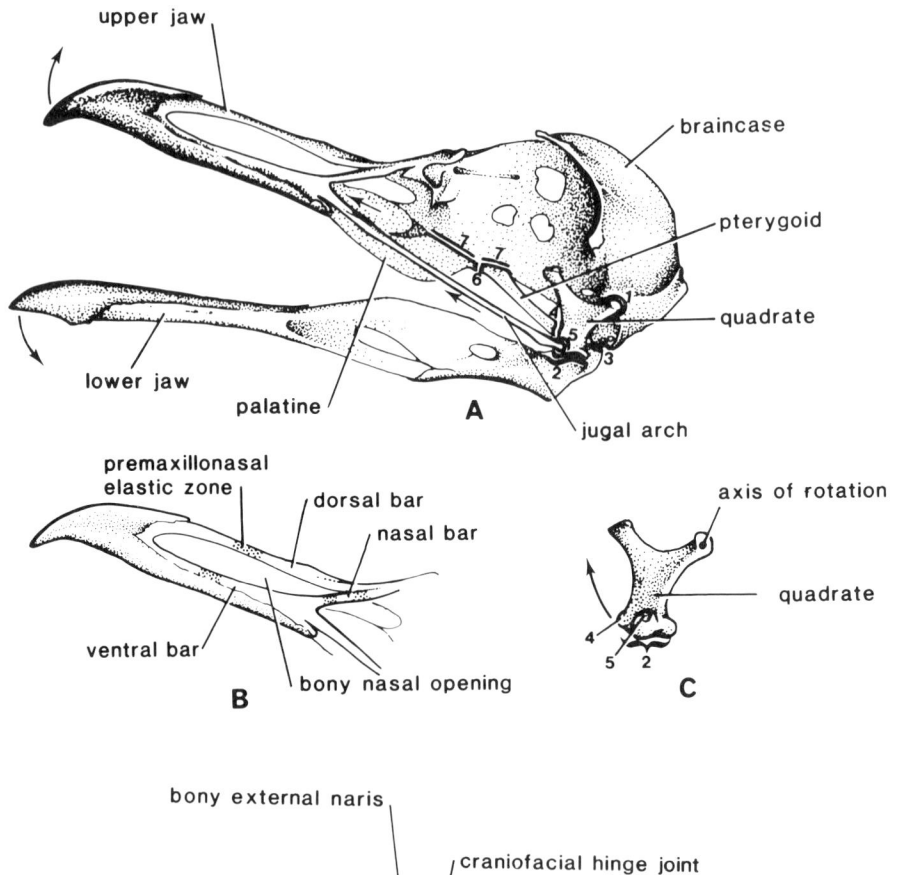

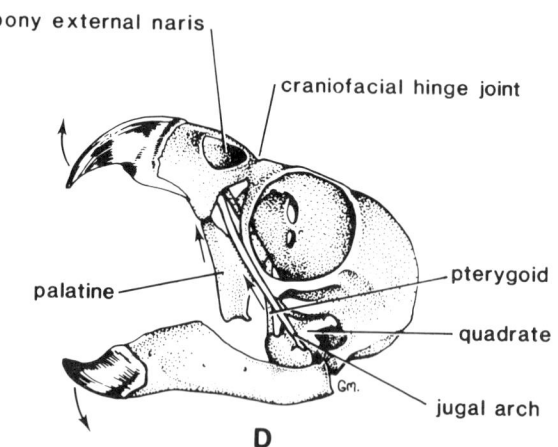

Fig 4–2 The kinetic avian upper jaw. In **A** the upper jaw is rhynchokinetic, the bones being arranged in the form of a framework (the parts of which are labelled in **B**) and the bony nasal opening being large and slitlike. Elastic zones, stippled in **B**, occur in the dorsal, nasal and ventral bars of the framework. In rhynchokinesis the quadrate bones rotate rostrally (as indicated by the arrow in **C**), the resultant forces being transmitted via the pterygoid and palatine bones and the jugal arches (straight arrows in **A**) to the rostral part of the upper jaw which is elevated at the premaxillonasal elastic zone. 1 to 7 in **A** and **C** are the synovial joints which are involved in kinesis. 1, quadratal-squamosal-otic; 2, quadratomandibular; 3, mandibulosphenoidal (present only in a few

genioglossus) elevate, protract and retract the tongue. Two muscles (dilator glottidis and constrictor glottidis) control the size of the *laryngeal glottis*. Several pairs of muscles are associated with the *larynx* and *trachea* (sternohyoideus, sternotrachealis, cleidotrachealis and tracheolateralis) (Figs 7–5 and 7–6). Although primarily tracheal muscles, these also play a part in oscillating the syrinx and trachea cranially and caudally in sound production. Some variable small muscles are associated directly with the *syrinx*. Oscine species usually have five pairs of syringeal muscles (tracheobronchialis dorsalis, brevis and ventralis, and the syringealis dorsalis and ventralis) (Fig 7–6), while suboscine passerines characteristically have four pairs of muscles (vocalis dorsalis and ventralis, and obliquus ventralis and lateralis). Non-passeriforms usually have only one pair of syringeal muscles (tracheolateralis) (Fig 7–5).

NECK AND TRUNK

Vertebrae

The vertebrae of the trunk are almost entirely fused together. To compensate for this rigidity, the cervical vertebrae are much more numerous and mobile than in mammals.

The number of vertebrae in the various regions of the vertebral column is uncertain. This is partly because the individual vertebrae tend to be masked where the bones are fused together, as in the synsacrum. It is also uncertain in the neck, since there is no reliable way of deciding which vertebrae in the caudal end of the neck are 'cervical' and which are 'thoracic'. However, according to a widely used convention thoracic vetebrae are identified as those which carry a complete rib. A complete rib consists of two parts, a dorsal (vertebral) component articulating with a vertebra, and a ventral (sternal) component which articulates, or almost articulates, with the sternum. According to this criterion, in the domestic fowl there are five or six thoracic vertebrae; these comprise four vertebrae which carry ribs articulating with the sternum, and one or two vertebrae which carry ribs articulating with the preceding rib rather than with the sternum.

The total number of vertebrae ranges from 39 to 64, the pygostyle counting as one vertebra. Passerines have fewest vertebrae, while the largest number

---

species); 4, quadratopterygoid; 5, quadratal-quadratojugal; 6, pterygopalatine; and 7, gliding joints between the pterygoid and palatine bones and the braincase. D, The prokinetic upper jaw of the parrot. In the prokinetic jaw the bones characteristically form a rigid triangular block and the bony nasal openings are oval. Bending of the upper jaw in prokinesis occurs at the junction of the upper jaw and braincase where there is a craniofacial elastic zone. Parrots are unusual among prokinetic species in that the craniofacial elastic zone is converted into a synovial joint, the craniofacial ginglymus joint (craniofacial hinge). Basically, prokinesis involves the same movements of the quadrates, pterygoids and palatine bones (straight arrows in **D**) as in rhynchokinesis. In **A, B** and **D** the upper and lower jaws are shown partly enclosed by their horny sheaths (rhamphothecae). Based on Bühler (1981).

occurs in ratites and swans. Five regions of the vertebral column are usually distinguished: cervical vertebrae, thoracic vertebrae, synsacrum (fused thoracic, lumbar, sacral and caudal vertebrae), free caudal vertebrae and the pygostyle (fused caudal vertebrae).

Cervical vertebrae. The number of cervical vertebrae ranges is typically about 14 or 15 with a maximum of 25 in swans. In the domestic fowl there are usually 16 cervical vertebrae. In parakeets there are said to be 11, and a number as low as eight has been claimed in some very small birds but this is believed to be an error (did some of these tiny bones get lost during preparation?). In birds generally the number is much greater and much more variable than in mammals (in which there are regularly seven with only a very few exceptions). The atlas articulates with the skull by a single occipital condyle. The atlas and axis are fused in the adults of some hornbills. In the domestic fowl the last cervical vertebra is fused to the first three thoracic vertebrae (Fig 4–4). The joints between the bodies of the other cervical vertebrae are synovial, with a thin ringlike cartilaginous meniscus. Unlike those of other vertebrates, the articular surfaces of the bodies of the cervical vertebrae are heterocoelous (saddle-shaped), conferring great mobility. The cranial articular surfaces are concave from side to side and convex vertically. These curves are reversed on the caudal articular surfaces. The arrangement of the joint surfaces allows free movement of the cranial and caudal thirds of the neck in the forward direction and free movement of the middle third in the backward direction. This produces the characteristic sigmoid shape of the neck. Most of the cervical vertebrae possess dorsal, ventral and lateral processes for the attachments of ligaments and the complex musculature of the neck. An unusual feeding adaptation of the cervical vertebrae occurs in darters. In these birds the neck is kinked, since the eighth cervical vertebra articulates with the seventh and ninth vertebrae in such a way that the eighth vertebra lies almost at right angles to the rest of the neck. By rapidly straightening the neck the bird is able to dart the head forwards a short distance with the minimum of effort to impale a fish on the bill. On the ventral surfaces of the cervical vertebrae is a groove or canal in which lies the single or paired internal carotid artery (Fig 14–2). In this position near the axis of ventral curvature of the neck, the artery is unaffected by cervical movements.

The large number of cervical vertebrae, the mobile atlanto-occipital articulation, and the saddle-shaped articular surfaces, give a long and very flexible neck which enables the beak to be used for grooming, and nest building as well as feeding, and compensates for the total commitment of the forelimb to flying. While most birds can turn the head upon the neck through 180 degrees and thus look backwards, the Wryneck is said to turn it through 360 degrees.

All the cervical vertebrae, except the atlas, carry recognizable vestiges of ribs (Fig 4–3). Typically, the last few cervical vertebrae carry ribs which are freely movable (but lack a sternal component). In the other cervical vertebrae the rib has become the costal process, pointing caudally like a spike and ankylosed to the transverse process (similar vestigial ribs occur in the synsacral vertebrae). The transverse foramen, where the root of the spike

SKELETOMUSCULAR SYSTEM

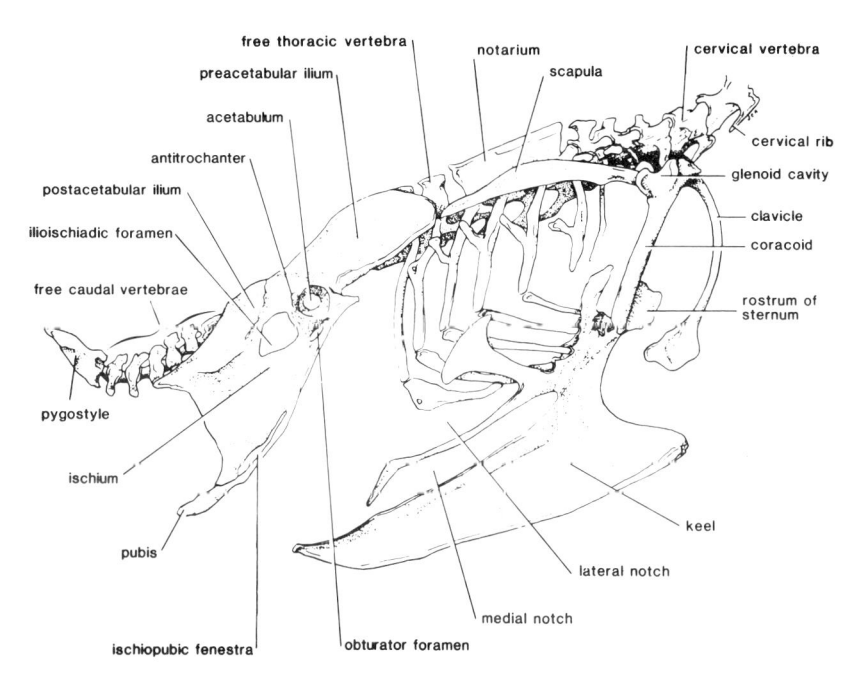

Fig 4–3 Trunk skeleton of the domestic fowl. The pointed structure labelled 'cervical rib' is an example of the vestigial immovable ribs which in this species are carried by all cervical vertebrae except the first and the last two. The two ribs which are directed ventrally at the cranial end of the thoracic cage are movable 'floating' cervical ribs.
From Raikow (1984), with kind permission of the publisher.

attaches to the body of the cervical vertebra, is formed between the vestigial tubercle and head of the rib; as in mammals the vertebral artery runs through this foramen.

Thoracic vertebrae. The first thoracic vertebra is the first vertebra to carry a pair of complete ribs which articulate with the sternum (Fig 4–3); at the caudal end of the series one or more may be absorbed into the synsacrum, but there is no reliable way of distinguishing the last thoracic from the first lumbar vertebra. It is generally stated that there are between three and ten thoracic vertebrae. In many birds the first two to five thoracic vertebrae are fused to make a single bone called the *notarium* (in the domestic fowl the notarium is formed by the last cervical vertebra and the first three thoracic vertebrae). A notarium is characteristically present in galliform (Figs 4–3 and 4–4) and columbiform birds, and in ibises and spoonbills among the ciconiiforms. The fusion of the bones in the domestic fowl starts about four months after hatching. Since little or no movement occurs between these fused thoracic vertebrae they provide a rigid base for the forces acting on the rib cage during flight. Caudal to the notarium are one or more mobile thoracic vertebrae (one in the domestic fowl), the last of these articulating with the synsacrum (Fig 4–3). The joints between the free thoracic vertebrae are heterocoelous (saddle-shaped) except in penguins, auks, gulls and some parrots in which they are opisthocoelous (centrum concave caudally, convex cranially). Judging by the way the mobile fourth vertebra in the domestic fowl seems to be singled out by disease (see below), it appears to be the weak point in the overall design of the vertebral column in this species. The median ventral crest on the more cranial of the thoracic vertebrae receives the attachment of the ventral longus colli muscle and is especially well developed in penguins, loons, and some anseriforms as an adaptation for movement of the neck under water. There are one or more thoracic vertebrae in the fused synsacrum (Fig 4–10). In

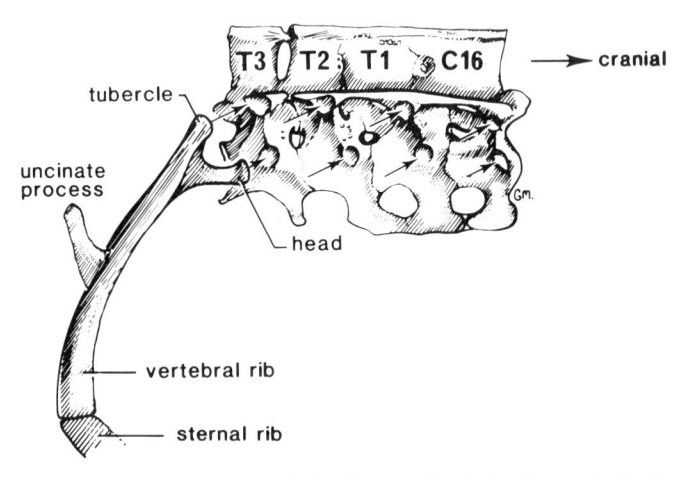

Fig 4–4 Drawing of the notarium and third true rib of the domestic fowl, right lateral view. In this species the notarium is formed by fusion of the last cervical vertebra (C16) and the first three thoracic vertebrae. The most cranial pair of arrows indicates the articulations of the last incomplete rib, thereby identifying the last cervical vertebra (C16). The next pair of articular facets carries the first complete rib, thereby identifying the first thoracic vertebra.

the domestic fowl, for example, the cranial end of the synsacrum bears one and sometimes two ribs which have both dorsal and ventral components (Fig 4–3), suggesting the presence of two thoracic vertebrae.

Synsacrum. This fused series of vertebrae (Fig 4–10) comprises about 10 to 23 thoracic, lumbar, sacral and caudal vertebrae, but the exact boundaries between the vertebrae of these regions cannot be readily identified. In the domestic fowl it contains between 15 and 16 vertebrae. Fusion starts in this species at about seven weeks and spreads caudally. The synsacrum is extensively fused to the ilium. It provides a rigid framework through which the body is supported by the limbs.

Free caudal vertebrae. Behind the synsacrum are five to eight free caudal vertebrae (usually six in the domestic fowl, but five in Figs 4–3 and 4–10). The joints between the vertebral bodies have a synovial cavity and an incomplete articular disc. They are classified as symphyses (fibrocartilaginous joints). The transverse processes of the free caudal vertebrae are well developed and provide attachments for the tail muscles.

Pygostyle. The pygostyle is a single flattened upturned bone (Figs 4–3 and 4–10), consisting of four to ten fused caudal vertebrae (four to six in the domestic fowl) and giving attachment to the innermost retrices. A pygostyle is absent in tinamous and most ratites. Although a similar bone is present in the Ostrich it is not homologous to the pygostyle of carinates. The pygostyle provides support for the tail feathers. Fitting closely against the ventral surface of the caudal vertebrae (including the pygostyle) there is a well defined fibroadipose mass, the *retrical bulb*; all the rectrices (except the central pair, 1a and 1b, in Fig 2–5) are implanted in the rectrical bulb. The central pair are attached to the pygostyle. The pygostyle and free caudal vertebrae are very well developed in woodpeckers and some other species which use the tail as a support in climbing. In *Archaeopteryx* there was no pygostyle, there being a long tail based on 20 caudal vertebrae instead.

Abnormalities of the vertebral column in the domestic fowl. In broilers *spondylolisthesis* (kinky back) is a condition in which the cranial end of the single free thoracic vertebra, T4, becomes displaced ventrally. Most cases are subclinical but the displacement can cause slight to severe damage to the spinal cord, leading to incoordination of the legs or complete paraplegia.

Scoliosis, a lateral deviation of the vertebral column usually in the thoracic region, is not uncommon in meat and laying birds and is probably genetic in origin. It seldom causes locomotor symptoms.

Kyphosis, or 'humpback', of the lumbosacral region occurs in rickets. Kyphosis also occurs through staphylococcal osteomyelitis of the vertebra T4 and its two adjacent vertebrae, possibly secondary to damage caused by shearing forces at this freely movable vertebra; the clinical signs can be similar to those of spondylolisthesis.

Fractures of the thoracic vertebrae, associated with paralysis, occur in osteomalacia ('cage layer fatigue'), a condition of battery-caged hens in which lack of exercise and a shortage of phosphorus appear to be involved.

Ribs

Birds have three to nine pairs of 'true' ribs. Pigeons have fewest ribs and swans have the most. Five or six pairs are present in the domestic fowl (Fig 4–3). Each of the ribs consists of a dorsal vertebral component or vertebral rib articulating with the thoracic vertebrae by a head and a tubercle (Fig 4–4), and a ventral sternal component or sternal rib articulating with the sternum (Fig 4–3). In some species, one or more of the sternal ribs (the last one or two in the domestic fowl) does not reach the sternum but articulates with the rib cranial to it (Fig 4–3). Caudal to the ribs there are frequently a variable number of 'floating' ribs (often consisting only of a short vertebral part) which may articulate with the cranial part of the ilium.

The *vertebral ribs* slope caudoventrally. The head and tubercle of each rib articulate with the vertebral body and transverse process respectively (Fig 4–4). Several of the vertebral ribs carry a large caudodorsal process, the *uncinate process* (Fig 4–4), which gives attachment to some of the trunk muscles and ligaments and presumably strengthens the thoracic wall. The processes are unusually long in powerful divers like guillemots and loons and help the thorax to resist the pressure of water in deep dives. The entire thoracic cage in diving birds is laterally flattened and lengthened, thus reducing the resistance to movement through the water. In birds generally uncinate processes develop independently from the ribs, and in penguins and kiwis they remain unfused with the ribs. Uncinate processes are absent in screamers.

The *sternal ribs* are ossified costal cartilages, homologous to the costal cartilages of mammals. The bony sternal ribs presumably help to resist the forces of dorsoventral compression exerted by the pectoral muscles during flight. They articulate with the vertebral ribs by a cartilaginous joint.

As already stated, the last few cervical vertebrae carry movable vertebral ribs but no sternal ribs. In the domestic fowl there are two such pairs of 'floating' cervical ribs (Fig 4–3), but some species have as many as four pairs. All other cervical vertebrae except the atlas bear vestigial immovable ribs (Fig 4–3).

Sternum

The sternum is much better developed in birds than in other vertebrates and is a fundamental factor in the classification of birds. In the 'ratites' the sternum is flat, i.e. raftlike. In the 'carinates', which includes the great majority of birds, it carries a prominent ventral median keel, the *carina* (Fig 4–3). The keel helps to strengthen the bone and also provides attachments for the powerful muscles of flight (especially for the supracoracoid which produces the upstroke of the wing). Consequently, the size of the keel tends to vary with the power of flight, being particularly well developed in hummingbirds and swifts and reduced in some flightless carinates (e.g. the Kakapo of New Zealand). A distinct keel is also present in penguins which use their wings as flippers for swimming. The sternum in diving birds tends to be much longer than in other species. In birds generally the caudal portion of the sternum is often deeply notched or perforated by large holes (windows) (Fig 4–3). The notches and windows are completed by fibrous membranes and represent areas of the sternum which have failed to ossify. They are generally reduced in strongly flying birds and well developed in poor flyers such as the domestic fowl in which they extend most of the length of the sternum. The cranial part of the sternum in swans and some cranes is excavated to form a cavity in which lie coils of the elongated trachea (see Chapter 7).

Trunk muscles

Epaxial and hypaxial groups of muscles act on the vertebral column. The epaxial vertebral muscles connect together the dorsal parts of the vertebrae; the hypaxial vertebral muscles connect together the ventral parts of the vertebrae. The muscles of the vertebral column are best developed in the neck and rump regions. The cervical muscles (e.g. rectus capitis, longus colli dorsalis and ventralis, and intertransversarii) are characteristically subdivided and overlapping (Fig 4–6), and this enables the neck to perform an enormous range of subtle movements (p. 52). One of the neck muscles, the complexus (Fig 4–6), overlies the base of the skull and the cranial part of the neck and is commonly known as the 'hatching muscle'; however, it is now doubtful if this muscle is directly concerned with 'pipping' of the shell.

The vertebral musculature is much reduced in the thoracic and synsacral regions where the vertebrae are extensively fused, but a complex group of muscles acts on the *tail*. They arise from the bones of the pelvic girdle, or from the synsacrum and the free caudal vertebrae. Without exception their insertions include attachments to the inter-rectrical elastic ligament which is a band of elastic tissue stretching transversely between and uniting the follicles of the rectrices; there are also many insertions on the free caudal vertebrae and the pygostyle. The follicles of all the rectrices except the central pair insert into the *rectrical bulb*. The rectrical bulb is capable of moving as a whole, and is also flexible enough to allow movement of the individual rectrices embedded in it. The rectrical bulb is itself enveloped by a thin coat of striated muscle, the muscle of the rectrical bulb, which arises from the ventral surface of the pygostyle and attaches over the entire surface of the bulb. A great variety of subtle tail movements is therefore possible, ranging from motility of the tail as a whole to movements of individual feathers. These movements play an important part during flight in controlling pitch and direction by moving the tail as an air or water brake, and in providing a variable tail surface for soaring. *Elevation* of the tail is brought about mainly by the levator caudae (Fig 4–6). This muscle is especially large in stifftail ducks such as the Musk Duck in which the tail is directed forward over the back during the sexual display known as 'whistle-kick'. *Depression* is caused by several muscles including the depressor caudae (Fig 4–6) and the pubocaudalis externus and internus. The larger number of muscles involved in lowering the tail indicates the relative importance of this action. *Lateral movements*, *tilting* and *rotation* of the tail are accomplished by the contraction of various combinations of the elevator and depressor muscles. The most extreme rotation occurs during preening when the tail may be turned about 90 degrees.

Spreading of the tail feathers is accomplished by the lateralis caudae muscle and bulbi retricium muscle; the spread tail is returned to its resting position by the recoil of the interrectrical elastic ligament interconnecting the shafts of the feathers, and by the contraction of the numerous small adductor rectricium muscles.

Of the muscles of respiration, the main *inspiratory muscles* are the intercostales externae and the costosternalis, and the main *expiratory muscles* are the intercostales internae and the abdominal muscles (see also Chapter 7, **Respiratory mechanics**).

The four *abdominal muscles* (obliquus externus abdominis, obliquus internus abdominis, transversus abdominis and rectus abdominis) are arranged essentially as in mammals, except that in some species including the domestic fowl, the rectus abdominis is represented only by a membrane.

THORACIC GIRDLE AND WING

Thoracic girdle

The thoracic girdle consists of the scapula, clavicle and coracoid bones (Fig 4–3). The *scapula* (Fig 4–3) is a long bladelike bone that is strongly attached to the ribs by ligaments and reaches caudally to the ilium in some species. It is longest in strong flying birds and very small in flightless birds like the Ostrich. It is extremely broad in penguins which use their wings for swimming. In ratites and frigate birds the scapula and coracoid are fused.

The *clavicles* (Fig. 4–3) generally fuse ventrally to form the *furcula* (wishbone). In many parrots and owls, however, they are united only by cartilage or fibrous tissue. In some parrots and ratites the clavicles are absent or very reduced. When present in ratites they remain separate from each other and are sometimes fused to the scapula or coracoid. The ventral part of the furcula is expanded into a median blade which is attached to the apex of the sternal keel by ligaments. In a number of procellariiform species, the blade forms a synovial articulation with the apex, and in pelicans, frigate birds and the Secretary Bird the two bones are fused. The angle of the furcula is widest in strongly flying birds. One of the functions of the furcula is to act as a transverse spacer bracing the wings apart. It also provides major attachments for the pectoral muscles which produce the downstroke of the wings.

The *coracoid* (Fig 4–3) is massive in most birds. By articulating firmly with the sternum the coracoids act as struts, holding the wings away from the sternum during flight. In this way they help the vertebral ribs and ossified sternal ribs to prevent the thoracic cage from collapsing because of the powerful contractions of the pectoral muscles during the downstroke of the wings. In gliding flight the coracoids, along with the other bones of the thoracic girdle and the ribs, suspend the sternum which, like a platform, bears the weight of the viscera.

The scapula articulates with the coracoid and clavicle. The space where the three bones meet is called the *triosseal canal* and through this canal passes the tendon of the supracoracoid muscle (Fig 4–7). In some species the canal is formed between the scapula and coracoid only and in others, e.g. hornbills and the Hoopoe, it is formed within the coracoid alone. The resulting change of direction of the tendon causes the upstroke of the wing (Fig 4–7). The scapula and coracoid contribute to the glenoid cavity which is shallow permitting great mobility of the wing. The glenoid cavity in birds (like those of bats and pterosaurs) is directed laterally, thus allowing the movements of abduction and adduction which are required for flight.

Skeleton of the wing

The skeleton of the wing characteristically consists of the humerus, radius and ulna, the carpal bones, a carpometacarpus and three digits (Fig 4–5). No strong correlation has been found between the relative proportions of the wing segments and the strength and method of flight. In birds specialized for

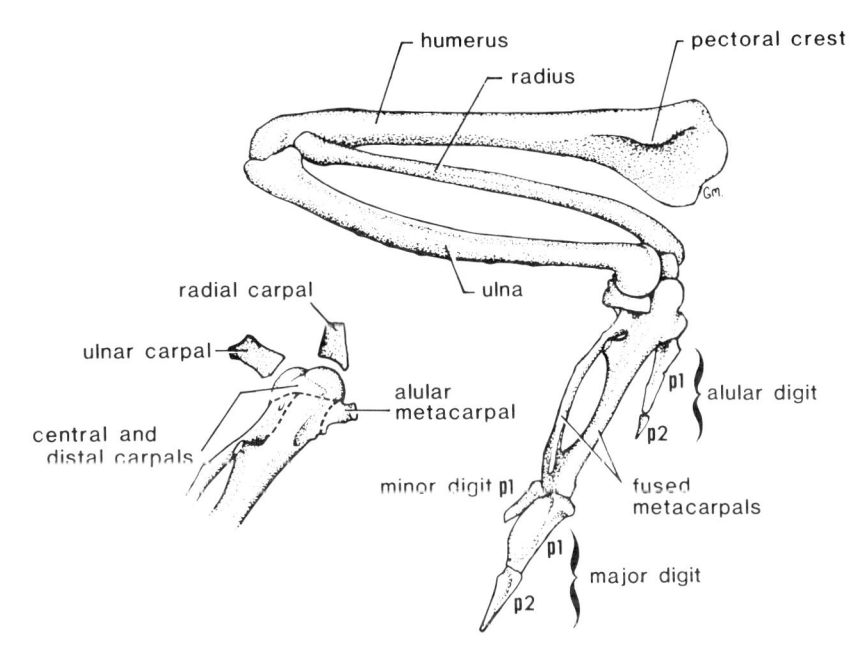

Fig 4–5 Lateral view of the bones of the right wing of the domestic fowl. On the left, the carpal bones and the carpometacarpus have been drawn separated from one another.

soaring, however, the wing is usually very long (p. 18), especially in the antebrachium and manus.

Humerus. When the wing is folded, the humerus lies close against the thoracic cage. In flight it rotates away from the body, the dorsal edge at rest becoming the trailing edge in flight. The humerus in most birds is a relatively short and broad bone (Fig 4–5), but in the albatross, a gliding bird, it is longer than the bones of both the forearm and hand. In hand-propelled divers such as penguins, diving petrels and auks, the humerus and the other bones of the wing are very flattened. In birds generally the proximal end of the humerus is expanded and has a well developed pectoral crest (Figs 4–5 and 4–7) for the insertion of the pectoralis muscle. On the medial surface of the greater tuberosity in many birds there is a large pneumatic foramen, through which the lateral diverticulum of the clavicular air sac enters the medullary cavity of the bone.

Radius and ulna. The ulna is more massive than the radius (Fig 4–5). The two bones are quite widely separated and form a unit which is slightly bowed along their length. These features increase the resistance to bending forces in the plane in which the bones lie. Both of the bones of the forearm are best developed in birds with wings that are specialized for soaring. The roots of the secondary flight feathers are anchored to the ulna by ligaments. In many species the points of attachment on the ulna take the form of raised quill knobs. Sesamoid bones may be present at the elbow joint. Flexion and extension of the elbow joint are restricted to the plane of the surface of the wing. Rotation of the two bones on each other is very limited.

Manus. The skeleton of the manus has undergone considerable simplification and adaptation to flight. In the adult only two free bones, the ulnar and radial carpal bones, remain to represent the proximal row of carpal bones (Fig 4–5). Several central and distal elements arise separately in the embryo; by the second or third week after hatching these have been condensed into two carpal bones, which are then absorbed into a composite carpometacarpus. The ulnar carpal bone is absent in some ratites. The carpometacarpus forms a rigid platform for the attachment of the primary flight feathers. The digits are reduced to only three: the alular digit, the major digit and the minor digit (Fig 4–5). The three metacarpal bones of these digits are fused into the carpometacarpus. Most commonly, there is one phalanx in the alular digit, two phalanges in the major digit and one phalanx in the minor digit. However, in many species including the domestic fowl (Fig 4–5), the alular digit has two phalanges. In kiwis and cassowaries, reduction has proceeded so far that only one digit remains, possibly the primordial third. The alular digit has some power of independent movement and forms the basis of the alula or 'bastard wing', which is raised during flight to prevent stalling at slow speeds like the slotted wing of an aircraft. In the Canada Goose the proximal end of the carpometacarpus in the region of the alula forms a knobby enlargement in sexually mature males which increases the effectiveness of the wing as a weapon. The knob has been used in the field to identify mature males in wintering populations.

To prevent flight, especially in ornamental ducks and large aquatic birds such as pelicans and cranes, *pinioning* is performed by cutting off the manus at the carpal joint. The optimum age for the operation is four to ten days after hatching and is generally restricted to one wing, thus producing an imbalance which makes flight impossible. In pheasant chicks, flight is prevented by brailing, a procedure in which the manus is temporarily held in the flexed position by means of a leather strap.

Muscles of the wing

Movements of the wing bones are traditionally described with the wing in the outstretched position as in gliding flight, the upper surface being dorsal and the lower surface ventral.

(a) Movements of the *humerus* are elevation (as in the upstroke of the wing), depression (as in the downstroke of the wing), protraction, retraction, dorsal rotation (when the leading edge of the wing is rotated dorsally) and ventral rotation (when the leading edge of the wing is rotated ventrally). The main *elevator* of the humerus is the supracoracoideus muscle, which extends upwards from the keel through the triosseal canal to insert on the dorsal side of the humerus (Figs 4–7 and 6–3). The supracoracoideus is well developed in species which are specialized for steep take-offs and for hovering. The main

Fig 4-6 The superficial muscles of the Budgerigar. patag = tensor propatagialis. The unlabelled muscle immediately ventral to the pseudotemporalis and adductor mandibulae is a further subdivision of the adductor mandibulae. A small muscle bundle is shown lying ventral to the caudal half of this subdivision of the adductor mandibulae and attaching to the caudal end of the lower jaw; this is a part of the pterygoid muscle. From Evans (1969), with kind permission of the publisher.

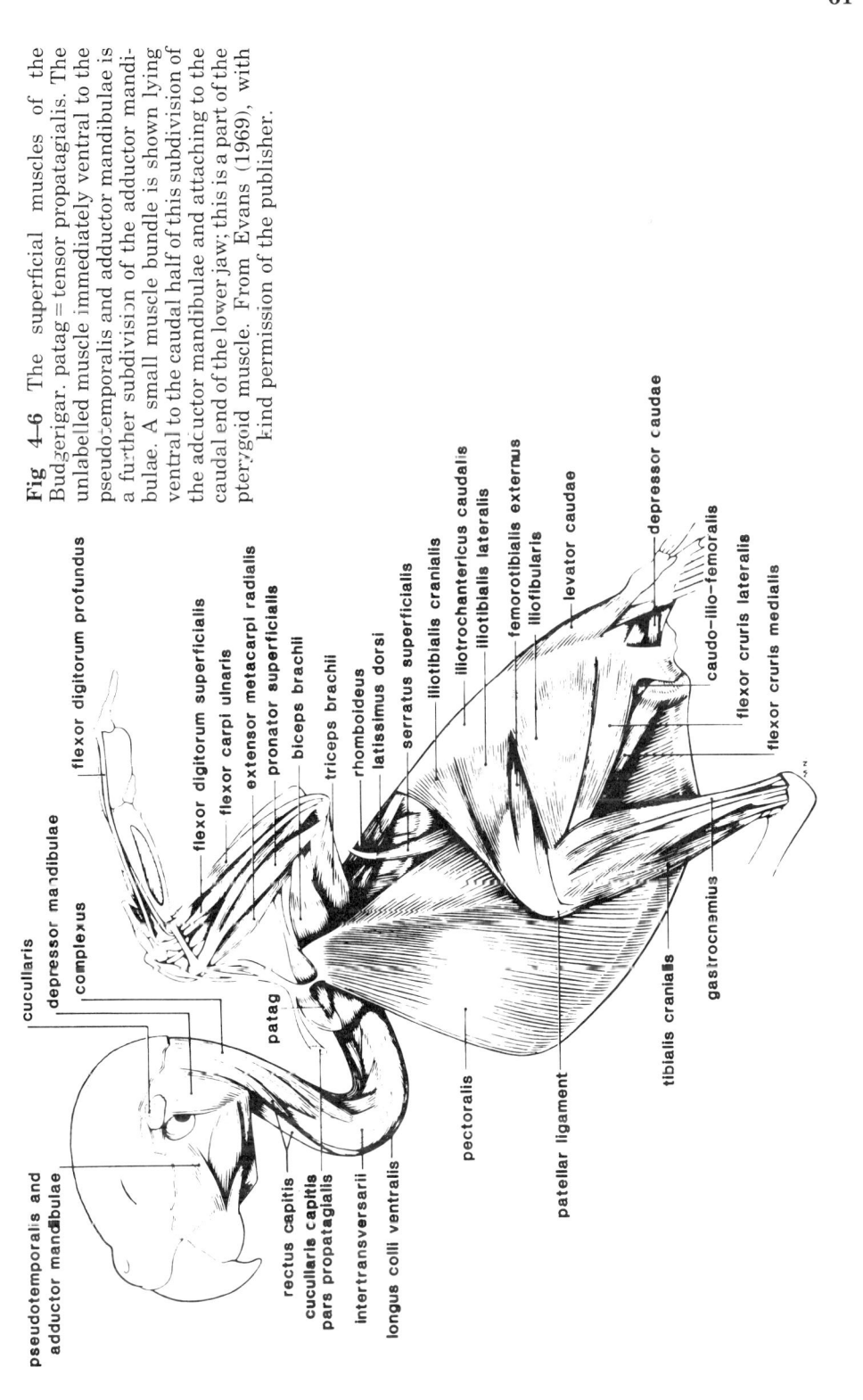

depressor of the humerus is the pectoralis muscle (Figs 4–6 and 6–3) which in most birds extends mainly from the clavicle and the coracoclavicular membrane, and to a lesser extent from the keel of the sternum, to the pectoral crest of the humerus (Fig 4–7). In birds generally, it forms about 15 per cent of the total body weight and in some species with high wing loading for their size it constitutes 21 per cent. In soaring birds, the pectoralis is divided into superficial and deep parts. The deep part (absent in non-soarers) is possibly a fast-acting muscle which can rapidly adjust the position of the wing relative to the wind to keep the wing in a horizontal position while gliding. In some birds the size of the keel appears to be correlated with the size of the

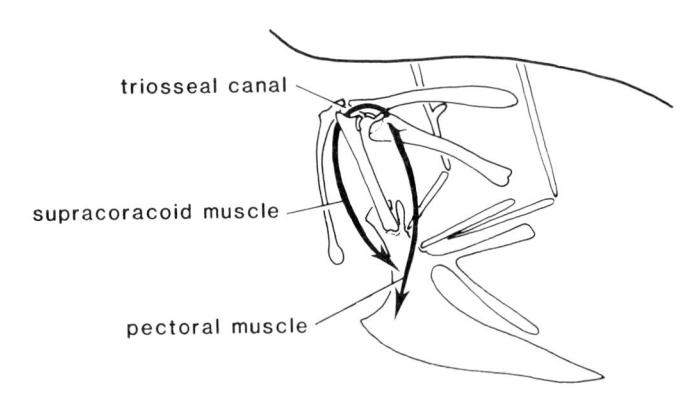

Fig 4–7 The lines of action of the flight muscles. The supracoracoid muscle sends its tendon through the triosseal canal, lifting the humerus. The pectoral muscle inserts directly on the pectoral crest of the humerus, causing the downstroke of the wing.

pectoralis; on the other hand, in a number of groups (e.g. ducks, grebes, loons and tinamous) the pectoral muscle arises from a median septum ventral to the keel (as in some species of bat), so that no correlation exists. Possibly in birds generally, the keel supports the pectoralis muscle to avoid pressure being exerted on the supracoracoideus while the latter is contracting. The relatively large size of the furcula in *Archaeopteryx* compared to that in modern birds suggests that in this species the pectoral muscle was also well developed and that powered flight was possible.

In the wing-propelled diving penguins the thrust for swimming is produced during both elevation and depression of the wing (unlike fast horizontal flapping flight in which the thrust is produced solely during depression) and the supracoracoid muscle in these birds is very large relative to the pectoralis. (In penguins the wing musculature in general is very reduced, many muscles being entirely lost or totally tendinous.) The pectoralis and supracoracoideus are very reduced in the terrestrial flightless birds.

Deep pectoral myopathy (Oregon muscle disease) is a necrotic condition of the supracoracoid muscle in adult turkeys and broiler breeders. It appears to be caused by excessive sudden activity of the muscle and the resulting accumulation of interstitial fluid (which is a characteristic of active skeletal muscle). The supracoracoideus cannot expand, being completely enclosed by a strong fascial coat which binds it to the keel of the sternum; consequently, when the pressure in the muscle rises through the accumulation of fluid its blood vessels are compressed and this results in ischaemic necrosis. The condition is rarely diagnosed in life. In the poultry processing plant it is identified in the whole carcase by inserting a light probe into the body cavity of the bird in a darkened room, the affected muscle casting a shadow.

In birds generally the humerus is weakly *protracted* by the variably developed coracobrachialis cranialis and deltoideus minor. In contrast, *retraction* is brought about by a number of large muscles including the latissimus dorsi, scapulohumeralis, subscapularis, subcoracoideus and deltoideus major. These muscles also have an important postural role by keeping the folded wing fixed against the side of the body. *Dorsal rotation* of the humerus is produced by the supracoracoideus, coracobrachialis and deltoideus. *Ventral rotation* is produced by the scapulohumeralis caudalis and the pectoralis.

(b) Movement at the *elbow joint* is mainly flexion and extension in the plane of the surface of the wing. The elbow joint is *extended* by the triceps brachii (Fig 4–6) which in some species has a sesamoid bone in its tendon. The main *flexors* of the elbow joint are the biceps brachii (Fig 4–6) and the brachialis. When the elbow joint is flexed some rotation of the antebrachium is possible, i.e. *dorsal rotation* by the supinator and *ventral rotation* by the pronator superficialis (Fig 4–6) and pronator profundus.

(c) Movements of the *carpal joint* are restricted to extension and flexion in the plane of the surface of the wing. Extension and flexion of the carpal joint are linked to extension and flexion of the elbow joint. The main *extensor* of the carpal joint, unfolding the distal part of the wing, is the extensor metacarpi radialis (Fig 4–6). The main *flexor* of the carpal joint is the flexor carpi ulnaris (Fig 4–6).

(d) Of the three *digits*, the alular digit is the most movable and is important in preventing stalling during flight. The digit may be moved away from the wing (*extended*) or pressed against the wing (*flexed*) and some *elevation* and *depression* is also possible. The movements are accomplished by four small intrinsic muscles (abductor alulae, flexor alulae, extensor brevis alulae and adductor alulae). The alular muscles are reduced in number in passerines. The major and minor digits move together. When the digits are extended, spreading of the primary flight feathers occurs; flexing of the digits folds the primary feathers. The digits are *extended* by a number of extrinsic muscles, e.g. the extensor digitorum communis and the extensor longus digiti majoris, as well as by several intrinsic muscles (e.g. the abductor digiti majoris). They are *flexed* by two small muscles only, the interosseous ventralis and the flexor digiti minoris. The actions of some 'flexor' muscles (e.g. flexor digitorum superficialis and flexor digitorum profundus) are complex and may prove to be extensor rather than flexor. The number of extensor muscles is greater than the number of flexors because the digits return to a flexed position, when the wing is folded, mainly by elastic recoil. Some of the digital muscles are also involved in a small amount of elevation and depression of the major digit, thus altering the angle of the wing surface (primary feathers).

(e) The *propagatium* is a triangular fold of skin on the leading edge of the brachium and antebrachium which increases the surface area of the wing (Fig 2–7). It is supported internally by a number of muscles, including the tensor propatagialis pars longa (Fig 4–6), the pectoralis pars propatagialis longus and the tensor propatagialis pars brevis. These muscles alter the shape of the propatagium according to the requirements of flight. The smaller fold of skin, the *metapatagium*, on the trailing edge of the brachium is tensed by the serratus superficialis pars metapatagialis.

(f) When the folded wing is spread, the manus (including its major and minor digits) is extended and the flight feathers unfurled. The mechanism producing the complex remigial movements involves a variety of structures (Fig 4–8).

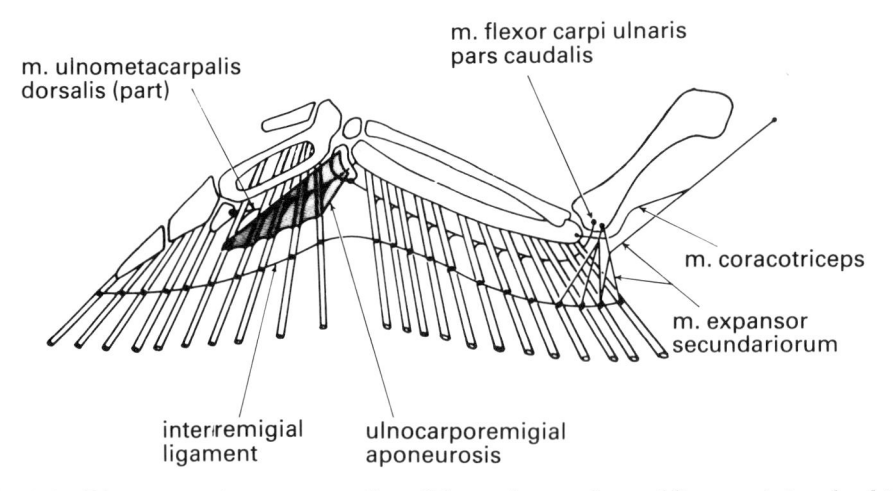

Fig 4–8 Diagrammatic representation of the main muscles and ligaments involved in coordinating the spreading and folding of the remiges in the domestic pigeon. From Raikow (1984), with kind permission of the publisher.

They include (i) an interremigial elastic ligament, which lies in the skin fold on the caudal edge of the forearm (postpatagium) and manus, and unites the follicles of the primary and secondary flight feathers; (ii) a triangular apneurotic sheet, the ulnacarporemigial aponeurosis, on the ventral side of the wing (visible through the skin in a small bird), which inserts on several of the proximal primaries; (iii) two striated muscles, one of which gives off tendon slips to the primaries (ulnometacarpalis dorsalis), and the other to the secondaries (flexor carpi ulnaris pars caudalis); and (iv) a smooth muscle, the expansor secundariorum, which arises from the distal end of the humerus and inserts on the bases of some of the most proximal secondaries, where it is closely associated with the interremigial ligament. When the wing is spread, extension of the digits causes the distal primary to swing outwards since it is firmly fixed to the manus. The remaining primaries, as well as the secondaries, follow mainly as a result of the pull of the elastic interremigial ligament. The elasticity of the ligament ensures that the feathers are not only swung distally but also diverge from each other. Spreading of the primaries is assisted by the ulnocarporemigial aponeurosis and the ulnometacarpalis dorsalis muscle, which resist the lateral movement of the proximal primaries. Spreading of the secondaries is assisted by the flexor carpi ulnaris pars caudalis and expansor secondariorum muscles, which resist the lateral movement of the secondaries. The expansor secondariorum also helps to spread the primaries via the interremigial ligament. Closing of the remiges when the wing is being folded is probably mainly passive, but contraction of the muscles attaching to the bases of the remiges may be involved. When the wing is folded the primary feathers lie parallel to each other, with their vanes overlapping, and below the secondaries (Fig 2–2).

PELVIC GIRDLE AND HINDLIMB

Pelvic girdle

The elongated pelvis consist of the ilium, ischium and pubis, which in the adult are partly fused together. Generally, the ilium is also fused to the synsacrum (Fig 4–10). In nearly all birds the pelvic girdle is incomplete ventrally, there being no articulation of the pelvic bones in the ventral midline (Fig 4–10). Presumably, this favours the passage of large fragile eggs through the pelvic canal. The arched shape of the pelvic bones enables birds to carry the weight of the body, which is usually light, in the bipedal standing position. However, in the Ostrich and rheas, which are heavy flightless birds a ventral symphysis is present and is either pubic or ischial. The pubic symphysis of the Ostrich may help to prevent compression of the viscera when the bird sits on the ground. In birds generally an articular facet, the antitrochanter, lies on the ilium and ischium just dorsal to the acetabulum (Fig 4–9) and articulates with the

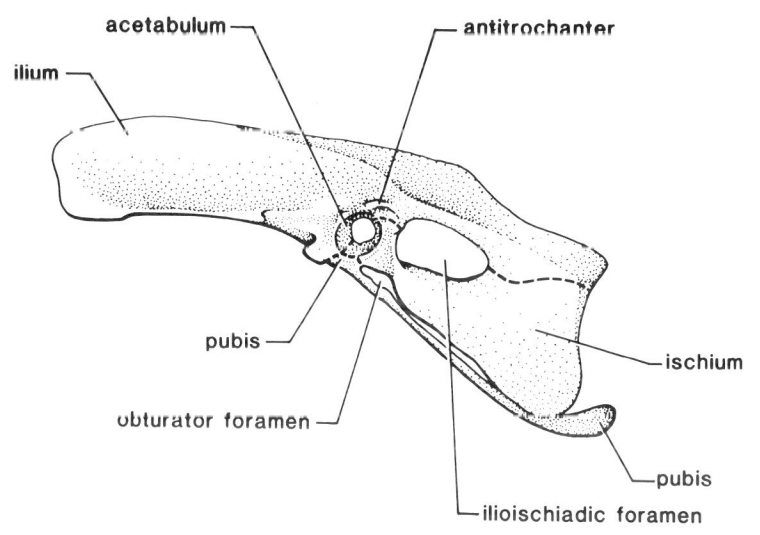

Fig 4–9 Left pelvic bones of the domestic fowl. Broken lines indicate the boundaries between the three pelvic bones.

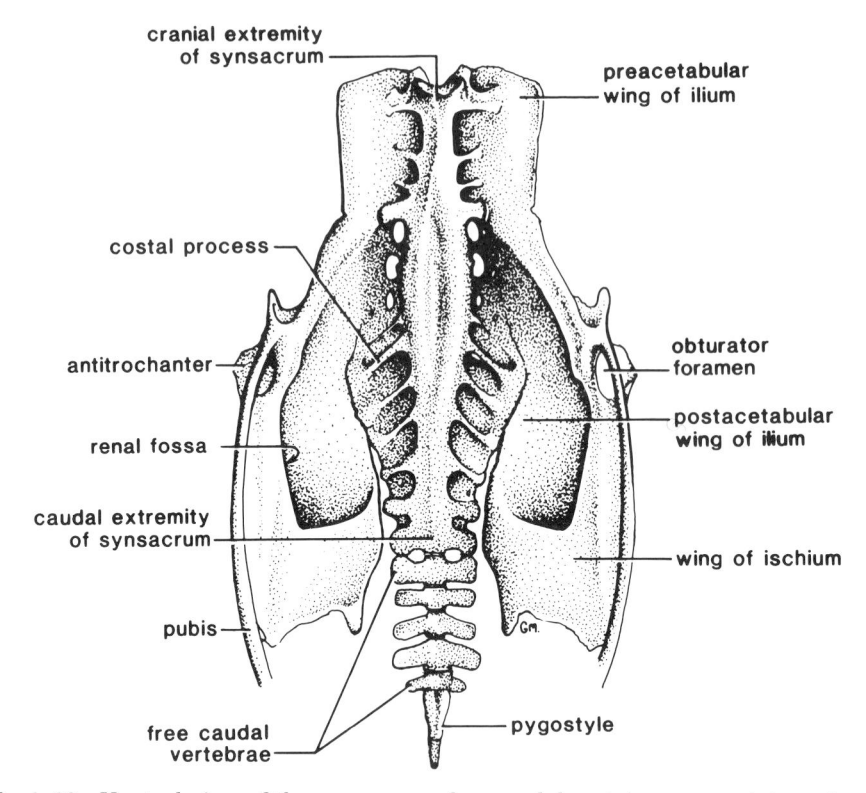

Fig 4–10 Ventral view of the synsacrum, free caudal vertebrae, pygostyle and pelvic bones of the domestic fowl.

antitrochanteric articular facet of the femur (Fig 4–11); this reinforces the weak adductor muscles. The acetabulum is deep and perforated in its centre. The ilioischiadic foramen transmits the ischiadic nerves and vessels and is much larger than the obturator foramen (Fig 4–9). It is open caudally in kiwis. The general shape of the pelvis is closely correlated with the mode of locomotion. In running and climbing birds, for example, the pelvis tends to be relatively wide whilst in the foot-propelled divers, such as loons and grebes, it is much longer and narrower.

Skeleton of the pelvic limb

As in the wing, there is considerable simplification of the skeleton of the pelvic limb, with fusion of some bones and elimination of others especially in the distal part of the limb. Characteristically, the pelvic limb comprises the femur, patella, tibiotarsal bone, fibula, the tarsometatarsus, a single free metatarsal bone and four digits (Fig 4–11).

Femur. In birds generally the femur is a stout and relatively short bone (Fig 4–11). Its distal end slopes cranially when the bird is standing so that the

SKELETOMUSCULAR SYSTEM

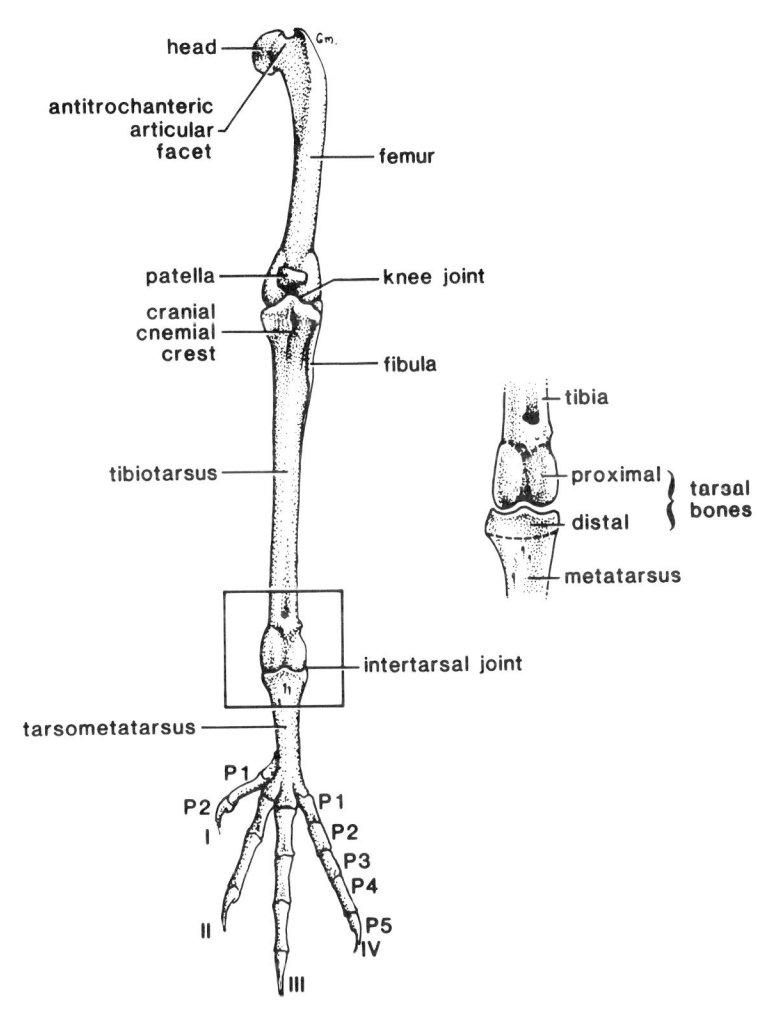

Fig 4–11 The left pelvic limb of a bird. On the right, magnified, are the bones forming the intertarsal joint. I, II, III and IV are the numbers of the digits. P = phalanx.

limbs as a whole are brought well forward on the body close to the centre of gravity of the bird.

Patella. A *patella* (Fig 4–11) is present in most birds and is usually large in aquatic species. It is associated with the tendon of the femorotibialis muscle and articulates with the patellar sulcus of the femur.

Tibiotarsus and fibula. The tibiotarsus is formed by fusion of the tibia to the proximal row of tarsal bones (Fig 4–11). The proximal row of individual tarsal bones can, however, be recognized for a few months after hatching. The hock (or true ankle) joint in birds is, therefore, an intertarsal joint between the tibiotarsus and tarsometatarsus. One of the tarsal bones remains separate in kiwis. The proximal extremity of the tibiotarsus carries the cranial cnemial crest (Fig 4–11) to which are attached muscles of the thigh and leg, including the femorotibialis muscle which extends the knee joint. In foot-propelled diving birds such as grebes and loons the tibiotarsus is elongated and lies almost parallel to the vertebral column, close to the body to which it is bound by skin. In these birds movements of the tibiotarsus are basically backwards and forwards with only a small amount of rotation at the knee. To extend the area for the attachment of the crural muscles the cnemial crest in divers is highly developed and projects cranial to the knee joint. In grebes the crest is fused to the patella, but in loons the tibiotarsus and patella remain separate, the latter being much reduced. The increased area of attachment of the crural muscles in darters and cormorants is provided by the patella. The trochlea on the caudal surface of the distal end of the tibiotarsus articulates with a block of fibrocartilage known as the tibial cartilage; the tendons of the gastrocnemius and superficial flexors pass over this cartilage and the tendons of the deep digital flexor muscles pass through it.

The *fibula* is typically reduced to a slender pointed rod extending about two-thirds of the way down the tibiotarsus to which it is usually fused (Fig 4–11). In penguins and darters, however, the fibula almost reaches the intertarsal joint. The reduction in length of the fibula in most birds limits the ability to rotate the leg. In the genetic abnormality 'creeper', which affects domestic fowl, the fibula is thickened and extends further distally than normal.

Within the intertarsal joint medial and lateral menisci are usually present, although the medial meniscus is absent or poorly developed in some species including the domestic fowl. Both menisci are absent in flamingos. In some species a sesamoid bone is present in the region of the intertarsal joint. Movements at the intertarsal joint are mainly flexion and extension.

Tarsometatarsus. This composite bone (Fig 4–11) is formed by the fusion of the distal row of tarsal bones to the three main metatarsal bones (of digits II, III and IV). In birds possessing four digits there is also a small movable metatarsal bone, distally and caudally. In penguins the metatarsals are incompletely fused. The tarsometatarsus in most birds is shorter than the tibiotarsus, but in long-legged birds like waders the two bones are nearly equal in length; this is essential in a long-legged bird if the centre of gravity of the bird is to remain above the toes when the bird is crouching. The tarsometatarsus in foot-propelled diving birds (e.g. grebes and loons) is laterally flattened to lessen the resistance to water during the recovery stroke. On the plantar aspect of the proximal extremity of the tibiotarsus is a process, the hypotarsus, which in most birds has grooves and high crests and is perforated by canals. Within the grooves and canals lie the digital flexor tendons. The distal part of the tarsometatarsus ends in three pulley-shaped trochleae. In males, and also in many females, a bony spur-core arises from the distal part of the medial surface of the tarsometatarsus.

Digits. In most birds, including the domestic fowl, digits I to IV are present (with two, three, four and five phalanges, respectively). The first digit is usually directed backwards, the other three forwards. For details of the variations in the number and arrangement of the toes see page 19.

Muscles of the pelvic limb

The actions of many of the muscles in the avian pelvic limb have not been conclusively established. The main muscle groups are as follows.

(a) Most of the muscles acting on the *hip joint* are 'one-joint' muscles (spanning only a single joint) extending between the pelvic girdle and femur, but some 'two-joint' muscles (spanning two joints) inserting on the tibiotarsus or fibula are also important. The main *extensors* of the hip joint are the pubo-ischio-femoralis and the iliofemoralis. In addition to retracting the femur during walking these muscles play an important part in posture, and also on landing when they act as shock absorbers opposing the femoral protractors. The main *flexor* of the hip joint is the iliotibialis cranialis (Fig 4–6). *Medial* and *lateral rotation* of the femur is brought about by muscles which extend from the pelvic girdle to the proximal end of the femur. The largest medial rotator is the iliotrochantericus caudalis (Fig 4–6), whilst the ischiofemoralis is the main lateral rotator. The rotator muscles are responsible for rotation of the hindlimb during the recovery phase of walking and for rotating the body to bring it over the supporting leg while the opposite limb is raised during recovery. These movements result in the waddling gait which is especially obvious in birds with short legs, such as ducks. *Abduction* and *adduction* movements of the hip joint are virtually non-existent in birds.

(b) Movements of the *knee joint* are mainly extension and flexion. The main *extensor* is the femorotibialis, within its tendon lies the patella (Fig 4–6). *Flexion* is produced principally by the iliofibularis (Fig 4–6) and the flexor cruris medialis and lateralis (Fig 4–6). *Medial rotation* of the tibiotarsus is brought about by the femorotibialis internus. In addition to these movements the crus may undergo some *lateral rotation, abduction* and *adduction*.

(c) Movements of the *intertarsal joint* are mainly flexion and extension. The main *flexor* is the tibialis cranialis (Fig 4–6) which is assisted by the extensor digitorum longus. The main *extensor* is the gastrocnemius (Fig 4–6), which in birds has three separate bellies and inserts by a single tendon on the hypotarsus. *Medial rotation* of the tarsometatarsus is brought about by the fibularis brevis.

In foot-propelled divers the gastrocnemius is greatly enlarged to produce the power stroke of the foot. The main muscle involved in the recovery stroke is the tibialis cranialis. In grebes and phalaropes, which have a lobate variety of foot, the tibiotarsus is rotated 90 degrees between the power and recovery strokes. Thus in the recovery stroke the medial side of the foot is directed vertically forwards so that resistance to the forward movement of the foot is reduced to a minimum. At the end of the recovery stroke the foot returns to its normal position in order to provide a large surface area for the backward movement of the power stroke. Other birds with lobed toes do not rotate their feet between the power and recovery strokes, and in the recovery stroke the lobes of the feet are folded against the sides of the toes.

(d) The main movements of the *digits* are flexion and extension and there is a certain amount of abduction and adduction. Extrinsic and intrinsic digital muscles are involved. The only extrinsic *extensor* muscle is the extensor digitorum longus which has an insertion on all the main digits. In addition there are a variable number of small intrinsic muscles each of which acts only on a single digit. In passerines the first digit is extended by the intrinsic extensor hallucis longus. It is aided by an elastic ligament which holds the distal interphalangeal joint extended when the flexor of the digit is relaxed. The digits are *flexed* mainly by several extrinsic 'superficial', 'intermediate' and 'deep' digital flexor muscles. The 'superficial' digital flexors (the musculi flexores perforati II, III and IV) arise from the femur and fibula just proximal and distal to the knee joint and their tendons pass distally through the tibial cartilage to end each on a single digit. The 'intermediate' digital flexors (musculi flexores perforantes et perforati digiti II and III) arise from the femur, tibia and fibula and end on the digits. Complex perforating relationships occur between the tendons of the superficial and intermediate digital flexor muscles. The superficial flexors are probably assisted by a large medial thigh muscle, the ambiens, which arises from the ilium or pubis: its tendon of insertion reaches the lateral side of the crus by passing through or laterally below the patellar ligament, and contributes to the superficial flexors of digits II, III and IV. The 'deep' digital flexors include the flexor hallucis longus which acts on digit I, and the flexor digitorum longus which acts on digits II, III and IV. The flexor hallucis longus is especially large in perching passerines. Variations in the arrangement and interconnections of the deep digital flexor tendons and the fibrous band, the vinculum, which unites them, have been used in taxonomy.

Perching in birds is assisted by the digital flexor tendons which cross the caudal surface of the intertarsal joint. When the bird crouches, flexion of the joint passively tenses the tendons and thus automatically clamps the digits around the perch. Further locking is provided by means of minute projections on the under surface of the digital flexor tendons at the metatarsophalangeal joints. These projections are forced, by the weight of the crouching bird, to interdigitate like a ratchet with ridges on the insides of the tendon sheaths. When the bird rises the intertarsal joints are extended; this removes the tension from the flexor tendons and allows the digits to be extended. The ambiens muscle is sometimes known as the 'perching muscle', but is absent in many species including the passeriforms which are often referred to as 'the perching birds'. In species which do have it, section of the muscle does not affect their perching ability.

Locomotion

Most passerine species are 'hopping' birds and have relatively short legs. The majority of non-passerine species, as well as some passerines, e.g. crows and starlings, are 'walking' and 'running' birds. This type of locomotion is best developed in the flightless ratites such as the Ostrich. Walking and running birds tend to have long legs. Some birds, such as larks and wagtails, hop when they are young but later adopt a walking form of locomotion. Pigeons and ducks have relatively short legs and are poor walkers with a waddling type of gait; they are much happier either in the air or in the water. In some swimming birds (e.g. loons) the legs are very far back on the body and therefore far from the centre of gravity (in terrestrial birds, the centre of gravity lies between the feet); consequently, these birds have difficulty in standing for a long time and move forwards by wriggling clumsily on their abdomens. In contrast to other birds, the leg bones in foot-propelled diving birds are tied into the body by skin, almost to the level of the intertarsal joint. Some birds, especially swifts which rarely walk, have extremely short legs which are only adequate for clinging.

Abnormalities of the hindlimb in the domestic fowl and turkey

Chondrodystrophy. The essential defect in this condition is a failure of the cartilaginous growth zone to produce sufficient cells for the normal longitudinal growth of long bones throughout the skeleton. Consequently, the long bones are abnormally short and are often bowed. On the other hand, appositional growth and mineralization are normal. Most severely affected are the rapidly growing bones such as the tibiotarsus and tarsometatarsus. Enlargement of the intertarsal joint occurs. If the condition is severe the gastrocnemius tendon slips from the intercondylar groove of the tibiotarsus, resulting in 'slipped tendon'. 'Perosis', 'enlarged hock disorder', and 'turkey syndrome '65' are all essentially similar chondrodystrophies, the most conspicuous changes being bony enlargements at the intertarsal joint. The cause may be one of various primary nutritional deficiencies in young birds (e.g. of zinc, nicotinic acid, etc.). Alternatively it may be a secondary nutritional deficiency, for example from congenital infection with *Mycoplasma meleagridis* (as in 'turkey syndrome '65') which apparently blocks the nutrition of the growth zones of long bones.

Rickets. Rickets may arise in young birds (from deficient vitamin D_3, or an imbalance of calcium or available phosphorus) and is an osteodystrophy in which there is impaired mineralization of osteoid tissue and epiphyseal cartilage. When bones are deformed it is the proximal end of the tibiotarsus which is most commonly affected. The growth zones are not always obviously widened. At first, the long bones are softer and more pliable, and later the rate of growth is reduced and deformities arise.

Tibial dyschondroplasia. Tibial dyschondroplasia is a condition in which the zone of hypertrophied cartilage cells in the proximal end of the tibiotarsal bone fails to become calcified because its blood supply is absent. The proximal end of the tibiotarsus may also be affected. Gross deformity or fracture may result. In poultry this abnormality occurs only in table breeds, but in ducks it is common.

It can occur at any age until the growth zone has become ossified. The cause is unknown.

Twisted leg deformity. In this condition, which is relatively common in table birds, the distal part of the tibiotarsus is twisted laterally and there is bending of the proximal part of the tarsometatarsus. Chondrodystrophic changes are absent. The cause is unknown.

Rotated tibia. Rotated tibia most frequently affects turkeys of 2 to 14 weeks of age and is characterized by rotation of the tibiotarsus through approximately 40°, so that the tarsometatarsus is directed laterally. The aetiology is not known.

Cage layer osteoporosis. This condition is seen in heavily-laying battery birds and results from excessive depletion of minerals producing weakening and deformity of the bones. Spontaneous fractures of the bones of the pelvic limb may occur.

Degeneration of the femoral head. Degeneration of the femoral head occurs in turkeys aged 10 weeks or more. The lesions are on the medial side of the head several centimetres below the growth plate, and consist of necrosis and replacement of cortical bone by dense fibrous tissue. Blood-stained fluid is present in the hip joint and the joint capsule is thickened. Degeneration of the femur also occurs in broilers aged 10 weeks or more and may be due to molybdenum deficiency.

Enlargement of the hock joint. Various nutritional deficiencies in turkey poults (e.g. of calcium, vitamin E, choline and nicotinic acid) cause swelling of the joint.

Ossification of tendons. In the domestic fowl ossification of tendons occurs on the dorsal and plantar tarsometatarsus, and variably on the caudal tibiotarsus, dorsal carpometacarpus and caudal antebrachium. The mineralization begins about 90 days after hatching and therefore after the age at which broilers are killed in the processing plant, but within the age range for layers. Ossification of the tendons also occurs in turkeys and is likely to be a much more serious commercial problem.

Pneumatic bones

In birds generally, including the domestic fowl, most of the vertebral, pelvic, sternal and costal bones are invaded by diverticula of the air sacs which replace the bone marrow. The limb bones vary greatly in their degree of pneumaticity in different species (see p. 135), only the humerus being pneumatic in the domestic fowl. The pneumatic spaces within the bones of the skull are derived from either the nasal cavity or the tympanic cavity (p. 46).

Ossification and growth of cartilage bones

As in mammals most bones are preformed in cartilage and the cartilaginous model is subsequently replaced completely by the process of ossification. The first bone to appear is laid down by the perichondrium, around the middle of the cartilaginous shaft. The cartilaginous core of the shaft is then removed, leaving the primary marrow cavity. Blood vessels and connective tissue invade the cartilage at the ends of the bone, occupying cartilage canals, many of which turn into the shaft and eventually join up with the marrow cavity. The cartilage cells in the walls of these canals hypertrophy and are removed. Endochondral bone replaces the cartilaginous walls of the canals. Ultimately, the cartilage is entirely converted into bone, except for a thin layer under the fibrocartilage at the articular surface. Usually secondary centres of ossification in bony epiphyses are absent (an exception is the proximal end of the tibiotarsus)

The bone grows in length through proliferation of cartilage cells in a wide growth zone. This zone seems less clearly marked off than the cartilaginous epiphyseal disk of mammals; also it is penetrated by blood vessels, whereas the mammalian disk is avascular. Growth in diameter is achieved as in mammals by appositional growth from the periosteum.

Medullary bone

The cortical bone of the long bones of birds of both sexes is similar to that of mammals. Medullary bone is labile bone which occurs only in female birds during the reproductive phase.

Structure. Medullary bone consists of interconnected spicules of bone resembling ordinary embryonic bone. At the approach of laying (about two weeks before its onset in the domestic fowl) the spicules grow out from the endosteal surface of the cortical bone, the spaces between them being occupied by blood sinuses and red marrow. In the domestic fowl these spicules grow throughout the laying period, eventually penetrating about 1 mm into the marrow cavity but seldom entirely filling it. The spicules contain osteocytes and are covered with variable numbers of osteoblasts and osteoclasts. There are no Haversian systems. Neither the collagen fibres nor the hydroxyapatite crystals are orientated, showing that the mechanical functions of medullary bone are unimportant. There is less collagen than in cortical bone. However, there is more chondroitin sulphate in the organic interfibrillar substance of medullary bone than of cortical bone (or alternatively, the chondroitin sulphate may be in a different state of polymerization). These differences in the interfibrillar substances are reflected in differences in staining reactions of decalcified sections (e.g. medullary bone stains more intensely with periodic acid-Schiff than cortical bone).

Factors controlling formation of medullary bone. The formation of medullary bone is induced by the combination of oestrogens and androgens.

This combination also induces a great increase in the retention of calcium and phosphorus from the alimentary tract, thus providing the necessary minerals for the calcification of the medullary bone. In the period immediately before laying the total weight of the skeleton increases by about 20 per cent.

Mobilization of medullary bone. Phases of formation and destruction of medullary bone alternate during the egg cycle. During the period of calcification of the egg shell much of the medullary bone is reabsorbed, the bony trabeculae becoming shorter, narrower and much less continuous. However, there seems to be great variation from bird to bird in the amount of medullary bone which remains as the egg passes through each part of the oviduct. Histologically, groups of numerous osteoblasts and osteoclasts seem to operate more or less simultaneously and side by side throughout the egg cycle, and the net movement of calcium into and out of the bone must depend on the relative activity of these bone-forming and bone-destroying cells. The extent to which the calcium needed for the shell comes from the skeleton rather than the diet seems uncertain. Apparently, pullets are normally in negative calcium balance for the first few weeks of lay, but otherwise the calcium-rich diet used in the poultry industry could well be a fully sufficient source of calcium for the shell in normal birds. However, the bird certainly draws on its medullary bone whenever absorption of calcium from the gut is inadequate. Perhaps medullary bone should be regarded as a reserve of mineral enabling the bird to smooth out large or small fluctuations in the calcium available from the diet.

Factors controlling mobilization of medullary bone. The reabsorption of medullary bone may be induced by a reduction in circulating oestrogens. Another possibility is that a fall in the diffusable calcium level of the plasma induces secretion of parathyroid hormone. The exact mechanism is not yet known.

Red and white muscles

The existence of red and white muscles in birds will be known to almost everyone, with the possible exception of vegetarians. There are two types of muscle fibre, red and white. The red muscle fibres contain relatively large amounts of myoglobin and it is this that causes the red colour. Compared with the white fibres, red fibres also tend to have other characteristics, including relatively more mitochondria, a higher content of lipid globules and greater vascularity. Red fibres use fat rather than glycogen as a source of energy, and this makes them more efficient than white fibres since weight-for-weight, fat releases more energy than carbohydrate. Because of all these characteristics red muscle fibres are adapted for sustained effort. It is likely, however, that the classical red and white fibres are the two extremes of a continuum of fibre types.

Most avian muscles contain a mixture of red and white fibres, the proportion depending on how prolonged the activity of the muscle must be. In the pectoral muscles of strong fliers like the pigeon, red fibres predominate, and they may even be the only type of fibre as in the pectoral muscle of the hummingbird which is perhaps the most metabolically active skeletal muscle among all the vertebrates and the ultimate in efficiency. The muscles of diving birds are deep red in colour indicating the presence of large amounts of myoglobin for storing oxygen, presumably for use during diving.

FLIGHT

Four basic types of wing can be identified: elliptical, broad soaring, high speed and long soaring. For a description of their characteristics see Chapter 2.

A bird's wing is a streamlined structure that is beautifully designed to perform the functions of both the wing and propeller of an aircraft. The upper

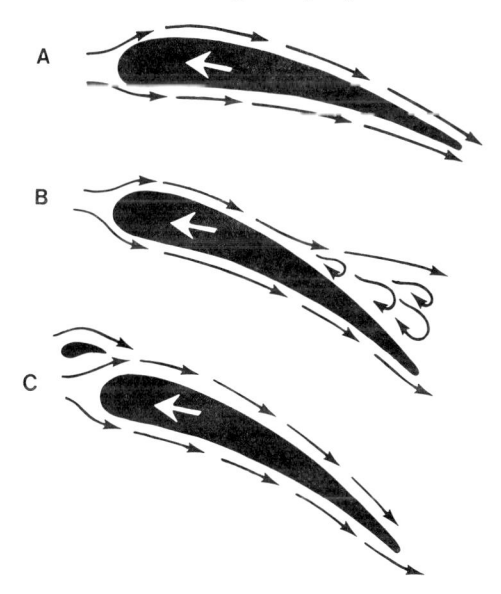

Fig 4–12 Wing showing deflection of air above and below the leading edge. In normal horizontal flight **A**, a zone of low pressure is created above the upper surface of the wing resulting in lift. During stalling **B**, when the angle of attack is too great, eddying occurs above the upper surface. In **C** the slot formed by the alula smooths the air flow over the upper surface and reduces eddying. From Spearman and Hardy (1984), with kind permission of the publisher.

surface of the wing is convex in cross-section, the lower surface is concave. The wing is thickest at its leading edge where the bones and muscles provide strength and rigidity. Caudally, it tapers to the flexible trailing edge formed by the flight feathers projecting backwards. As the aerofoil moves forwards two forces are generated, a lift force acting upwards and a drag force opposing the forward motion. The lift and drag forces are proportional to the square of the speed. Sustained flight is possible if there is sufficient speed to generate a lift force equal to the weight of the body. During flight air flowing past the leading edge travels further and therefore faster over the upper convex surface of the wing than the air passing over the lower concave surface (Fig 4–12A). This creates a fall in pressure above the wing, while the pressure on the under surface remains almost constant. The difference in the pressure above and below the wing results in lift. If the leading edge of the wing is tilted upwards slightly, i.e. increasing the angle of attack, the negative pressure above the wing is increased thus creating more lift. When the angle of attack is too great, the air stream separates from the upper surface, turbulence develops, and the wing stalls (Fig 4–12B). A serious source of drag on the wing results from air on the under surface flowing over the trailing edge into the area of low pressure on the upper surface, thus creating turbulence. This phenomenon is greatest at the wing tip and is known as the tip vortex. Drag due to tip vortex is lessened if the length of the wing is increased so that as large as possible an area of air between the wing tips is undisturbed.

In order to delay stalling, which may occur when the speed falls too low or the angle of attack is raised above 20°, the wings of many birds have slots or spaces that are formed by separation of the tips of the primary feathers. The size of the slots is usually further increased by the presence of notches on the feather vanes. These slots let through a part of the air and maintain the smooth stream which is necessary on the upper surface of the wing for lift. An additional slot is provided by the alula when it is drawn forward away from the rest of the manus (Fig 4–12C). The alula and wing slots are prominent in the elliptical and broad-soaring types of wing. Wing slots are generally absent in the high-speed wing and the long-soaring wing, although in the latter the alula is often large.

In *flapping flight* the wings undergo a screwlike motion, downwards and forwards and then backwards and upwards. Propulsion in fast flapping flight results solely from the downstroke of the wing. The upstroke is entirely a passive action, the wing being lifted by the pressure of the air and not by muscle contraction. Much of the thrust is usually provided by the flexible wing tips, the forces on them causing them to bend upwards and twist at an angle to the wing as a whole. At the end of the downstroke the tips of the wings are level with the bill. However, in birds with the relatively inflexible elliptical type of wing, the twisting effect necessary for propulsion is provided by air passing through slots between the primaries, causing the individual feathers to rotate and each to act like a twisted wing. The medial part of the wing in fast, level flapping flight provides much of the lift in both the upstroke and downstroke. On the upstroke the primaries separate to allow easy passage of air. The ratio of the mass of the pectoralis to that of the supracoracoideus in fast flyers such as terns, is about 10:1. In the *slowly ascending and descending flight* of small woodland birds lift and propulsion are generated in a similar manner to that in fast, level flapping flight. However, in the *steeply ascending and descending flight* of strong flyers like pigeons, ducks, hawks and pheasants, propulsion is usually produced during the upstroke in which rapid rotation at the shoulder joint and extension of the elbow joint produces a violent backward flick of the primary feathers resulting in forward motion. The downstroke in these birds is an active process involving muscular activity and provides much lift, but little propulsion. In birds with steep take-offs the supracoracoideus is relatively well developed, the ratio of the mass of the pectoralis to that of the supracoracoideus being about 6:1.

As well as assisting the wings in steering and braking, the *tail* when depressed and fanned out at low speeds has an important role in providing additional wing area. This results in air being sucked down over the central region of the wing so that lift is increased. The additional lift is particularly valuable during landing and take-off when controlled flight at extremely low speeds is required.

In *hovering flight*, as in hummingbirds, the body is held almost vertical in position and the wings beat actively backwards and forwards up to 200 times a second producing lift in both strokes. On the forward stroke the wing moves with the leading edge slightly angled providing lift but no thrust. On the backward stroke the wing rotates almost 180° at the shoulder joint so that the leading edge is now turned backward, again providing lift but no thrust. In hummingbirds the ratio of the mass of the pectoralis to that of the supracoracoideus is only 2:1.

Further reading

Aulie, A. (1983) The fore-limb muscular system and flight. In *Physiology and Behaviour of the Pigeon* (Ed.) Abs, M. New York and London: Academic Press.

Baumel, J.J. (1979) Osteologia. In *Nomina Anatomica Avium* (Ed.) Baumel, J.J., King, A.S., Lucas, A.M., Breazile, J.E. & Evans, H.E. London and New York: Academic Press.

Baumel, J. J. (1979) Arthrologia. In *Nomina Anatomica Avium* (Ed.) Baumel, J.J., King, A.S., Lucas, A.M., Breazile, J.E. & Evans, H.E. London and New York: Academic Press.

Bellairs, A. d'A. & Jenkin, C.R. (1960) The skeleton of birds. In *Biology and Comparative Physiology of Birds* (Ed.) Marshall, A.J. Vol. 1. New York and London: Academic Press.

Berger, A.J. (1952) The comparative functional morphology of the pelvic appendage in three genera of Cuculidae. *Am. Midl. Nat.*, 47, 513–605.

Berger, A.J. (1960) The musculature. In *Biology and Comparative Physiology of Birds* (Ed.) Marshall, A.J. Vol. I. New York and London: Academic Press.

Bock, W.J. (1964) Kinetics of the avian skull. *J. Morph.*, 114, 1–42.

Bock, W.J. (1974) The avian skeletomuscular system. In *Avian Biology* (Ed.) Farner, D.S. & King, J.R. Vol. IV. New York and London: Academic Press.

Bühler, P. (1970) Schädelmorphologie und Kiefermechanik der Caprimulgidae (Aves) *Z. Morph. Okol. Tiere*, 66, 337–399.

Bühler, P. (1981) Functional anatomy of the avian jaw apparatus. In *Form and Function in Birds* (Ed.) King, A.S. & McLelland, J. Vol. 2. London and New York: Academic Press.

Cracraft, J. (1979) The functional morphology of the hindlimb of the domestic pigeon, *Columba livia. Bull. Amer. Mus. Nat. Hist.*, 144, 171–268.

Evans, H.E. (1969) Anatomy of the budgerigar. In *Diseases of Cage and Aviary Birds* (Ed.) Petrak, M.L. Philadelphia: Lea and Febiger.

Feduccia, A. (1975) Aves osteology. In *Sisson and Grossman's The Anatomy of the Domestic Animals* (Ed.) Getty, R. Vol. 2, 5th Edn. Philadelphia: Saunders.

Fisher, H.I. (1946) Adaptations and comparative anatomy of the locomotor apparatus of New World vultures. *Am. Midl. Nat.*, 35, 545–727.

Fisher, H.I. & Goodman, D.C. (1955) The myology of the Whooping Crane, *Grus americana. Ill. Biol. Monogr.*, 24, 1–127.

Fujioka, T. (1959) On the origins and insertions of the muscles of the thoracic limb in the fowl. *Jap. J. Vet. Sci.*, 21, 85–95.

Fujioka, T. (1962) On the origins and insertions of the muscles of the head and neck in the fowl. Part 1. Muscles of the head. *Jap. J. Vet. Sci.*, 25, 207–226.

George, J.C. & Berger, A.J. (1966) *Avian Myology*. New York and London: Academic Press.
Hartman, F.A. (1961) Locomotor mechanisms of birds. *Smithson. misc. Collns*, 143, 1–49.
Harvey, E.B., Kaiser, H.E. & Rosenberg, L.E. (1968) *An Atlas of the Domestic Turkey (Meleagris gallopayo). Myology and Osteology*. US Atomic Energy Commission, Division of Biological Medicine, US Government Printing Office, Washington D.C.
Hudson, G.E. & Lanzillotti, P.J. (1964) Muscles of the pectoral limb in galliform birds. *Am. Midl. Nat.*, 71, 1–113.
Hudson, G.E., Lanzillotti, P.J. & Edwards, G.D. (1959) Muscles of the pelvic limb in galliform birds. *Am. Midl. Nat.*, 61, 1–67.
Jollie, M.T. (1957) The head skeleton of the chicken and remarks on the anatomy of this region in other birds. *J. Morph.*, 100, 389–436.
King, A.S. & King, D.Z. (1979) Avian morphology: general principles. In *Form and Function in Birds* (Ed.) King, A.S. & McLelland, J. Vol. 1. London and New York: Academic Press.
Klemm, R.D. (1969) Comparative myology of the hind limb of procellariiform birds. *Sth. Ill. Univ. Monogr., Sci. Ser.*, 2, 1–269.
Pennycuick, C.J. (1972) *Animal Flight*. London: Arnold.
Pennycuick, C.J. (1975) Mechanics of flight. In *Avian Biology* (Ed.) Farner, D.S. & King, J.R. Vol. V. New York and London: Academic Press.
Raikow, R. (1970) Evolution of diving adaptations in the stifftail ducks. *Univ. Calif. Publs Zool.*, 94, 1–52.
Raikow, R. (1984) Locomotor system. In *Form and Function in Birds* (Ed.) King, A.S. & McLelland, J. Vol. 3. London and New York: Academic Press.
Riddell, C. (1981) Skeletal deformities in poultry. In *Advances in Veterinary Science and Comparative Medicine* (Ed.) Cornelius, C.E. & Simpson, C.F. Vol. 25. New York and London: Academic Press.
Spearman, R.I.C. & Hardy, J.A. (1984) Integument. In *Form and Function in Birds* (Ed.) King, A.S. & McLelland, J. Vol. 3. London and New York: Academic Press.
Taylor, T.G., Simkiss, K. & Stringer, D.A. (1971) The skeleton: its structure and metabolism. In *Physiology and Biochemistry of the Domestic Fowl* (Ed.) Bell, D.J. & Freeman, B.M. Vol. 2. London and New York: Academic Press.
Van Tyne, J. & Berger, A.J. (1976) *Fundamentals of Ornithology*. 2nd Edn. New York: Wiley.
Vanden Berge, J.C. (1970) A comparative study of the appendicular musculature of the order Ciconiiformes. *Am. Midl. Nat.*, 84, 289–364.
Vanden Berge, J.C. (1975) Aves myology. In *Sisson and Grossman's The Anatomy of the Domestic Animals* (Ed.) Getty, R. Vol. 2, 5th Edn. Philadelphia: Saunders.
Vanden Berge, J.C. (1979) Myologia. In *Nomina Anatomica Avium* (Ed.) Baumel, J.J., King, A.S., Lucas, A.M., Breazile, J.E. & Evans, H.E. London and New York: Academic Press.
Weymouth, R.D., Lasiewski, R.C. & Berger, A.J. (1964) The tongue apparatus in hummingbirds. *Acta anat.*, 58, 252–270.
Young, J.Z. (1981) *The Life of Vertebrates*. 3rd Edn. Oxford: Clarendon Press.
Zusi, R.L. (1967) The role of the depressor mandibulae muscle in kinesis of the avian skull. *Proc. U.S. natn. Mus.*, 123, 1–28.
Zusi, R.L. & Storer, R.W. (1969) Osteology and myology of the head and neck of the Pied-billed Grebes (*Pobilymbus*). *Misc. Publs Mus. Zool. Univ. Mich.*, 139, 1–49.
Zweers, G.A. (1974) Structure, movement and myography of the feeding apparatus of the Mallard (*Anas platyrbynchos* L.). *Neth. J. Zool.*, 24, 323–457.
Zweers, G.A. (1982) The feeding system of the pigeon. *Adv. Anat. Embryol. cell Biol.*, 73, 1–108.

Chapter 5

COELOMIC CAVITIES

The coelomic cavities are probably similar in birds generally, but the following account is based on the domestic fowl since this species has been relatively extensively investigated. Sixteen distinct and separate cavities are enclosed within the body wall of the adult. Eight are cavities of the air sacs (see p. 131) and the remaining eight are the cavities of the coelom proper, as listed below. Of the latter the first five are peritoneal cavities formed by peritoneal partitions and are not represented in mammals; the remaining three are pleural and pericardial, and are formed essentially as in mammals.

1. Left ventral hepatic peritoneal cavity
2. Right ventral hepatic peritoneal cavity
3. Left dorsal hepatic peritoneal cavity
4. Right dorsal hepatic peritoneal cavity
5. Intestinal peritoneal cavity
6. Left pleural cavity
7. Right pleural cavity
8. Pericardial cavity

The peritoneal partitions

The five peritoneal cavities within the avian coelom (1 to 5 above) are formed by five sheets of peritoneum which form partitions within the peritoneal coelom (see below). One of these sheets (the combined dorsal and ventral mesentery) is present in mammals in a simpler form; the other four (the left and right components of the posthepatic septum, and left and right hepatic ligaments) do not occur in mammals.

The combined dorsal and ventral mesentery. Together these two mesenteries form a continuous vertical sheet from the dorsal to the ventral body wall, as far caudally as the gizzard (Figs 5–1 and 5–2). Caudal to this level there is only a dorsal mesentery. (In mammals, the ventral mesentery is much less extensive, being confined to the extreme cranial and caudal regions of the abdomen, near the liver and bladder.)

The left and right sheets of the posthepatic septum. These two double-layered peritoneal sheets, one on the left side and one on the right, extend craniocaudally from near the last two thoracic vertebrae to the caudal wall of the peritoneal cavity (Figs 5–2 and 6–8). Cranially and ventrally, the two sheets unite in the midline. The gizzard is situated between the two layers of the left sheet (Figs 5–2 and 6–8). Ventrally, the dorsal mesentery joins the inner (medial) layer of the left sheet where the latter passes over the gizzard (Fig 5–2). The outer (lateral) layer of the left sheet and that of the right sheet

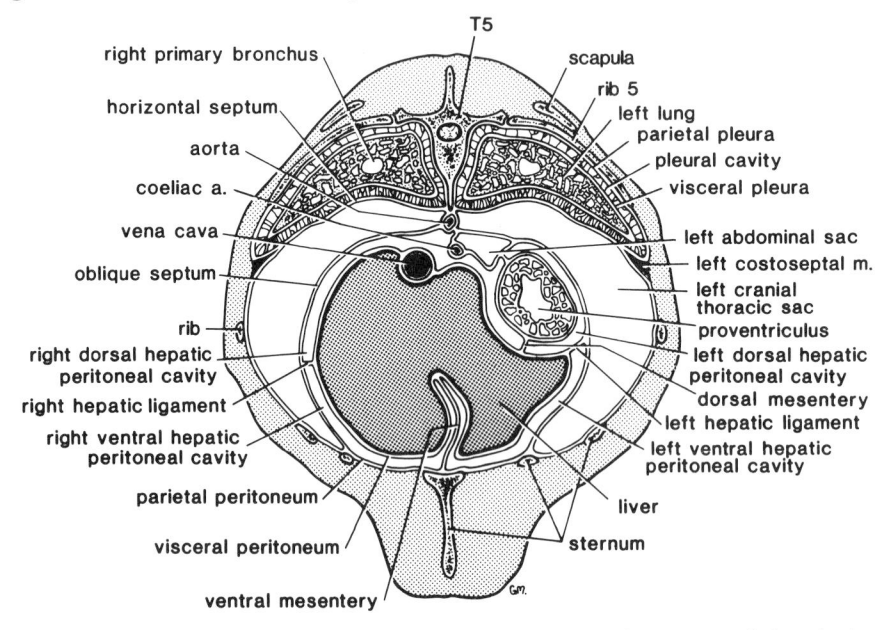

Fig 5–1 Transverse section through the fifth thoracic vertebra just caudal to the heart in a young domestic fowl, cranial view. Redrawn from Campana (1875).

face laterally into the left and right ventral hepatic peritoneal cavities, respectively (Fig 5–2). A large quantity of fat is stored between the two peritoneal layers of the posthepatic septum in the normal adult domestic fowl, constituting the principal peritoneal fat depot.

The posthepatic septum divides the peritoneal cavity into three principal cavities: (a) a midline cavity, called the intestinal peritoneal cavity; and (b) two lateral cavities which enclose the liver. Each lateral cavity is further subdivided by the left and right hepatic ligaments into the left and right dorsal and left and right ventral hepatic peritoneal cavities (see below).

The left and right hepatic ligaments. Each lies entirely cranial to the posthepatic septum, parallel with the ground (Fig 5–1). The left ligament is on the left side of the dorsal and ventral mesentery; the right ligament is on the right side of the dorsal and ventral mesentery. Each is continuous with other peritoneal sheets (laterally with the parietal peritoneum of the oblique septum as described in Chapter 7, in the midline with the dorsal and ventral

COELOMIC CAVITIES 81

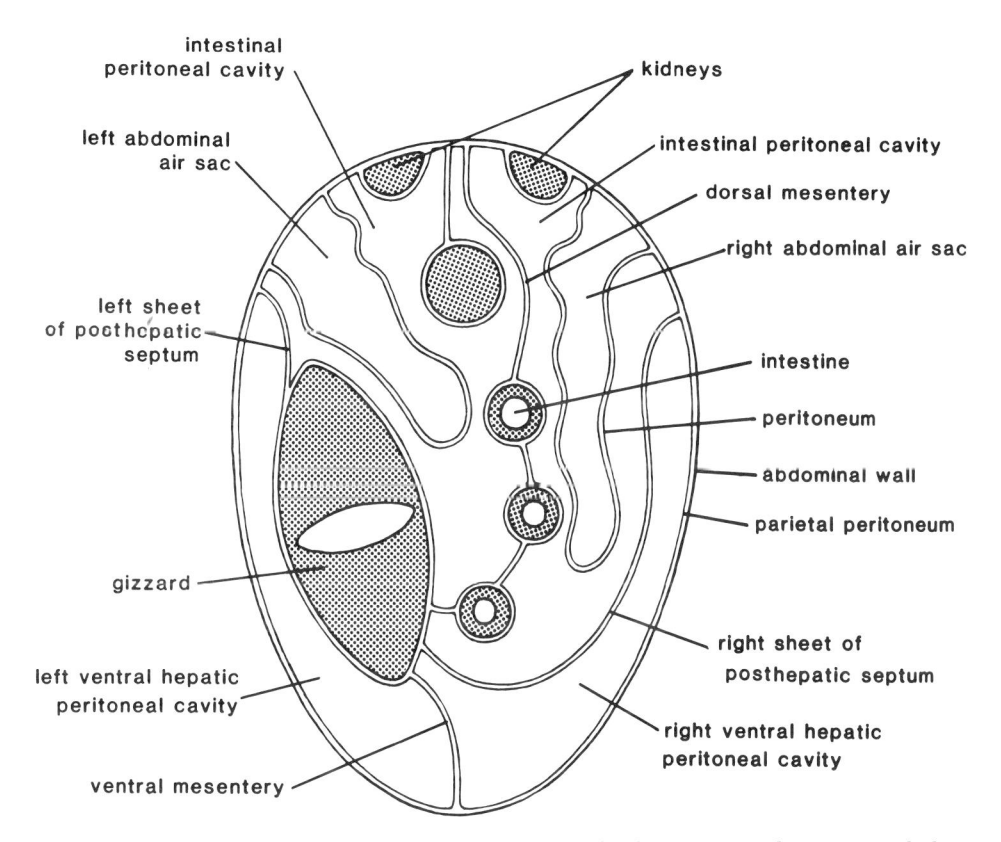

Fig 5–2 Diagrammatic transverse section through the peritoneal cavities of the bird, caudal view. The diagram shows three of the five peritoneal cavities, i.e. the intestinal peritoneal cavity and the left and right ventral hepatic peritoneal cavities. The left and right dorsal hepatic peritoneal cavities, which are much smaller than the other three, are not shown. Note that the left and right ventral hepatic peritoneal cavities enclose the intestinal peritoneal cavity, so that an incision anywhere in the abdominal wall at the level of this transverse section can only enter either the left or the right ventral hepatic peritoneal cavity; at a more caudal level of the body immediately cranioventral to the vent, a ventral incision will enter the intestinal peritoneal cavity. From McLelland (1979), with kind permission of the publisher.

mesentery, caudally with the posthepatic septum, and cranially with the parietal peritoneum).

The left hepatic ligament divides the left hepatic peritoneal cavity into dorsal and ventral sub-compartments. The right hepatic ligament similarly divides the right hepatic cavity. Thus the left and right hepatic ligaments create four cavities, i.e. the dorsal and ventral left hepatic peritoneal cavities and the dorsal and ventral right hepatic peritoneal cavities.

The peritoneal cavities

The intestinal peritoneal cavity. This is a single midline elongated cavity, enclosed between the left and right hepatic cavities (Fig 5–2). It extends from the liver cranially to the vent caudally. It does not reach the body wall, except caudally, and only in the caudal region can it be directly entered by simply incising the abdominal wall.

Various structures are suspended by mesenteries within the intestinal peritoneal cavity. These include the intestines, and the gonads and their ducts.

The hepatic peritoneal cavities. The left and right *ventral hepatic peritoneal cavities* are large elongated blind cavities extending along the lateroventral body wall from the liver to the caudal body wall (Figs 5–1 and 5–2). The left lobe and right lobe of the liver protrude into the left and right ventral hepatic cavities respectively, but otherwise nothing is suspended within these two cavities.

The left and right *dorsal hepatic peritoneal cavities* are much smaller than the two ventral hepatic peritoneal cavities and the intestinal peritoneal cavity. They are dorsal and cranial in position in the abdomen (Fig 5–1). During dissection or postmortem examination they can be entered through the cavity of the cranial thoracic air sac, by penetrating the oblique septum near the midline (Fig 5–1). At this level, the combined dorsal and ventral mesentery, with the proventriculus suspended between them, form a median partition between the left and right cavities. In theory, these dorsal hepatic peritoneal cavities can also be entered by cutting the hepatic ligament (Fig 5–1), but this ligament is very delicate and is not an easy landmark to identify. The craniodorsal regions of the left lobe and right lobe of the liver project into the left and right cavities respectively, but nothing else is suspended within them. The right dorsal hepatic peritoneal cavity is blind, but the left dorsal hepatic peritoneal cavity connects directly with the intestinal peritoneal cavity. With this one exception, each individual peritoneal cavity is blind and has no connections of any kind.

By separating the intestinal peritoneal cavity from the hepatic peritoneal cavities (except for the single connection which has just been mentioned) the posthepatic septum plays an important role in restricting the spread of certain diseases. For instance secondary tumours which spread across the coelom from the ovary or oviduct are often confined to the intestinal peritoneal cavity and left dorsal hepatic peritoneal cavity. Remnants of ovulated ova that have failed to enter the infundibulum of the oviduct (internal laying) are found in the intestinal peritoneal cavity, and it is in this cavity that egg peritonitis begins. Not surprisingly, the ascites (dropsy) which may arise from egg peritonitis is more often found in the intestinal peritoneal cavity than elsewhere. Peritonitis caused by a foreign body (e.g. shot wounds in wild birds) is often restricted to one peritoneal compartment. On the other hand in a systemic disease which causes peritonitis the inflammation may involve all of the peritoneal cavities to varying degrees. In the domestic fowl hernias of the intestine or oviduct into the ventral hepatic peritoneal cavities, i.e. via a rupture of the posthepatic septum, have been described.

The pericardial and pleural cavities

The pleural cavity is described on page 136. The pericardial cavity is essentially similar to that of mammals. However, in birds the lungs are so far dorsal that they no longer enclose the heart. In the absence of a diaphragm the liver extends on either side of the heart, and the parietal pericardium becomes continuous with the peritoneal partitions of the coelom.

Further reading

Beddard, F.E. (1896) On the oblique septa ('diaphragm' of Owen) in the passerines and in some other birds. *Proc. zool. Soc. Lond.*, 1896. 225–231.

Bittner, H. (1925) Beitrag zur topographischen Anatomie der Eingeweide des Huhnes. *Z. Morph. Okol. Tiere*, 2, 785–793.

Butler, G.W. (1889) On the subdivision of the body-cavity in lizards, crocodiles and birds. *Proc. zool. Soc. Lond.*, 1889. 452–474.

Campana, A. (1875) *Anatomie de l'Appareil Pneumatique pulmonaire, etc., chez le Poulet*. Paris: Masson.

Duncker, H.-R. (1978) Coelom-Gliederung der Wirbeltiere—Funktionelle Aspekte. *Verh. anat. Ges., Jena*, 72, 91–112.

Duncker, H.-R. (1979) Coelomic cavities. In *Form and Function in Birds* (Ed.) King, A.S. & McLelland, J. Vol. 1. London and New York: Academic Press.

Goodchild, W.M. (1970) Differentiation of the body cavities and air sacs of *Gallus domesticus* post mortem and their location *in vivo*. *Br. Poult. Sci.*, 11, 209–215.

Goodrich, E.S. (1930) *Studies on the Structure and Development of Vertebrates*. Reprinted 1958. New York: Dover Publications.

Huxley, T. H. (1882) On the respiratory organs of *Apteryx*. *Proc. zool. Soc. Lond.*, 1882. 560–569.

Kern, D. (1963) *Die Topographie der Eingeweide der Körperhöhle des Haushuhnes (Gallus domesticus) unter besonderer Berücksichtigung der Serosa- und Gekroseverhaltnisse*. Vet. Med. Diss. Giessen.

McLelland, J. (1979) Pericardium, pleura et peritoneum. In *Nomina Anatomica Avium* (Ed.) Baumel, J.J., King, A.S., Lucas, A.M., Breazile, J.E. & Evans, H.E. London and New York: Academic Press.

McLelland, J. & King, A.S. (1970) The gross anatomy of the peritoneal coelomic cavities of *Gallus domesticus*. *Anat. Anz.*, 127, 480–490.

McLelland, J. & King, A.S. (1975) Aves coelomic cavities and mesenteries. In *Sisson and Grossman's The Anatomy of the Domestic Animals* (Ed.) Getty, R., Vol. 2, 5th Edn. Philadelphia: Saunders.

Petit, M. (1933) Peritoine et cavité péritonéale chez les oiseaux. *Rev. vét., Toulouse*, 85, 376–382.

Poole, M. (1909) The development of the subdivisions of the pleuroperitoneal cavity in birds. *Proc. zool. Soc. Lond.*, 1909, 210–235.

Chapter 6

DIGESTIVE SYSTEM

ORAL CAVITY AND PHARYNX (OROPHARYNX)

The relationship between the pharynx and the oral and nasal cavities differs from that in mammals. The soft palate is absent and therefore the pharynx is not divided into nasal and oral parts. Also, there is nothing comparable to the glossopalatine arch (cranial pillar of the soft palate, anterior pillar of the fauces) which in some species of mammal forms a constriction, and therefore constitutes a distinct boundary between the oral cavity and pharynx. Thus together the oral cavity and pharynx of birds form one common cavity which is often called the oropharynx. However, it has been suggested on embryological grounds that the homologous boundary between the oral and pharyngeal cavities lies between the choanal opening and the infundibular cleft (Fig 6–1).

Lips and teeth

These are absent in all living birds and are replaced functionally by the cutting edge (tomia) of the horny beak (see Chapters 2 and 3). Teeth, however, were present in the earliest known fossil birds, *Archaeopteryx* (Fig 1–2).

Choana and palate

The choana is a median fissure in the palate connecting the oropharynx to the nasal cavity (Figs 6–1 and 7–1). In most species the *palate* is ridged both lateral and rostral to the choana.

The arrangements of the ridges of the palate is especially well developed in seed-eating passerines which use them to remove the shell. The shell of dicotyledonous seeds is removed by cutting while that of monocotyledenous seeds is removed by crushing. In the Fringillidae, all of which feed on dicotyledonous seeds, the seed is wedged in the groove of the palate formed between the lateral palatine ridge and the edge of the upper bill. Elongated seeds lie with their long axis parallel to the bill, round seeds with their seam longitudinally orientated. The shell is cut by rapid rostro–caudal movements of the sharp-edged lower bill, the seed being supported by the point of the tongue. Usually, the seed is transferred to the opposite side of the mouth and the procedure repeated. With especially hard seeds as many as twenty such transfers may be required. The shell is removed by the bird inserting its tongue between the shell and the kernel and the seed being rotated by the tongue. In species which feed on monocotyledonous seeds (e.g. the Emberizidae and Ploceidae) the ridge rostral to the choana is expanded into a buttress-like prominence. Elongated seeds are positioned transversely to the bill and close to the prominence against which they are crushed by upward and downward movements of the lower mandible. Round seeds, in contrast, are placed in the

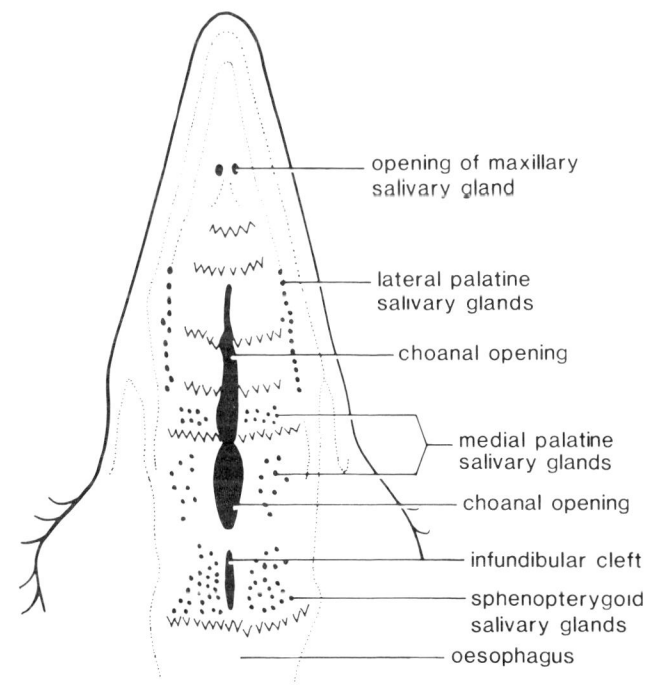

Fig 6-1 The roof of the oropharynx of the domestic fowl. Six rows of caudally pointing papillae are shown. There are numerous openings of salivary glands.

lateral groove of the palate and crushed by up-and-down movements of the blund-edged lower mandible.

In the Hawfinch which is capable of crushing especially large seeds, like those of cherries and olives, the seed is positioned between a pair of knoblike projections on the caudal part of the palate and corresponding projections on the lower bill. Consequently, the strain of cracking the large seeds is evenly distributed between both sides of the jaw.

Infundibular cleft

The infundibular cleft is a midline slitlike opening that is common to the right and left pharyngotympanic (Eustachian) tubes and is situated immediately caudal to the choana (Fig 6-1). In contrast to mammals the orifice of the cleft in birds is not guarded by apposable folds; consequently, the great changes in atmospheric pressure which some birds (notably the Bar-headed Goose, which ascends more than 9000 m) experience during descent from high altitudes cannot passively close the orifice due to the development of unequal pressures on the two sides of the tympanic membrane, as tends to happen in man. Lymphatic tissue ('pharyngeal tonsil') is abundant in the wall of the infundibular cleft.

Tongue

The tongue is supported by the hyobranchial (hyoid) apparatus (see Chapter 4) and displays an enormous variety of dietary adaptations. Basically, three forms of adaptation can be distinguished.

(a) *Adaptations for collecting food.* The tongues in this category are long narrow and protrusible, and function as probes, spears, brushes and capillary tubes. The tongue of woodpeckers is capable of being thrust out of the mouth a considerable distance to procure insects and the sap of trees. Frequently, the apical part of the tongue is armed laterally with barbs and the dorsal surface roughened by numerous minute spinous papillae (Fig 6–2Be). The exceptionally well-developed hyobranchial apparatus of woodpeckers is described on page 49. The tongues of most flower-frequenting species (e.g. flowerpeckers, sunbirds, honeyeaters and nectar-feeding warblers) are also usually highly protrusible and capable of being thrust in and out of a flower to obtain nectar, pollen and small insects. Depending on the diet the rostral portion of the tongue of flower-frequenting species may be split, curled, fringed or tubular (Fig 6–2Bc).

(b) *Adaptations for manipulating food.* Tongues in this category are non-protrusible. Examples include the short thick fleshy tongues of seed- and nut-eaters such as parrots and finches; the thick rasplike tongues of birds of prey (Fig 6–2Bd) in which the rostral portion is often extremely hard and rough and in eagles and vultures sometimes curled or troughlike; the tongues of many fish-eaters such as penguins, mergansers and shearwaters which carry numerous stiff sharp papillae designed for holding slippery prey (Fig 6–2Ba and –Bb); and the tongues of anseriforms which, with the lamellae of the bill, form an organ either for filtering solid food particles from water as in the Mallard, or for cropping grass as in the Muscovy Duck.

Adaptations of the tongue in ducks, geese and swans include the modification of the rostral portion to form a scooplike process and the presence on the lateral margins of a double row of overlapping bristles which are interspersed caudally with a number of toothlike processes. These bristles and processes interdigitate with the lamellae of the bill. On the dorsal surface of the tongue is a shallow median groove while at the root there is a fleshy eminence or cushion. In straining ducks the tongue is first depressed allowing water to fill the dorsal groove. The tongue is then pressed against the palate so that the water is ejected sideways between the bristles which retain solid food particles. As a cropping organ a part of the tongue, e.g. the processes at the edge of the

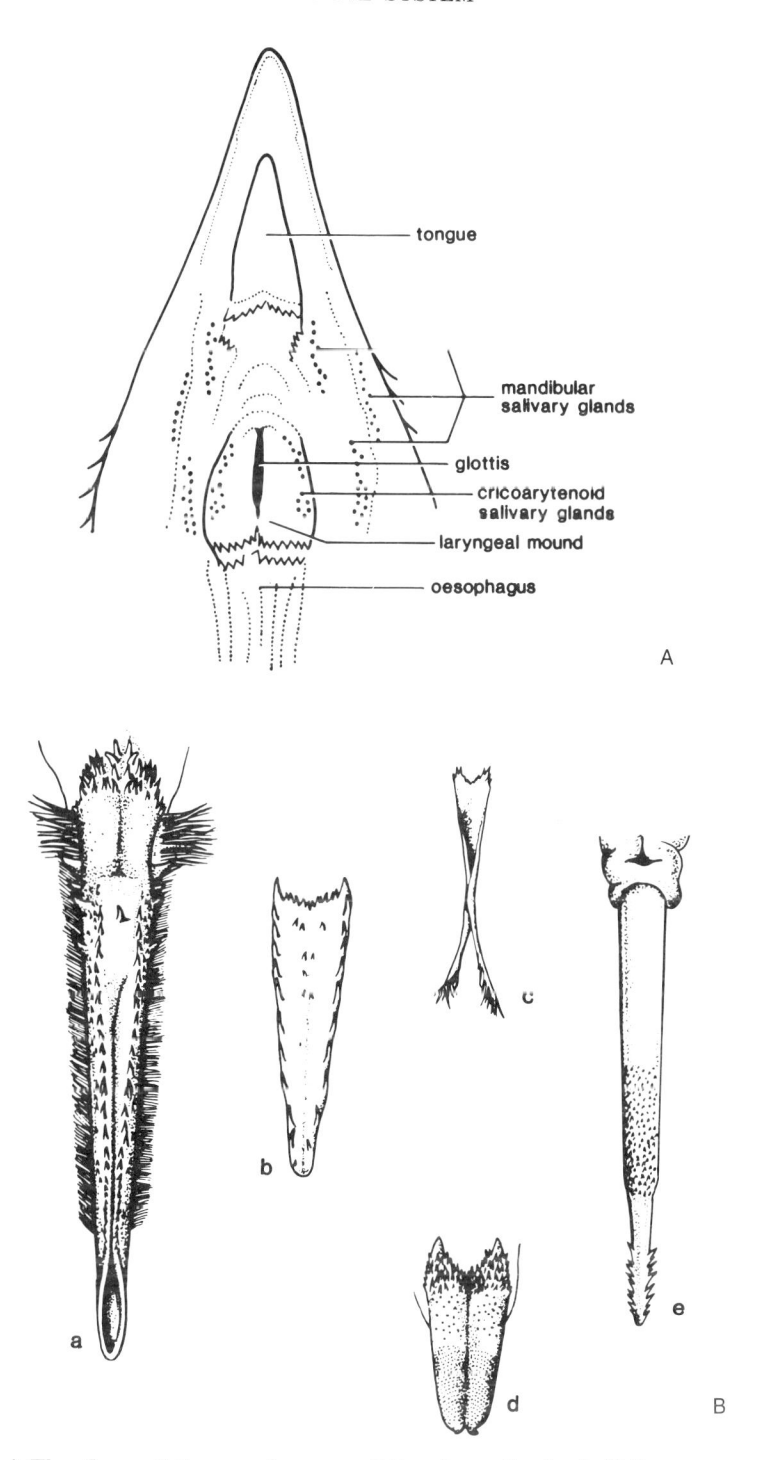

Fig 6–2 A The floor of the oropharynx of the domestic fowl. B Tongues. a, Red-breasted Merganser; b, Sooty Shearwater; c, Bananaquit; d, American Kestrel; e, White-headed Woodpecker. From Gardner (1926).

dorsal groove in the Whooper Swan or the lateral bristles in the Canada Goose, is converted into a coarse cutting margin which functions to hold and tear the vegetation.

(c) *Adaptations for swallowing.* Again, tongues in this category are non-protrusible, and frequently have as their main feature, as in the domestic fowl, caudally-directed papillae which are designed to move the bolus caudally and prevent regurgitation (Fig 6–2A). Also in this category are the rudimentary tongues of many species (e.g. pelicans) which swallow large items of food whole and quickly.

Except in parrots, there are no true intrinsic muscles in the avian tongue, although some of the hyoid muscles attach to its caudal part; movement of the tongue in birds is mainly due to the great mobility of the hyobranchial apparatus. In the domestic fowl and some other species the epithelium of the ventral surface of the tongue undergoes a process of hard keratinization similar to that of the beak. Folding of the tip of the tongue ('curled tongue') is sometimes seen in chicks and turkey poults, and is thought to be due to the feeding of dry mashes of fine physical consistency during the first few weeks after hatching.

The distribution and functional significance of *taste buds* in birds is dealt with in Chapter 16.

Laryngeal mound

The prominent laryngeal mound (Fig 6–2A) carries the narrow slitlike opening into the glottis of the larynx. There is no epiglottis. In nearly all species of bird the mound has several rows of strong backward-pointing cornified papillae. These help in the raking movements of the larynx during swallowing which are known to occur in the domestic fowl (see **Swallowing**), and the typical presence of such papillae suggests that similar raking by the laryngeal mound may occur in many groups of birds.

Papillae

Numerous caudally-directed papillae are distributed, either irregularly or in transverse rows, on the roof of the oropharynx, along the edges of the choana and infundibular cleft, on the tongue and on the laryngeal mound (Figs 6–1 and 6–2A).

Salivary glands

Salivary glands are best developed in birds which have a relatively dry type of diet like that of seed- or insect-eaters, and least developed in birds with a well-lubricated diet such as fish-eaters. Salivary glands are totally absent in a few species including the Great Cormorant. The salivary glands of the domestic fowl are well developed and comprise the maxillary, palatine and sphenopterygoid glands in the roof of the oropharynx; glands in the angle of the mouth and the cheeks; and the mandibular, lingual and cricoarytenoid glands in the floor of the oropharynx. Many of the large number of gland openings in the roof and floor are visible with the naked eye (Figs 6–1 and 6–2A). As in most birds, the glands of the domestic fowl secrete only mucus, the main function of which is to act as a lubricant in swallowing. Recently, however, the glands of the House Sparrow have also been shown to secrete amylase.

In woodpeckers the large mandibular glands below the tongue secrete a sticky fluid which coats the tongue and allows it to act as a probing 'lime-stick' to trap insects. In swifts the mandibular glands secrete an adhesive glycoprotein which is used in nest-building; the nests of some of the cave swiftlets of South-east Asia (*Collocalia* species) are made up entirely of this secretion and are the source of the edible nests ('bird's nest soup') which are an expensive and popular delicacy in that part of the world. In the Grey Jay parts of the mandibular glands are greatly enlarged and secrete abundant mucus which is used by the birds to make boli and then to stick the boli to conifer needles and other parts of trees, thus providing a store of food when the ground is covered with snow.

Oral sacs

The oral cavity of a few species contains one or more ventral diverticula. In some birds (e.g. the males of several species of bustard) the sacs are inflated with air during the breeding season to act as display chambers. In other birds such as nutcrackers they are concerned with carrying food. In pelicans, the entire floor of the mouth is enormously enlarged as an adaptation for catching fish.

The lining of the oral cavity and pharynx

The surface of the oropharynx is lined by a stratified squamous epithelium which is keratinized in regions subject to abrasion, notably the caudally-directed papillae in the domestic fowl. The mouth lining in the chicks of some species, especially passerines, is brightly coloured and has a distinctive pattern of markings. Conspicuous flanges may also be developed on the bill. When the nestling opens its mouth the appearance of the lining, combined with the exaggerated gaping, forms a begging display which stimulates the parent to feed the chick. The distinctive appearance of the oral cavity disappears when the chick can feed itself and no longer requires the help of the parent.

Swallowing

Swallowing is a complicated process, which in the domestic fowl has been shown by cineradiography to involve the following stages.

A pellet of food is seized by the beak and moved by the tongue to the palate, where it is held by the sticky mucous secretion of the salivary glands. The choanal opening is reflexly closed by muscular action to prevent the food from entering the nasal cavity.

Rapid rostrocaudal movements of the tongue then roll the bolus caudally into the caudal part of the oropharynx, with the aid of the caudally directed papillae on the tongue and the roof of the oropharynx. The infundibular cleft and the glottis are reflexly closed to prevent food from entering the pharyngotympanic infundibulum and the cavity of the larynx.

Within the caudal part of the oropharynx, the propulsive action of the tongue on the food is continued by similar movements of the laryngeal mound; darting rostrocaudal movements of the mound, aided by the caudally-directed papillae and by the presence of much sticky saliva, rake the food towards the oesophagus. Food may accumulate for a short while just caudal to the laryngeal mound before passing down the oesophagus by peristalsis.

In drinking, fluid enters the mouth by rapid rostrocaudal movements of the tongue and accumulates in the floor of the oropharynx, caudal to the tongue. Simultaneously, the larynx moves rostrally to lie close to the choana. The head is then raised and the fluid flows around the larynx and into the oesophagus, mainly by gravity. Other methods of drinking which are utilized by birds include the lapping of parrots, the sucking mechanism of hummingbirds in which the tubular tongue is thrust rapidly in and out of the liquid, and the procedure used by pigeons in which the beak is immersed throughout the entire swallowing process.

OESOPHAGUS

The avian oesophagus is thin-walled and distensible with a relatively greater diameter than that of mammals. Most of the cervical part lies on the right side of the neck, the opposite side to that of mammals (Fig 6–3). The internal surface of the oesophagus is longitudinally folded, thus increasing the distensibility of the tube. Both the diameter of the oesophagus and the development of the folds tend to be greater in species such as hawks, owls and cormorants, which take in large items of food or store food throughout the whole length of the oesophagus. It is narrowest and least folded in species such as swifts which feed on very small items of food. The oesophagus is lined by an incompletely keratinized stratified squamous epithelium with subepithelial mucous glands, the latter being generally more numerous in the thoracic oesophagus.

Crop

Just cranial to the thoracic inlet in many species, the oesophagus enlarges to form the crop or ingluvies (Figs 6–3, 6–8 and 14–2). A common misconception is that all birds have a crop, but in fact it is absent in many groups including, for example, gulls and penguins. When present the crop varies in its external appearance (Fig 6–4). Where it is very large, as for example in galliform species (including the domestic fowl and turkey), parrots and a number of passerines, it usually takes the form of a ventral or lateral saclike diverticulum

DIGESTIVE SYSTEM 91

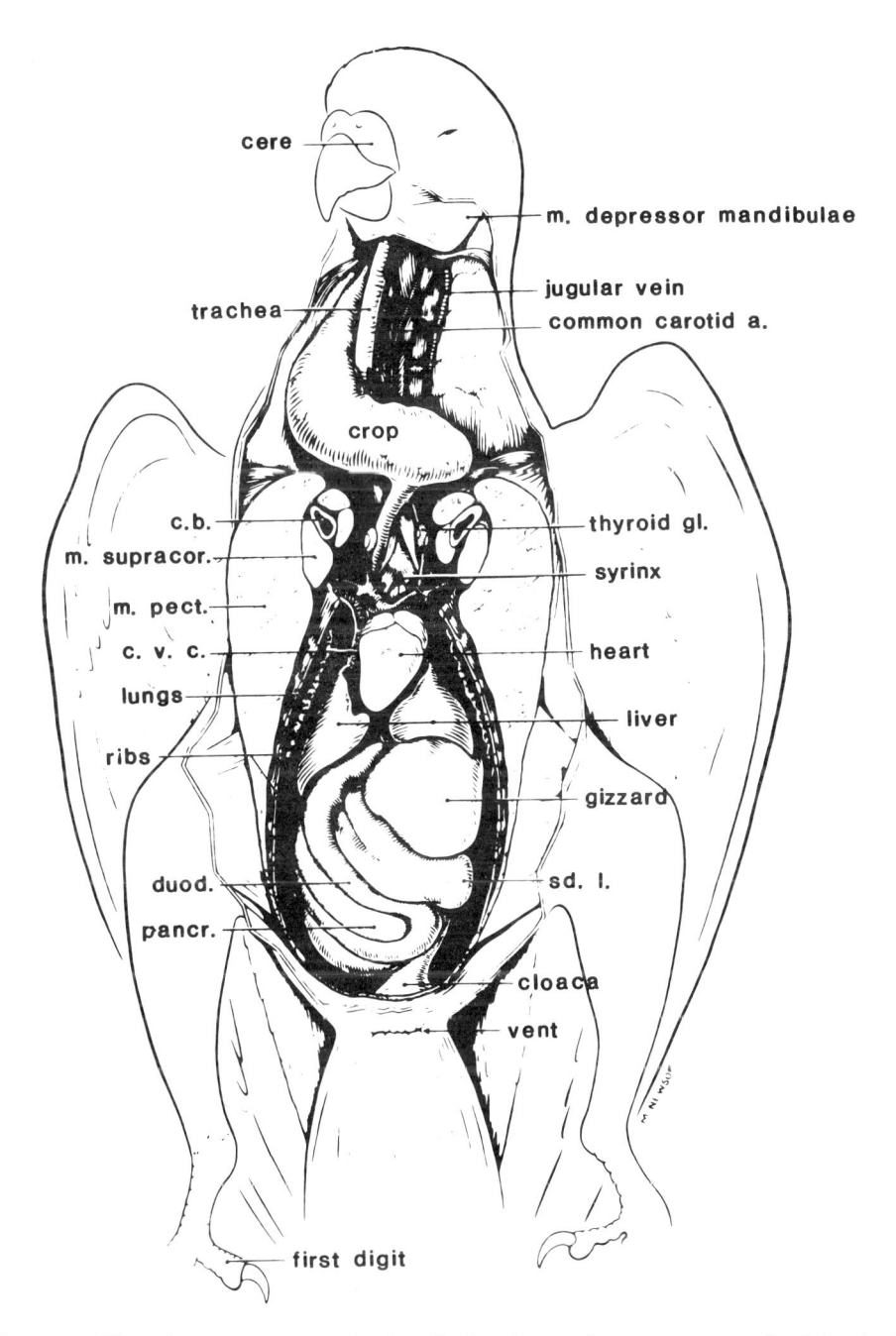

Fig 6-3 The digestive tract of the Budgerigar; the sternum and abdominal wall have been removed. a. = artery; c.b. = coracoid bone; c.v.c. = caudal vena cava; duod. = duodenum; m. pect. = pectoral muscle; m. supracor. = supracoracoid muscle; pancr. = pancreas; sd.l. = supraduodenal loop. From Evans (1969), with kind permission of the publisher.

(Figs 6–4b and –4c). In contrast to other species, the crop in parrots and the Budgerigar is orientated transversely across the neck (Fig 6–3). The crop of pigeons is especially well developed and is divided into two large lateral sacs (Fig 6–4d). The massively developed crop of the Hoatzin is unusual in having not only a cervical component but also two thoracic components; the cervical component is especially large and suppresses the development of the cranial part of the sternal keel. In contrast to these large forms of crop, that of many birds, including ducks and geese and a number of songbirds, is a simple fusiform widening of the oesophagus which is often difficult to identify (Fig 6–4a). The crop epithelium in the domestic fowl and turkey resembles that of

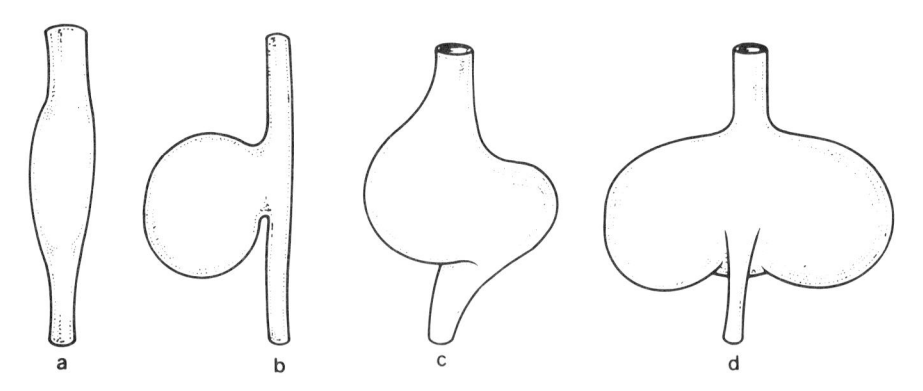

Fig 6–4 Crops. **a**, Great Cormorant; **b**, Peafowl; **c**, Budgerigar; **d**, domestic pigeon. Ventral views except **d**, which is a dorsal view. **a** and **b** from Pernkopf and Lehner (1937), with kind permission of the publisher; **c** and **d** from McLelland (1979), with kind permission of the publisher.

the oesophagus, except that mucous glands are absent. In these species the crop is strongly attached to the underlying skin and thus can easily be seen and palpated externally. In order to remove the gastrointestinal tract in the postmortem evisceration of poultry in slaughter plants, the oesophagus is transected at a variable point caudal to the crop. Since the bird is not fed between two and eight hours before death, there is usually no contamination of the carcase with food resulting from the cut. In the domestic fowl impaction of the crop frequently results from the ingestion of large quantities of bulky or dry feed. Impaction of the crop by fur or other roughage can also occur in raptors if the moisture content of the diet is inadequate.

Oesophageal sac

The oesophageal sac is an inflatable croplike diverticulum or bilaterally symmetrical expansion of the cervical oesophagus which is possessed by only a small number of species (e.g. the Sage Grouse, the Great Bustard and the Prairie Chicken). The enlargement in the majority of such species occurs only in the male, and is used during courtship to produce mating calls or in 'showing off'.

Transport of food

Movement of food is aided by the lubricating mucous secretion of the oesophageal glands. In the domestic fowl peristaltic waves occur in the cervical oesophagus at intervals of 15 seconds and in the thoracic oesophagus at intervals of about 50–55 seconds. In some species such as penguins and pigeons, retroperistaltic waves in the oesophagus regurgitate food which has been stored in the oesophagus or crop, for feeding to the young. In Budgerigars kept alone in cages, such regurgitation of food may be part of a 'courtship feeding' display which the birds perform to themselves in front of a mirror.

Site of physical digestion

When the gizzard is full, food may be stored in the oesophagus either in the crop as in the domestic fowl, or throughout the whole length of the oesophagus as in species which have no crop (e.g. penguins and gulls). In a well-developed type of crop like that of the domestic fowl, the stored food undergoes softening and swelling. Entry of food to the crop inhibits its contraction for a while. Food is later moved caudally from the crop by powerful contractions of both the crop and the opposing oesophageal wall. When the gizzard is empty, food in the oesophagus travels directly to the stomach, the entrance to the crop being closed by contraction of the longitudinal muscle layer of the oesophagus. In the Hoatzin the highly muscular crop is the main site for the mechanical treatment of leaves on which the bird feeds, the gizzard (the main organ of physical digestion in other birds) being extremely reduced in size.

Site of chemical digestion

In most birds the oesophagus does not appear to play any part in chemical digestion.

Crop milk and other nutrients

'Crop milk' is fed by pigeons and doves to their young. It is produced by a desquamation of fat-laden cells of the proliferated stratified squamous epithelium lining the very well-developed crop of both sexes. The production of the milk is controlled by the pituitary hormone, prolactin. Proliferation of the crop epithelium begins at about the sixth day of incubation, and secretion starts at about the sixteenth day of incubation lasting until about two weeks after hatching. The response of the crop to intramuscular or intradermal injections of prolactin is the most frequently used method for bioassay of this hormone. The composition of the crop milk resembles that of mammalian milk, being very rich in fat and protein, i.e. fat 6.9–12.7 per cent, protein 13.3–18.6 per cent, ash 1.5 per cent and water 65–81 per cent; unlike mammalian milk, however, it lacks carbohydrates and calcium. The chicks receive crop milk during the first few days after hatching but later are fed increasing amounts of other types of food.

In the Greater Flamingo the merocrine oesophageal glands of both sexes produce seasonally a red-coloured nutritive juice which is regurgitated and fed to the young. A nutritive fluid is also formed by the male Emperor Penguin and fed to the chick, but this fluid is produced by desquamation of the oesophageal epithelial cells.

STOMACH

The stomach lies in the left dorsal and left ventral regions of the thoraco-abdominal cavity in birds generally (Figs 6–3 and 6–9). However, very little of

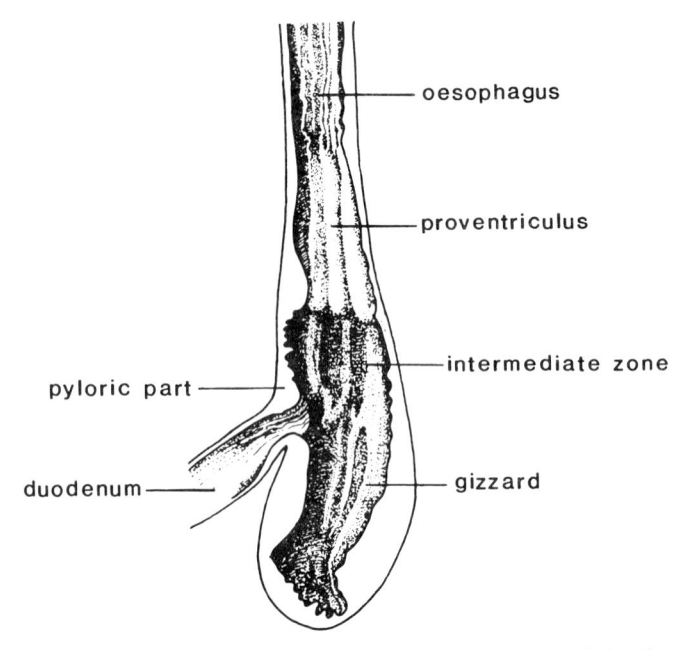

Fig 6–5 Stomach of a gull. From Pernkopf and Lehner (1937), with kind permission of the publisher.

the stomach is seen when the bird is first opened ventrally, since apart from the liver the abdominal organs are largely concealed by the fat-laden post-hepatic septum (Fig 6–8). The stomach consists of a cranial *proventriculus* or glandular part and a caudal *gizzard* or muscular part (Figs 6–5, 6–6 and 6–7). Lying between the proventriculus and gizzard is the *intermediate zone* (Figs 6–5 and 6–7). The *pyloric part* of the stomach joins the gizzard to the duodenum (Figs 6–5 and 6–6). The proventriculus secretes gastric juice. The gizzard is the site of gastric proteolysis and in many species it is also the organ of mechanical digestion. In some species, the proventriculus and the gizzard have an important role in the storage of food.

Depending on diet two basic types of stomach can be distinguished. One type is characteristic of birds like piscivores and carnivores, which feed on relatively large soft items of food and therefore require a stomach that is adapted more for storage than physical digestion. Typically in these birds the stomach is thin-walled and saclike, the junction between the proventriculus and gizzard often being difficult to identify externally (Fig 6–5). The other type of stomach characterizes birds like insectivores, herbivores and granivores (including domestic birds) which feed on relatively indigestible items of food and therefore require a stomach that is adapted for the physical phase of digestion. In these birds, typically the gizzard is massively developed with an especially thick muscle tunic, and the proventriculus and gizzard are easy to distinguish externally (Fig 6–6). Forms of stomach that are intermediate

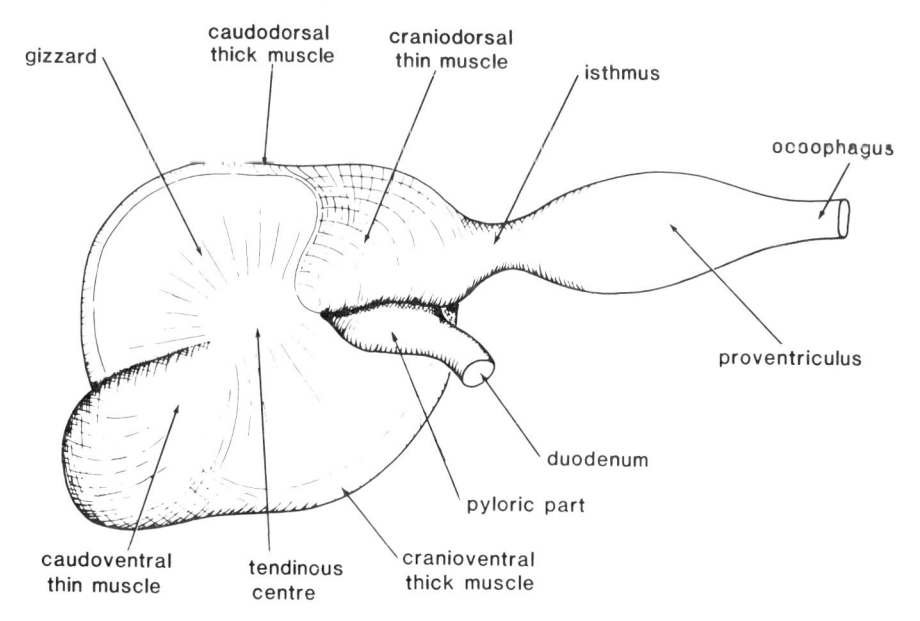

Fig 6 6 Right (medial) aspect of the stomach of the domestic fowl. From McLelland (1975), with kind permission of the publisher.

between those described above are characteristic of many groups including frugivores and testacivores (shellfish-eaters). Reduction of the gizzard, culminating in an inconspicuous band or a vestigial diverticulum, tends to occur in fruit-eating species such as tanagers.

Proventriculus

The oesophagus continues into the proventriculus without any distinct macroscopic boundary. Generally, the internal surface of the proventriculus lacks the large folds that are characteristic of the oesophagus, but such folds are usually present in fish- and meat-eaters (Fig 6–5). Opening into the proventriculus and visible to the naked eye, are the orifices of the gastric glands. In some species, including the domestic fowl, a series of conspicuous papillae project into the lumen of the proventriculus, the glands opening at the tops of the papillae (Fig 6–7). In most species including the domestic fowl, the glands are distributed throughout the proventriculus, but sometimes they are restricted to longitudinal tracts as in owls, to circular patches as in the Ostrich, or to separate diverticula as in the Anhinga. In birds generally, as well as in the domestic fowl, the main ducts of the unilobular or multilobular gastric glands are lined by very tall columnar mucous cells which discharge their mucus after feeding. The glandular alveoli drain into the central cavity of each lobule and contain only one type of cell which has ultrastructural features similar to both the parietal (acid-secreting) and the peptic (enzyme-secreting)

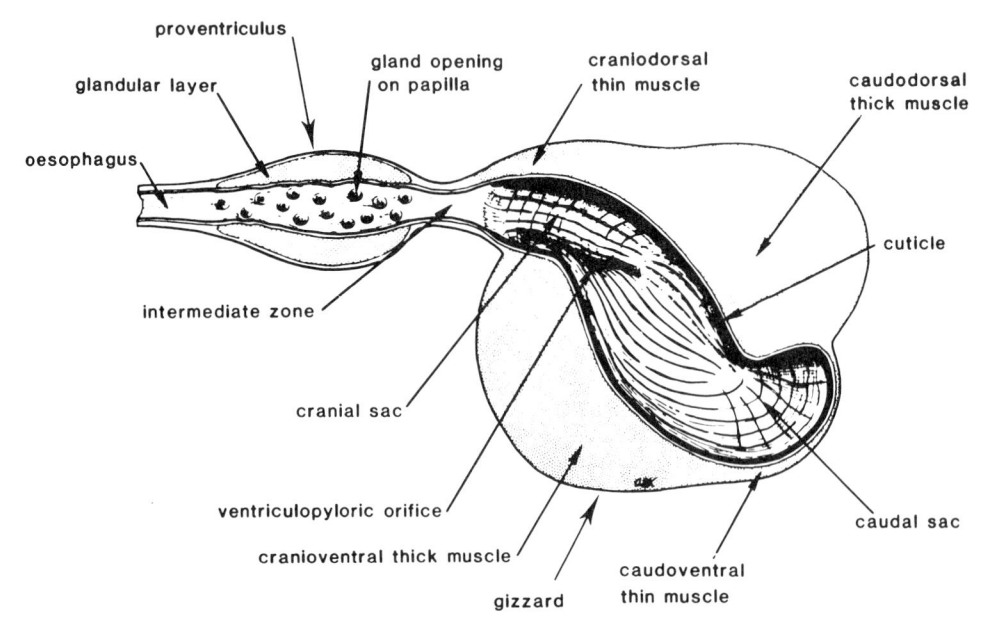

Fig 6–7 Interior of the stomach of the domestic fowl. From McLelland (1975), with kind permission of the publisher.

cells of the mammalian stomach. Therefore, as in other vertebrates except mammals, hydrochloric acid and pepsin are produced in birds by one type of cell, an oxynticopeptic cell. These cells vary from cuboidal to columnar according to their functional activity, but their luminal ends tend to project freely, giving a characteristic serrated appearance. They contain large spherical granules which diminish markedly in number within half an hour of feeding and are then restored in about six hours. Regurgitation of the gastric juice in poultry may be responsible for some cases of 'sour crop', a condition in which the crop is often filled with a large volume of foul-smelling fluid. The epithelium of the glandular alveoli of the proventriculus also contains endocrine cells (p. 105). Contrary to popular opinion, the pink 'stomach oil' shot out by petrels as food for the chick or in defence does not come from glands but is of dietary origin. Enlargement of the proventriculus related to diet has occasionally been described in poultry.

Gizzard

In species which feed on relatively soft food (e.g. carnivores and piscivores), the gizzard is a roundish sac often indistinguishable from the proventriculus, with a muscle tunic which is very poorly developed and of uniform thickness (Fig 6–5). In groups feeding on digestively resistant items of food, however, (insectivores, herbivores and granivores) the gizzard is thickened and biconvex; most of the wall consists of smooth muscle arranged in four semi-autonomous masses which attach to extensive aponeuroses (the right and left tendinous centres) (Figs 6–6 and 6–7). This muscle is derived from the original circular muscle layer of the stomach, the outer longitudinal layer being lost. The smooth muscle of the gizzard is very rich in myoglobin, its concentration being several times greater than in the striated muscles of the leg and heart. Relative to the long axis of the gizzard the muscles are asymmetrically arranged and include dark-coloured caudodorsal and cranioventral thick muscles, and lighter-coloured caudoventral and craniodorsal thin muscles. This asymmetrical arrangement results in rotatory as well as crushing movements when the gizzard contracts. Internally, the organ is lined by a simple columnar epithelium onto which, via crypts, the tubular glands of the lamina propria open (in the domestic fowl five to eight glands discharge into each crypt). The best developed glands occur in the highly muscular form of gizzard. The glands and crypts are lined mainly by protein-secreting chief cells. Chief cells in the base of the glands migrate towards the surface where they degenerate and become sloughed off. The epithelium of the gizzard also contains endocrine cells (p. 105).

Resting on the surface of the epithelium in the highly muscular form of gizzard is a hardened membrane, the *cuticle* (koilin layer) which is a carbohydrate–protein complex and not keratin as once believed. In the domestic fowl (Fig 6–7), the cuticle, like the muscle tunic, is asymmetrically developed, being thickest opposite the caudodorsal and cranioventral thick muscles. The cuticle is composed of a scaffolding of interconnecting vertical rods embedded in a horizontal matrix. The vertical rods are secreted by the glands of the lamina propria. The secretion of each gland hardens within the lumen as a filament. The filaments of all the glands opening into a crypt unite to form a vertical rod. The vertical rods project slightly beyond the surface of the membrane as the dentate processes. Lateral branches of the rods unite with those of neighbouring rods to confer great mechanical strength on the membrane. The horizontal matrix is a secretion of the cells of the crypts and surface epithelium. In contrast to the vertical rods, it does not harden immediately but only after it has spread over the surface of the epithelium and around the vertical rods. The hardening of the horizontal matrix is believed to be due to a fall in its pH as a result of diffusion through the membrane of hydrochloric acid from the proventriculus. Trapped within the horizontal matrix are the desquamated cells of the surface epithelium. The surface of this hard cuticle is constantly being worn away by grinding movements of the organ. The cuticle has many longitudinal parallel folds and is usually brown, green or yellow as a result of regurgitation of bile pigments from the duodenum. Despite the extraordinary toughness of the cuticle localized ulcerations extending into the glands of the lamina propria are not uncommon in chicks up to five weeks old when fed on certain diets, and the incidence becomes much greater with diets in which histamine is released by bacterial spoilage. During the postmortem processing of poultry in the slaughter plant the cuticle is removed mechanically. In species with a less muscular form of gizzard (Fig 6–5) a cuticle is also present, although it is relatively soft, the division into vertical rods and horizontal matrix being indistinct. In some of these stomachs, the cuticle may have an almost homogeneous appearance. In certain fruit-eating pigeons (*Ducula* species) the cuticle contains rows of hard pointed conical projections which are used to crush hard fruits such as nutmegs. Massive shedding and excretion of the cuticle occurs periodically in some species (e.g. the Common Magpie and the Common Starling), and in some hornbills the shed cuticle is regurgitated as a sac filled with seeds which the male bird feeds to the female sitting on the nest.

Intermediate zone

The intermediate zone (Figs 6–5 and 6–7) is a variably developed region between the proventriculus and gizzard which has a microscopic structure intermediate between that of the two major stomach compartments. Compound glands are absent, and in the domestic fowl the internal surface is relatively smooth and without papillae (Fig 6–7). Sometimes, as in the domestic fowl, the zone is separated from the proventriculus by a constriction or isthmus (Fig 6–6).

Pyloric part

The pyloric part of the stomach connects the gizzard to the duodenum. Its degree of development varies. In its smallest form, as in the domestic fowl, it is a narrow, lighter coloured region, about 0.5 cm in length and barely distinguishable from the rest of the digestive tract (Fig 6–6). Its mucosa is intermediate in histological structure between that of the gizzard and duodenum. Many endocrine cells are found here (p. 105). The pyloric part of the stomach in a number of species (especially those which take in large amounts of water with their food, e.g. the Great Cormorant and the Gray Heron) is a distinct expansion of the digestive tract. Its greatest development occurs in darters, where hairlike papillae, either alone or in combination with a large conical process which projects into the opening from the gizzard, appear to act as a valvular apparatus delaying the passage of hard material into the duodenum while allowing water to pass through. In general, however, the function of the pyloric part of the stomach is not known.

Digestion in the stomach

The *proventriculus* produces hydrochloric acid and pepsin. However, gastric proteolysis takes place mainly in the gizzard where a low pH is maintained.

The main function of the *gizzard* in herbivorous and granivorous species (such as the domestic fowl) is to triturate the food in preparation for gastric proteolysis. This is achieved by powerful asymmetrical contractions of the muscles, aided by the asymmetrically developed tough internal cuticle and the presence of grit in the lumen. In the domestic fowl the gizzard develops substantial pressures of 100–200 mmHg. The gizzard in species feeding on soft food plays a much less important part in physical digestion and functions instead as a storage organ where gastric juice can act.

Radiographic studies in young turkeys have demonstrated a complex cycle of gastric contraction (lasting 17–24 seconds) in which food is propelled in alternate directions between the proventriculus and gizzard, the sequence of

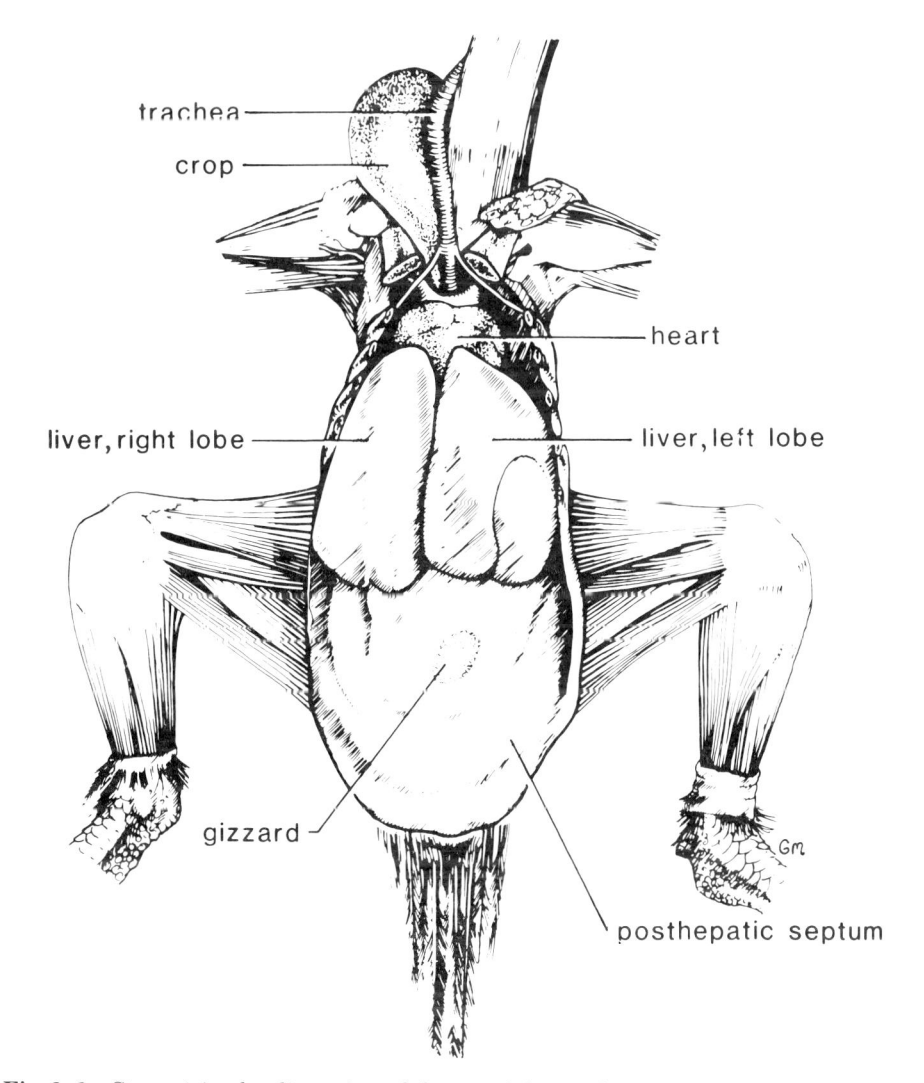

Fig 6–8 Stage 1 in the dissection of thoracoabdominal cavity of a 3-week-old female domestic fowl. Ventral view—the sternum and abdominal wall have been removed.

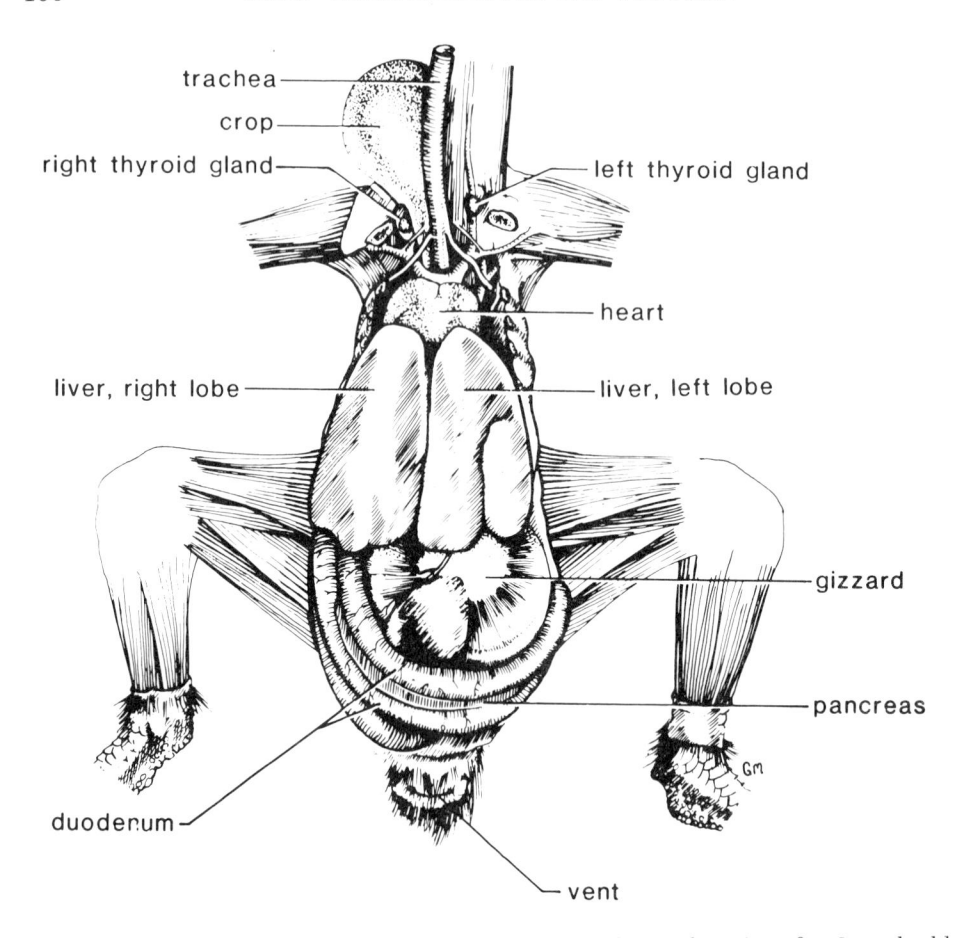

Fig 6–9 Stage 2 in the dissection of the thoracoabdominal cavity of a 3-week-old female domestic fowl. Ventral view—the posthepatic septum has been removed.

contraction being gizzard thin muscles, duodenum, gizzard thick muscles and proventriculus. In strigiform and falconiform species stomach motility is also concerned with arranging the indigestible portions of the stomach contents, i.e. bones, teeth and claws, into *pellets* which are then regurgitated. The contents of these pellets provide valuable clues to ornithologists investigating the feeding habits of birds producing them. The contraction sequence of the stomach in strigiform and falconiform species is different from that of the turkey, being proventriculus, isthmus, gizzard and duodenum.

When sharp foreign bodies such as nails or wire are ingested by poultry these sometimes penetrate the wall of the gizzard, the contractions of the powerful gizzard muscles often forcing the object through the wall. Impaction of the gizzard with food is sometimes a problem in chickens feeding on excessively long grass. The gizzard may also be impacted in raptors by a hard ball of fur. Perforation or impaction of the gizzard by foreign bodies is fairly common in the Ostrich. In Mute Swans which feed on the bottom of rivers the gizzard contents frequently include the small lead weights which are used by anglers to sink their fishing lines and are lost or thrown away. These weights are easily ground down by the gizzard and the lead is absorbed in the small intestine resulting invariably in lead poisoning (indeed postmortem examinations have shown that three-quarters of the dead swans from the Norfolk Broads die from this cause).

INTESTINAL TRACT

Small intestine

Duodenum. The duodenum in the majority of species is a narrow U-shaped loop on the right surface of the gizzard, with a proximal descending part and a distal ascending part held together by a narrow fold of mesentery (Figs 6–3 and 6–9). In a number of species (e.g. the White-tailed Sea Eagle and the Jackass Penguin) the duodenal loop is thrown into a series of secondary folds. Sometimes the duodenal loop is twisted, as in the Black Stork. The duodenum

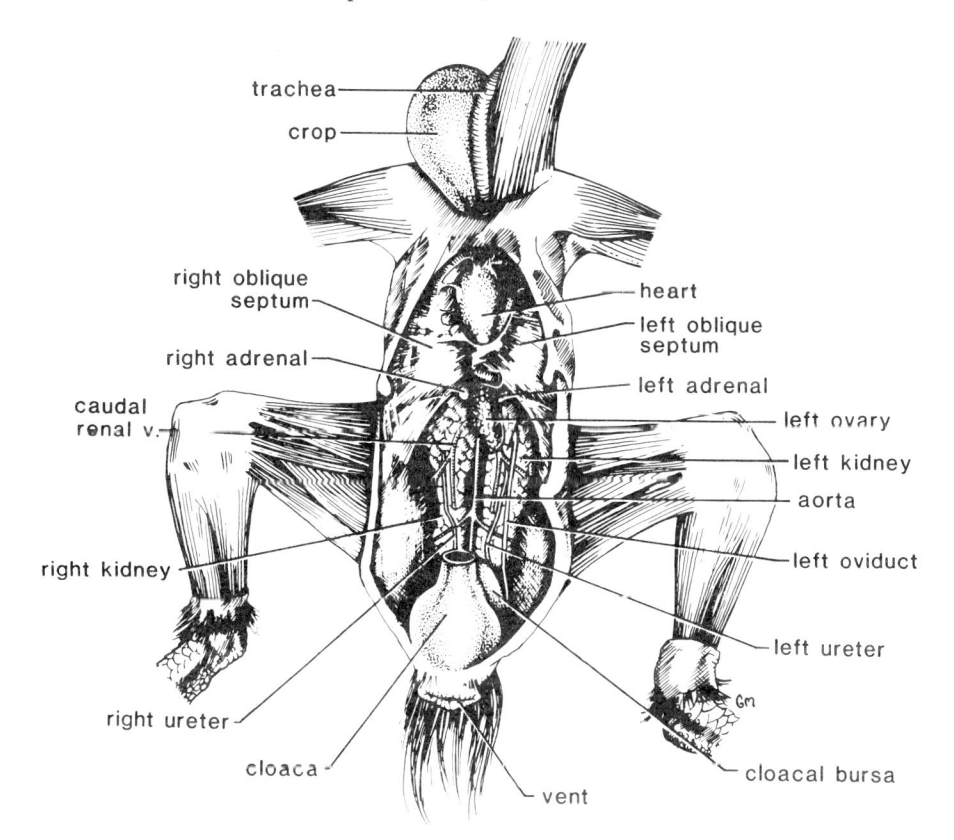

Fig 6–10 Stage 3 in the dissection of the thoracoabominal cavity of a 3-week-old female domestic fowl. Ventral view—the liver, stomach and intestines have been removed.

of a few species (e.g. the Northern Fulmar and the Gannet) is a compound structure consisting of more than one loop. Since the duodenum is always the most ventral part of the intestinal mass it is generally easy to identify in a postmortem examination. Duodenal (Brunner's) glands are absent, but mucous secretion is provided by goblet cells.

Bile and pancreatic ducts. These often open near to each other into the distal end of the ascending part of the duodenum opposite to the cranial part of

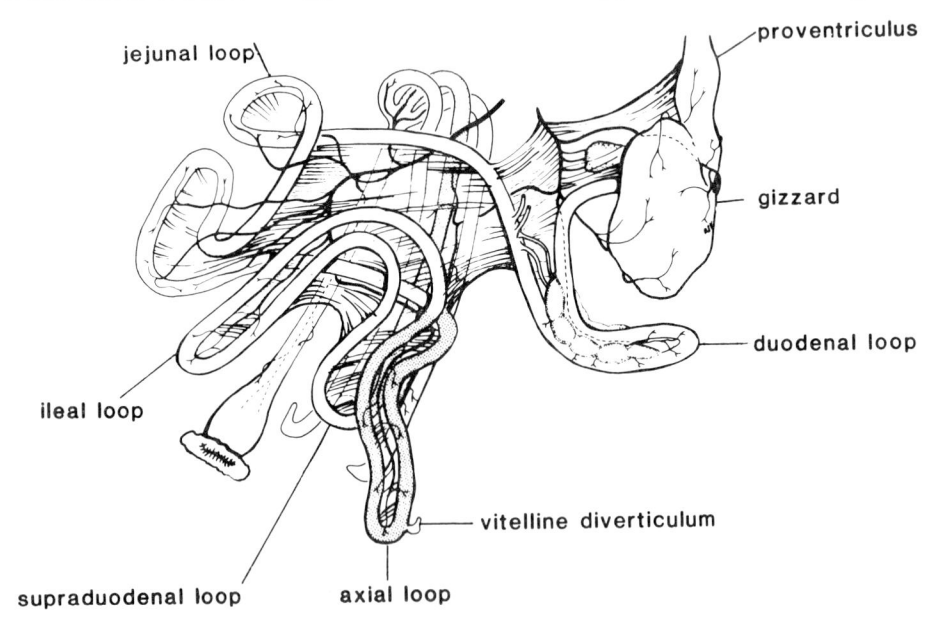

Fig 6–11 The gastrointestinal tract of the domestic goose. The duodenum, jejunum and ileum form a series of narrow U-shaped loops, as in many birds. The supraduodenal loop is the most distal loop of the ileum. The axial loop carries the vitelline diverticulum, opposite the distal end of the cranial mesenteric artery; the diverticulum marks the boundary between jejunum and ileum. The two pancreatic ducts are visible entering the distal end of the duodenal loop, and distal to them are the two ducts from the liver. On the right are the glandular and muscular parts of the stomach; these and the coeliac and cranial mesenteric arteries are as labelled in Fig 6–12.

the gizzard (Figs 6–11 and 6–12). In the domestic fowl there are two main ducts from the liver (a common hepatoenteric duct and a cysticoenteric duct) and three main ducts from the pancreas. In the domestic duck and goose there are usually only two pancreatic ducts. In these domestic birds the pancreatic and bile ducts open together on a papilla situated in an ampulla-shaped swelling of the duodenum. Sometimes, as in the Herring Gull, the openings of the pancreatic ducts are widely separated.

Jejunum and ileum. In the majority of species the jejunum and ileum are arranged in a number of narrow U-shaped loops at the edge of the long dorsal mesentery in the right part of the abdominal cavity (Fig 6–11). The pattern of these intestinal loops has some importance in taxonomy. The *vitelline (Meckel's) diverticulum* (Figs 6–11 and 6–12) is the short blind remnant of the yolk sac and yolk duct and lies at a point opposite the distal branches of the cranial mesenteric artery; it can be used to divide the jejunum from the ileum. At hatching the yolk sac in the domestic fowl weighs about 5.5 g, but by the 10th day the yolk has been resorbed and the sac converted into scar tissue. The duct of the yolk sac in the domestic fowl opens into the intestine on a small papilla. The *axial loop* of intestine carries the vitelline diverticulum and therefore has both jejunal and ileal components. The *supraduodenal loop* of intestine is usually the most distal loop of the ileum and lies immediately dorsal to the

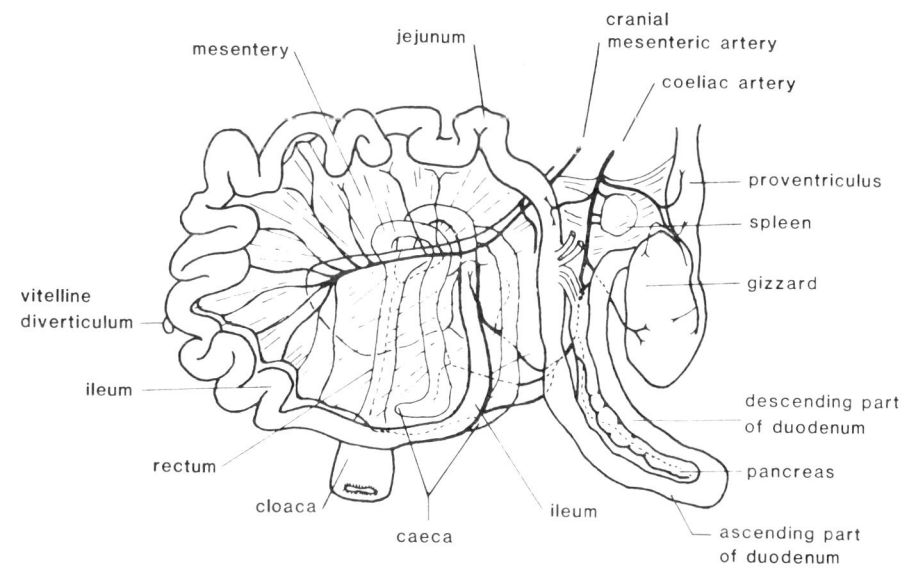

Fig 6–12 The gastrointestinal tract of the domestic fowl. The jejunum and ileum are arranged in short garlandlike coils, not U-shaped loops. Two pancreatic ducts and two ducts from the liver are visible entering the distal end of the ascending part of the duodenum. There is often a third pancreatic duct. Based on Grau (1943).

duodenum. In a relatively few species, such as pelicans, penguins and falcons, part of the most distal region of the ileum is arranged into a small *supracaecal loop*.

In another small group of species, including the domestic fowl and turkey (but not the duck and goose), the jejunum and ileum are not arranged into loops but instead form short garlandlike coils (Fig 6–12).

Large intestine

The large intestine generally consists of paired caeca and a short straight intestine which is probably homologous to the mammalian rectum.

Caeca. In most groups of birds right and left caeca arise from the rectum at the junction with the ileum (Fig 6–12). The form and size of the caeca vary greatly

(Fig 6–13). In the domestic species the caeca are exceptionally large and are a dark green colour, and are therefore easily identifiable on a postmortem examination. Because of their great length they are bent over backwards to point caudally. Very rarely in chickens, a caecal anomaly is seen in which only one tube is present, the distal part of which is divided. In owls the distal part of the caeca is enormously expanded (Fig 6–13c). In a few species including the Ostrich, the caeca are sacculated (Fig 6–13e). The two caeca in the Ostrich open at a common orifice. The caeca of passerine species are small. Whilst the caeca are widely reported to be reduced in number in some species including parrots, swifts and pigeons, rudimentary organs in many of these birds have

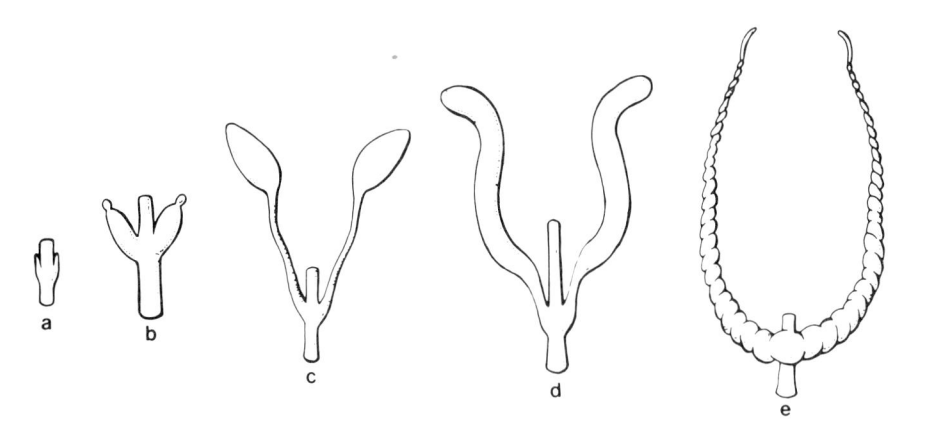

Fig 6–13 Caeca. **a**, Sparrow-hawk; **b**, Marabou; **c**, Barn Owl; **d**, Helmet Guineafowl; **e**, Ostrich. From Maumus (1902).

been described. In the domestic species, lymphatic tissue is well developed in the proximal part of each caecum forming the *caecal tonsil*. Each of the paired caeca in the domestic fowl has a sphincter where it opens at the ileorectal junction.

Rectum. The ileum continues through a sphincter into the straight rectum which lies in the dorsal part of the abdominal cavity (Fig 6–12). In a very small number of species (e.g. rheas) the rectum is looped or folded. At the junction of the ileum and rectum in the domestic fowl and turkey there is a caudally-directed mucosal fold, the *ileorectal valve*.

Folds, villi and intestinal crypts

The intestinal folds and villi vary widely in their form and distribution between species, different patterns of relief often being present in the same bird. In the domestic fowl only villi occur. In the Emu and a small number of other species folds are present containing a submucosal core. All intestinal folds and villi increase the surface area for absorption. The crypts (of Lieberkühn) extend between the folds and villi as simple slightly coiled tubular glands within the lamina propria. Unlike the crypts of mammals those of birds occur in the large intestine as well as in the small intestine. The intestinal epithelium consists of chief cells, goblet cells and endocrine cells. The chief cells have an absorptive function, their surface area being increased by the presence of a brush border. Goblet cells occur throughout the intestine, but become progressively more numerous towards the cloaca. The *endocrine cells* form, with those of the stomach and pancreas, a diffuse endocrine organ (p. 209). As in mammals different types of gastrointestinal endocrine cell have been identified, including cells believed to secrete enteroglucagon, gastrin, somatostatin, secretin, vasoactive intestinal polypeptide, neurotensin and avian pancreaticopolypeptide. Unlike the folds and villi of mammals, those of birds have no lacteals but possess instead a well-developed capillary network.

Digestion in the intestines

Chemical digestion and absorption of food take place in the small intestine. There is evidence that in the domestic fowl the cloaca and rectum may have the ability to reabsorb water from ureteral urine (p. 184) and this may be especially important in the water economy of desert birds. In the caeca of galliform species including the domestic fowl, the breakdown of food (especially of cellulose) by symbiotic bacteria takes place, food reaching the caeca by antiperistaltic movements of the rectum. The major function of the caeca in these birds may be to separate the intestinal contents into a nutrient-rich fluid fraction which enters the caeca for digestion, and an indigestible fibrous fraction which is rapidly extruded without entering the caeca. Recently, however, it has been shown that bacterial fermentation in the caeca supplies only a small fraction of the total energy requirements of birds, so the real importance of the caeca in the domestic fowl is in doubt. The 'caecal' droppings of the domestic fowl, in contrast to those of rectal origin, are dark brown and glutinous, and voided only once or twice each day.

Pancreas

The pancreas is a pale red or yellow lobulated organ lying in the dorsal mesentery between the two limbs of the duodenum (Figs 6–3, 6–9 and 6–12). It has three lobes: dorsal, ventral and splenic. The dorsal and ventral lobes are usually united by parenchymatous bridges which vary in extent. These connections are missing, however, in the domestic duck and pigeon, so that these two lobes remain separate. The small splenic lobe extends cranially from either the dorsal or ventral lobe. The pancreas drains by one, two or three ducts (three in the domestic fowl) which generally open into the distal part of the ascending duodenum close to the bile ducts (Fig 6–12). The exocrine part of the pancreas is a compound tubulo-acinar gland as in the mammal. The acini are lined by zymogenic columnar cells. Avian pancreatic juice contains enzymes, similar to those of mammals, which are important in the chemical phase of digestion in the small intestine. It is the major source of amylase, and also contains lipase, and proteinases including trypsin. The high pH of the pancreatic juice provides the necessary environment for the optimum activity of the pancreatic enzymes. The endocrine part of the pancreas is described in Chapter 12.

Liver

The liver is composed of right and left lobes which join cranially in the midline (Fig 6–14). In contrast to the other abdominal viscera, the liver is visible when the ventral body wall is removed since it is not covered by the fat-laden serosal posthepatic septum (Figs 6–3 and 6–8). In most species, including the domestic

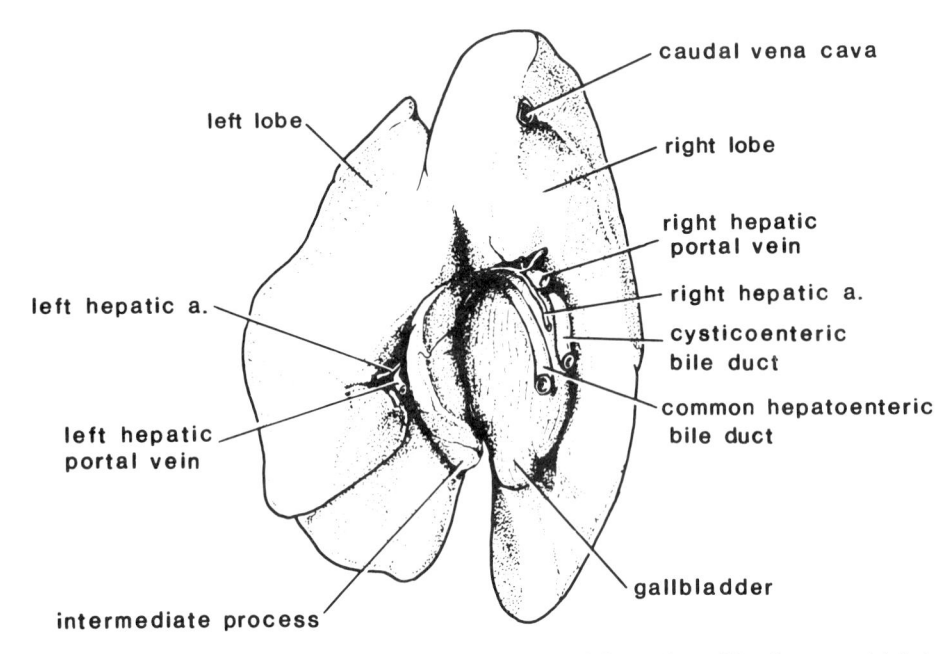

Fig 6–14 Liver of the domestic fowl, visceral (caudal) surface. The fissure which is shown running obliquely across the lower (ventral) region of the left lobe separates its dorsal part (above the fissure) from its ventral part (below the fissure). From McLelland (1979), with kind permission of the publisher.

fowl, the right lobe of the liver is larger than the left. The left lobe is subdivided in the domestic fowl and turkey into dorsal and ventral parts (Fig 6–14). However, in many species, including most passerines, it is the right lobe which is subdivided. In the domestic birds, at least, there are one or more intermediate processes which project from the visceral surface immediately ventral to the hilus (Fig 6–14). The cranioventral part of each liver lobe surrounds the apex of the heart. The caudal vena cava passes through the dorsal part of the right lobe (Fig 6–14). Much of the visceral surface of the liver carries impressions of the adjacent viscera. A *gallbladder* occurs in most species and lies on the visceral surface of the right lobe (Fig 6–14). Amongst the species in which a gallbladder has been reported to be absent are the majority of pigeons, many parrots and the Ostrich. The gallbladder in some woodpeckers, toucans and barbets is exceptionally long and in a few species may even extend caudally to the level of the cloaca. At hatching the liver has a yellow colour caused by the pigment carried with the lipids from the yolk to the liver in the late stages of incubation. At a variable time after hatching (8–14 days in the domestic fowl), the liver assumes its characteristic dark red colour.

During the evisceration of poultry carcases in the slaughter plant, the liver along with the gizzard, intestinal tract, heart and spleen is suspended outside the abdominal cavity for inspection. Afterwards, the liver (minus gallbladder), heart and gizzard form a major part of the giblets. Since the gallbladder is frequently cut accidentally in the postmortem processing of poultry carcases, bile is one of the most common sources of contamination of the carcase.

The liver consists of continuous and anastomosing sheets of hepatocytes. In many birds, as in man and other mammals, these sheets are one cell thick with a sinusoid on either side. In the domestic fowl, however, the sinusoids are separated by sheets which are two cells thick, as they are mainly in lower vertebrates. The sheets are perforated by lacunae, and the parenchyma therefore resembles a sponge. Perilobular connective tissue is absent. Consequently, the classic mammalian type of hepatic lobule, with the efferent (hepatic) vein at its centre and the portal tracts consisting of the afferent vessels (portal veins and hepatic arteries) and bile ducts at its perimeter, is difficult to identify histologically in birds.

Each lobe of the liver is drained by a bile duct. In the domestic fowl the hepatocystic duct drains bile from the right lobe to the gallbladder, while the common hepatoenteric duct drains bile from both lobes to the duodenum (Fig 6–14). When a gallbladder is absent, a branch of the bile duct drains the right lobe; this branch, the right hepatoenteric duct, opens directly into the duodenum.

Further reading

Aitken, R.N.C. (1958) A histochemical study of the stomach and intestine of the chicken. *J. Anat.*, 92, 453–466.

Beddard, F.E. (1911) On the alimentary tract of certain birds and on the mesenteric relations of the intestinal loops. *Proc. zool. Soc. Lond.*, 1, 47–93.

Berkhoudt, H. (1977) Taste buds in the bill of the Mallard (*Anas platyrhynchos* L). *Neth. J. Zool.*, 27, 310–331.

Calhoun, M. H. (1954) *Microscopic Anatomy of the Digestive System*. Ames: Iowa State College Press.

Clarkson, M.J. & Richards, T.G. (1971) The liver with special reference to bile formation. In *Physiology and Biochemistry of the Domestic Fowl* (Ed.) Bell, D.J. & Freeman, B.M. Vol. 2. London and New York: Academic Press.

Duke, G.E., Evanson, O.A., Redig, P.T. & Rhoades, D.D. (1976) Mechanism of pellet egestion in Great-horned Owls (*Bubo virginianus*). *Am. J. Physiol.*, 231, 1824–1829.

Dzuik, H.E. & Duke, G.E. (1972) Cineradiographic studies of gastric motility in turkeys. *Am. J. Physiol.*, 222, 159–166.

Eglitis, I. & Knouff, R. A. (1962) An histological and histochemical analysis of the inner lining and glandular epithelium of the chicken gizzard. *Am. J. Anat.*, 111, 49–66.

Elias, H. & Bengelsdorf, H. (1952) The structure of the liver of vertebrates. *Acta anat.*, 14, 297–337.

Evans, H.E. (1969) Anatomy of the Budgerigar. In *Diseases of Cage and Aviary Birds*, (Ed.) Petrak, M.L. Philadelphia: Lea and Febiger.
Fenna, L. & Boag, D.A. (1974) Filling and emptying of the galliform caecum. *Can. J. Zool.*, 52, 537–540.
Gadow, H. (1889) On the taxonomic value of the intestinal convolutions in birds. *Proc. zool. Soc. Lond.*, 1889, 303–316.
Gardner, L.L. (1926) The adaptive modifications and the taxonomic value of the tongue in birds. *Proc. U.S. natn. Mus.*, 67, article 19.
Gorham, F.W. & Ivy, A.C. (1938) General function of the gall bladder from the evolutionary standpoint. *Field Mus. Natur. Hist. Publ. Zool. Ser.*, 22, 159–213.
Grau, H. (1943) Artmerkmale am Darmkanal unserer Hausvögel. *Berl. tierärztl. Wschr.*, 23, 176–179.
Griminger, P. (1983) Digestive system and nutrition. In *Physiology and Behaviour of the Pigeon* (Ed.) Abs, M. New York and London: Academic Press.
Harrison, J.G. (1964) Tongue. In *A New Dictionary of Birds* (Ed.) Thomson, A.L. London: Nelson.
Hickey, J.J. & Elias, H. (1954) The structure of the liver of birds. *Auk*, 71, 458–462.
Hill, K.J. (1971) The structure of the alimentary tract. In *Physiology and Biochemistry of the Domestic Fowl* (Ed.) Bell, D.J. & Freeman, B.M. Vol. 1. London and New York: Academic Press.
Hill, K.J. (1971) The physiology of digestion. In *Physiology and Biochemistry of the Domestic Fowl* (Ed.) Bell, D.J. & Freeman, B.M. Vol. 1. London and New York: Academic Press.
Hill, K.J. (1983) The physiology of the digestive tract. In *Physiology and Biochemistry of the Domestic Fowl* (Ed.) Freeman, B.M. Vol. 4. London and New York: Academic Press.
Hill, K.J. & Strachan, P.J. (1975) Recent advances in digestive physiology of the fowl. In *Symp. Zool. Soc. Lond.*, No. 35 (Ed.) Peaker, M. London and New York: Academic Press.
Hill, W.C.O. (1926) A comparative study of the pancreas. *Proc. zool. Soc. Lond.*, 1, 581–631.
Hodges, R.D. (1972) The ultrastructure of the liver parenchyma of the immature fowl (*Gallus domesticus*). *Z. Zellforsch. mikrosk. Anat.*, 133, 35–46.
Hodges, R.D. (1974) *The Histology of the Fowl*. London and New York: Academic Press.
Hodges, R.D. & Michael, E. (1975) Structure and histochemistry of the normal intestine of the fowl. III. The fine structure of the duodenal crypt. *Cell Tissue Res.*, 160, 125–138.
Humphrey, C.D. & Turk, D.E. (1974) The ultrastructure of normal chick intestinal epithelium. *Poult. Sci.*, 53, 990–1000.
Maumus, J. (1902) Les caecums des oiseaux. *Annls Sci. nat. (Zool.)*, 15, 1–148.
McLelland, J. (1975) Aves digestive system. In *Sisson and Grossman's The Anatomy of the Domestic Animals* (Ed.) Getty, R. Vol. 2, 5th Edn. Philadelphia: Saunders.
McLelland, J. (1979) Digestive system. In *Form and Function in Birds* (Ed.) King, A.S. & McLelland, J. Vol. 1. London and New York: Academic Press.
McLelland, J. (1979) Systema digestorium. In *Nomina Anatomica Avium* (Ed.) Baumel, J.J., King, A.S., Lucas, A.M., Breazile, J.E. & Evans, H.E. London and New York: Academic Press.
Menzies, G. & Fisk, A. (1963) Observations on the oxyntico-peptic cells in the proventricular mucosa of *Gallus domesticus*. *O. Jl microsc. Sci.*, 104, 207–215.
Michael, E. & Hodges, R.D. (1973) Structure and histochemistry of the normal intestine of the fowl. I. The mature absorptive cell. *Histochem. J.*, 5, 313–333.
Mitchell, P.C. (1896) On the intestinal tract of birds. *Proc. zool. Soc. Lond.*, 1896, 136–159.
Mitchell, P.C. (1901) On the intestinal tract of birds, with remarks on the valuation and nomenclature of zoological characters. *Trans. Linn. Soc. Lond. (Zool.)*, 8, 173–275.

Miyaki, T. (1973) The hepatic lobule and its relation to the distribution of blood vessels and bile ducts in the fowl. *Jap. J. vet. Sci.*, 35, 403–410.

Miyaki, T. (1978) The afferent venous vessels to the liver and the intrahepatic portal distribution in the fowl. *Zbl. Vet. Med. C. Anat. Histol. Embryol.*, 7, 129–139.

Paik, Y.K., Fujioka, T. & Yasuda, M. (1974) Comparative and topographical anatomy of the fowl. LXXVIII. Division of pancreatic lobes and distribution of pancreatic ducts. *Jap. J. vet. Sci.*, 36, 213–229.

Pernkopf, E. & Lehner, J. (1937) Vorderdarm. Vergleichende Beschreibung des Vorderdarm bei den einzelnen Klassen der Kranioten. In *Handbuch der vergleichende Anatomie der Wirbeltiere* (Ed.) Bolk, L., Göppert, E., Kallius, E. & Lubosch, W. Vol. III. Berlin and Vienna: Urban and Schwarzenberg.

Purton, M.D. (1969) Structure and ultrastructure of the liver in the domestic fowl, *Gallus gallus*. *J. Zool.*, 159, 273–282.

Saito, I. (1966) Comparative anatomical studies of the oral organs of the poultry. IV. Macroscopical observations of the salivary glands. *Bull. Fac. Agri. Univ. Miyazaki*, 12, 110–120.

Simic, V. & Jankovic, N. (1959) Ein Beitrag zur Kenntnis der Morphologie und Topographie der Leber beim Hausgeflügel und der Taube. *Acta Vet., Beogr.*, 9, 7–34.

Sturkie, P.D. (1976) Alimentary canal: anatomy, prehension, deglutition, feeding, drinking, passage of ingesta and motility. In *Avian Physiology* (Ed.) Sturkie, P.D. New York: Springer-Verlag.

Sturkie, P.D. (1976) Secretion of gastric and pancreatic juice, pH of tract, digestion in alimentary canal, liver and bile, and absorption. In *Avian Physiology* (Ed.) Sturkie, P.D. New York: Springer-Verlag.

Toner, P.G. (1963) The fine structure of resting and active cells in the submucosal glands of the fowl proventriculus. *J. Anat.*, 97, 575–583.

Toner, P.G. (1964) The fine structure of gizzard gland cells in the domestic fowl. *J. Anat.*, 98, 77–86.

Ziswiler, V. & Farner, D.S. (1972) Digestion and the digestive system. In *Avian Biology* (Ed.) Farner, D.S. & King, J.R. Vol. II. New York and London: Academic Press.

Zweers, G.A. (1982) The feeding system of the pigeon (*Columba livia* L.). *Adv. Anat. Embryol. cell Biol.*, 73, 1–108.

Zweers, G.A., Gerritsen, A.F. Ch. & van Kranenburg-Voogd, P.J. (1977) Mechanics of feeding of the Mallard (*Anas platyrhynchos* L; Aves, Anseriformes). In *Contributions to Vertebrate Evolution* (Ed.) Hecht, M.K. & Szalay, F.S. Vol. 3. Basel: Karger.

Chapter 7

RESPIRATORY SYSTEM

NASAL CAVITY

Nostrils

The nostrils and the adjacent regions inside the nasal cavity are particularly variable among the different species of birds. The nostrils are often located at the base of the beak (Fig 2–1), being placed dorsally, ventrally or laterally, and uniquely are situated at the tip of the beak in kiwis. They may be screened by feathers as in crows and grouse. They may even be entirely closed by secondary overgrowth of cornified cells as in the Gannet so that breathing is through the mouth (a permanent gap is present at the corner of the mouth in the Gannet) probably as an adaptation for diving head first into the sea from a great height. A keratinized flap, the *operculum*, is situated dorsal to the nostrils in some species including the domestic fowl and turkey (Fig 7–1A). In the Wryneck it lies ventral to the nostril. In tapaculos the operculum is movable. The operculum in procellariiforms (the tubinares or tubenoses, e.g. petrels) is converted into a tube situated on the dorsal aspect of the bill.

Nasal septum

The nasal septum is usually partly bony and partly cartilaginous, and in many species including the domestic fowl it separates the right and left nasal cavities completely (Figs 7–1A and –1B). In some species, e.g. ducks and grebes, it is perforated rostrally and thus connects the two nasal cavities. Within the nasal septum in a few species including shrikes there are bony spaces forming the *septal sinus* which are continuous with the nasal cavity and are lined by mucociliary or non-ciliated epithelium. In wading and aquatic species as well as in kiwis and land kingfishers a crescentic mucosal fold, the *nasal valve*, rises from the roof of the nasal cavity at the caudal end of the middle concha. When in the operative position the valve extends ventrally from the roof of the nasal cavity, and in aquatic birds then deflects water away from the olfactory epithelium.

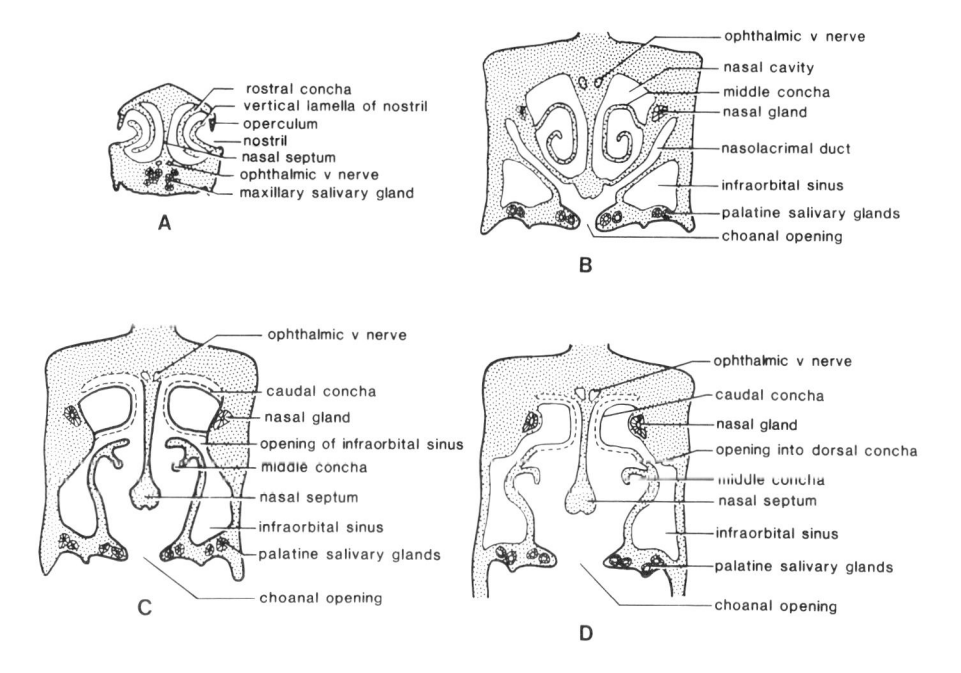

Fig 7–1 Transverse sections of the nasal cavity of the domestic fowl in rostrocaudal sequence, A being the most rostral. The olfactory mucosa is distributed in the region of the broken line in C and D. v nerve = trigeminal nerve. Redrawn from McLelland, Moorhouse and Pickering (1968), with kind permission from the editor of *Acta Anatomica*.

Nasal conchae

The nasal conchae are generally cartilaginous but may be partly or entirely bony. They increase the surface area over which the inhaled air must pass. In most birds there are three conchae: rostral, middle and caudal.

Rostral nasal concha. The rostral concha is a simple or convoluted fold which in transverse section may appear simple, branched, T-shaped or scroll-like (Fig 7–1A). It lies in the rostral *vestibular region* of the nasal cavity and is lined by stratified squamous epithelium. In some species (e.g. the Sulidae, apodiforms and quail) this concha is absent. In others, including the domestic fowl, an additional *vertical lamella* of cartilage (Fig 7–1A) arises from the ventral border of the nostril and acts to deflect the air stream. The rostral concha is highly vascular.

Middle nasal concha. This is the largest concha and fills the middle *respiratory region* of the nasal cavity (Fig 7–1B). In transverse section it is usually scroll-like with one-half to two turns (one-and-a-half in the domestic fowl), its cavity being continuous with the nasal cavity. In the Emu it is much branched. The epithelial lining is mucociliary, with alternating lines of ciliated cells and intraepithelial mucous glands. On its lumen side it is moderately vascular. This concha appears to be present in birds generally.

Caudal nasal concha. The caudal concha is usually a hollow mound projecting from the lateral wall of the caudally situated *olfactory region* of the nasal cavity (Fig 7–1C and –1D). Unlike the other conchae, its cavity connects with the infraorbital sinus, but not the nasal cavity (Fig 7–1D). The surface area is much increased in a few species by scroll-formation, and in kiwis by very extensive transverse folds. The epithelial lining of the nasal surface of the concha is olfactory. This concha is a fairly constant structure in birds, but is occasionally absent as in some falconiforms and swifts.

Septal concha. This small additional concha is unique to petrels. It arises from the nasal septum and interdigitates with the caudal concha. Its epithelial lining is olfactory.

Infraorbital sinus

The infraorbital sinus is a spacious triangular cavity under the skin in the lateral region of the upper jaw rostroventral to the eye (Fig 7–1B, –1C and –1D). Its walls consist almost entirely of soft tissues. The lumen of the sinus has two exits, both situated in its dorsal wall. One leads into the cavity of the caudal nasal concha (Fig 7–1D) and the other into the nasal cavity (Fig 7–1C). The latter opening lies immediately ventral to the caudal concha; in birds generally this opening is sited at the ventral border of the olfactory epithelium, suggesting that the sinus may somehow be involved in the olfactory mechanism. The sinus is lined by a mainly stratified squamous epithelium rostrally and by a ciliated columnar epithelium with a few mucous glands caudally. The sinus is absent in only a few species (e.g. some cormorants). Infection and swelling of the sinus is a relatively common condition in the domestic fowl and turkey.

The nasal gland

In most birds the nasal gland (formerly known sometimes as the salt gland) consists of a lateral lobe and a medial lobe, each with its own duct and ostium in the vestibular region of the nasal cavity. The gland secretes a hypertonic 5 per cent solution of sodium chloride in marine birds enabling these birds to drink sea water (a 3 per cent salt solution). It also has an osmoregulatory function in some desert species (e.g. the Ostrich as well as some falconiforms), enabling these birds to remain in water balance despite either limited fresh water and substantial water losses, or high levels of sodium chloride in the diet. However, in the large majority of terrestrial as opposed to marine birds, including all of the many passeriform species, the gland does not secrete salt, even though many of these species live in arid environments and some feed in regions of coastal rocks and sand. In the Herring Gull the gland is a crescent-shaped structure lying in a depression on the skull. Histologically the gland in this species consists of blind-ending secretory tubules radiating out from a central canal. The gland is supplied with blood via branches of the internal ophthalmic artery. These give rise to capillaries which extend between and parallel to the secretory tubules in such a manner that the flow of blood in the capillaries is in the opposite direction to the flow of secretion in the tubules; i.e., a countercurrent exchange system exists between the blood and the cells of the secretory tubules. Perfusion of the gland with blood and secretion by the gland are increased by stimulation of cholinergic vasodilator fibres in the palatine branch of the facial nerve. In the domestic fowl and closely related species only the medial lobe is present, with its duct and ostium. In such species the caudal part of the nasal gland then lies over the dorsal aspect of the eyeball and continues rostrally in the lateral wall of the nasal cavity (Fig 7–1B, –1C and –1D). Its single duct opens by a vertical slit in the nasal septum, level with the rostral concha.

Functions of the nasal cavity

Olfaction. Olfaction is an important function of the nasal cavity in birds generally and is dealt with in Chapter 16.

Filtration. As in mammals, the nasal cavity filters airborne particles by means of the mucous carpet secreted by the epithelium of the middle concha. The carpet is swept by cilia through the choanal opening and into the oropharynx where it is swallowed.

Water and heat economy: thermoregulation. The nasal cavity plays an important role in the *economy of water*. The inspired air becomes saturated with water vapour as it traverses the upper respiratory tract, this water being acquired by evaporation from the mucus covering the upper airways. Since the amount of water vapour which can be held in saturated air increases greatly as the temperature rises, the volume of water which is added to the inspired air as it reaches body temperature is substantial. If this saturated air was to be exhaled at about body temperature (as in man) nearly all the water which had been added would be lost. On the other hand, cooling of the exhaled air would reduce this loss of water, since condensation would occur as the temperature of the air falls. Such cooling does occur in birds, as in many mammals. During inhalation the walls of the nasal cavity are cooled by the air passing over them and by evaporation from their surfaces, the part nearest the nostrils being coolest and the temperature gradient rising caudally. During exhalation the warm air from the lungs, which is at body temperature, is cooled as it passes over the cool nasal walls. This cooling of the exhaled air causes condensation and thus reduces the loss of water. The amount of water which is recovered can be over 70 per cent if the ambient temperature is near room temperature and about 50 per cent if the ambient temperature is about 30°C, the better recovery rate at lower ambient temperatures being due to the greater cooling of the nasal walls.

This saving of water is of critical importance in the water balance of birds in desert habitats where there is no drinking water; the food of these birds may contain only a small amount of free water and the rest has to be metabolic water derived from oxidation of foodstuffs. Also critical is the saving of water in migrating birds. The main factor which limits the duration of non-stop long distance migrational flight is the supply of energy which can be taken on board at the beginning of the journey. This energy comes from stored fat reserves, which may account for about half of the total body weight. Oxidation of fat supplies both energy and oxidation water. Some water must always be lost at each expiration, but because of cooling and condensation the migrating bird recovers enough to remain in water balance throughout its journey.

The functional significance of the scrolls, laminae and projections of the nasal conchae is now apparent. The nasal cavity is a heat exchanger. To function efficiently it combines a narrow passageway with great surface area.

The nasal heat exchanger is also useful in *heat economy*. Energy has to be expended to warm and humidify the inspired air at low ambient temperatures, but most of this heat is recovered by condensation during expiration. This economy would be particularly important for birds living in very cold habitats. On the other hand the heat exchanger can be used to get rid of excess body heat. Essentially, this is achieved by evaporative cooling. There is the difficulty, however, that much of the excess heat which is shed during inspiration is reacquired during expiration. Some is lost, however, and the total amount that can be lost per unit time can be increased by panting. Most birds can pant, and some reinforce this by fluttering movements of the wall of the oropharynx, called gular fluttering. However, it is impossible to avoid the loss of increased quantities of water when excess heat is removed by panting. Consequently, birds also employ air-cooling. Their feathers should severely restrict this form of heat dissipation, and in cold environments they certainly do. However, birds are able to get rid of excess heat by radiation, conduction and convection, by exposing their legs and the undersurface of the wings where feathers are absent or reduced in quantity. The high body temperature of birds also favours the direct transfer of heat to the air. Even in hot environments many birds can escape to lower ambient temperatures by ascending to high altitudes.

LARYNX

The laryngeal skeleton consists of four cartilages (Fig 7–2) which become partly ossified. The *cricoid cartilage* is a median structure shaped like a sugar scoop. Caudally the left and right sides of the scoop (the cricoid wings) curve dorsally; each wing articulates in the dorsal midline with the small median comma-shaped *procricoid cartilage*, and makes a gliding contact with the body of the arytenoid cartilage. In the Raven and some species of crow each wing of the cricoid forms an articulation with the body of its cartilage. The *arytenoid cartilages* are paired, each being shaped like a tuning fork with the two prongs pointing caudally. The ventral limb of the fork is the body of the arytenoid cartilage, which articulates caudally with the procricoid and cricoid cartilages. The dorsal limb is the caudal process. In the Common Crow this process forms a synovial articulation with the body of the cartilage. The thyroid and epiglottic cartilages are absent. The *glottis* is the narrow slit between the two arytenoid cartilages (Fig 6–2A). The arytenoid cartilages are abducted and adducted by a dilator and constrictor muscle (dilator glottidis, constrictor glottidis) which thus regulate the aperture of the glottis. The muscles and cartilages form the prominent *laryngeal mound* (Fig 6–2A). The mucosal lining consists of a pseudostratified epithelium with unicellular mucous glands, intraepithelial mucous alveoli and projecting ridges carrying cilia.

As in mammals the main function of the larynx is to prevent the entrance of extraneous material into the lower respiratory tract, by reflex constriction of

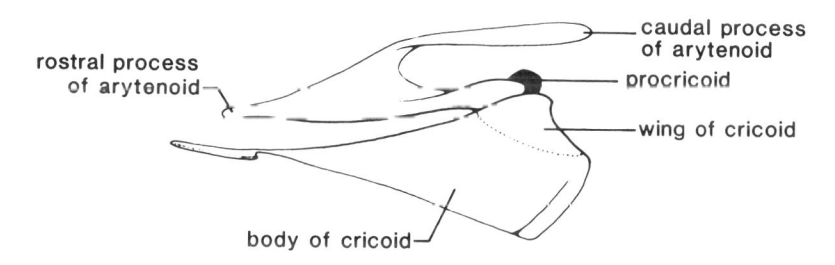

Fig. 7–2 Lateral view of the cricoid and procricoid cartilages and the left arytenoid cartilage of the domestic fowl. From King (1979), with kind permission of the publisher.

the glottis. The avian larynx plays no part in generating the voice but may modulate it. In crowing, the larynx of the domestic cock slides up and down the neck almost to the thoracic inlet, possibly functioning like the slide of a trombone. The raking action of the laryngeal mound in swallowing in the domestic fowl is mentioned in Chapter 6.

TRACHEA

The trachea is formed from a series of *tracheal cartilages* all of which in passerines and some large species are ossified. The form of these rings is rather constant in birds generally. Each is complete and not C-shaped as in mammals. The shape is like a signet ring, the broad part forming the left and right halves alternately of successive rings (Fig 7–3). Each broad part overlaps externally the narrow parts of the two adjacent rings. In the domestic fowl there are about 120 cartilages which progressively decrease in diameter caudally.

In all passerines and in the domestic fowl the trachea starts in the midline, passes to the right side of the neck, and then returns to the midline to enter the thoracic inlet (Figs 6–3 and 14–2). In species from several orders the trachea is peculiarly elongated into coils which lie between the skin and pectoral muscles as in curassows and spoonbills, or within an excavation in the sternum as in swans and cranes. In the Painted Snipe the trachea is long and convoluted in the female but short and straight in the male. In penguins the trachea is divided in the cranial end of the neck into left and right tubes by a median cartilaginous septum which is continuous with the tracheal rings. A similar septum occurs in some procellariiforms. An inflatable, saclike diverticulum, the *tracheal sac*, opens from the trachea in the Ruddy Duck and the Emu. In the male White-winged Scoter as well as in the males of some other anseriforms the caudal part of the trachea is expanded to form the *tracheal bulb*. Histologically, the mucosal lining of the trachea resembles that of the larynx.

The *tracheal muscles* are closely applied to the trachea and have diverse attachments on the sternum, clavicle, syrinx, trachea and hyobranchial apparatus. Their arrangement has not been reliably established in birds

Fig 7–3 Dorsal view of the middle of the trachea of an adult domestic fowl to show the shape of the tracheal cartilages and the way they overlap. From McLelland (1965), with kind permission of the editor of the *Journal of Anatomy*.

generally. In most species there are two such muscles, the tracheolateralis (Figs 7–5 and 7–6) and the sternotracheal (Figs 7–5 and 7–6). In many ducks, however, a third tracheal muscle, the cleidotrachealis, extends from the clavicle to the trachea (Fig 7–5). In the domestic fowl the sternohyoideus has an attachment to the cranial end of the trachea as well as to the larynx. Since all these muscles act on, and sometimes attach to, the syrinx they can also be regarded as extrinsic syringeal muscles.

As was noted in Chapters 2 and 4, the specialization of the forelimbs for flight compels the bird to use its bill for a relatively wide range of functions, including not only the manipulation of foodstuffs but many other activities such as grooming and nest building. These functions require a long neck (about × 2.7 longer than a mammal of the same body weight). This increased length increases the resistance to air flow in the trachea, but this is compensated by an increased tracheal radius (about × 1.3) compared to mammals. Consequently, the resistance to tracheal airflow is much the same in a bird and mammal of similar body weight. However, the dead space of the long wide avian trachea is about four times greater than in a mammal of comparable body size. This in turn is compensated by a much slower rate of breathing (about one-third of that of a comparable mammal) and a much greater tidal volume (about four times as high as in a comparable mammal).

SYRINX

The syrinx occurs at the junction of the end of the trachea with the beginning of the left and right primary bronchi. Its detailed structure is exceedingly variable among the avian species. The classical subdivision into tracheobronchial, tracheal and bronchial types of syrinx in different groups of birds is supposed to reflect the derivation of the cartilages of the syrinx from either the trachea or the primary bronchi. However, this depends on correctly identifying the tracheobronchial junction, and it is doubtful whether this is entirely practicable for birds generally in the present state of knowledge of this region. Nevertheless, in spite of these uncertainties the majority of birds are described as having a *tracheobronchial syrinx* since it is based on both tracheal and bronchial elements. A true *tracheal syrinx* does occur in the suboscine superfamily Furnarioidea (the so-called 'tracheophonae') and in some Ciconiidae, and is based solely on tracheal elements. A *bronchial syrinx* occurs in some cuculiforms and owls and is based purely on bronchial elements.

In most birds the syrinx is composed of a number of variably ossified cartilages, and vibrating soft structures. These hard and soft structures are combined to form both the median part of the syrinx cranially, and the divided part caudally. There are also the muscles of the syrinx.

The skeletal components of the syrinx

The syringeal cartilages (Figs 7–4A and –4B) are commonly ossified. The *tympanum* is the main component of the median part of the syrinx and is a direct continuation of the trachea. Its consists of a cylinder formed from the close apposition or fusion of two or more complete tracheal syringeal cartilages. A tympanum is usually present in the tracheobronchial type of syrinx and sometimes in the tracheal type. It is generally absent in the bronchial type of syrinx. Four such fused or firmly attached tracheal syringeal cartilages can be recognized in the tympanum of the domestic fowl (Fig 7–4B). In this species and many others the tympanum has a slightly increased diameter. The *pessulus* is a wedge-shaped cartilage, its blade lying dorsoventrally so that it divides the airway vertically. It is absent only rarely, for example, in larks and certain ratites. In pigeons the pessulus is represented by connective tissue. Caudal to the tympanum there is generally a number of thin flattened C-shaped tracheal syringeal cartilages (four in the domestic fowl, Fig 7–4B) which are typically attached to the pessulus at one end (to the ventral end in the domestic fowl) and free at the other end; in some species they are attached to the pessulus at both ends. The basis of the divided part of the syrinx is formed by the bronchial syringeal cartilages which are usually paired, incomplete and C-shaped. Three bronchial syringeal cartilages occur in songbirds (Fig 7–4A) and in the domestic fowl (Fig 7–4B), while five or six are present in the Herring Gull. The first bronchial syringeal cartilage in the domestic fowl is attached at both ends to the pessulus, the second is attached at its ventral end to the first cartilage, and the third is free at both ends; the latter free cartilage differs from the ordinary C-shaped cartilages of the

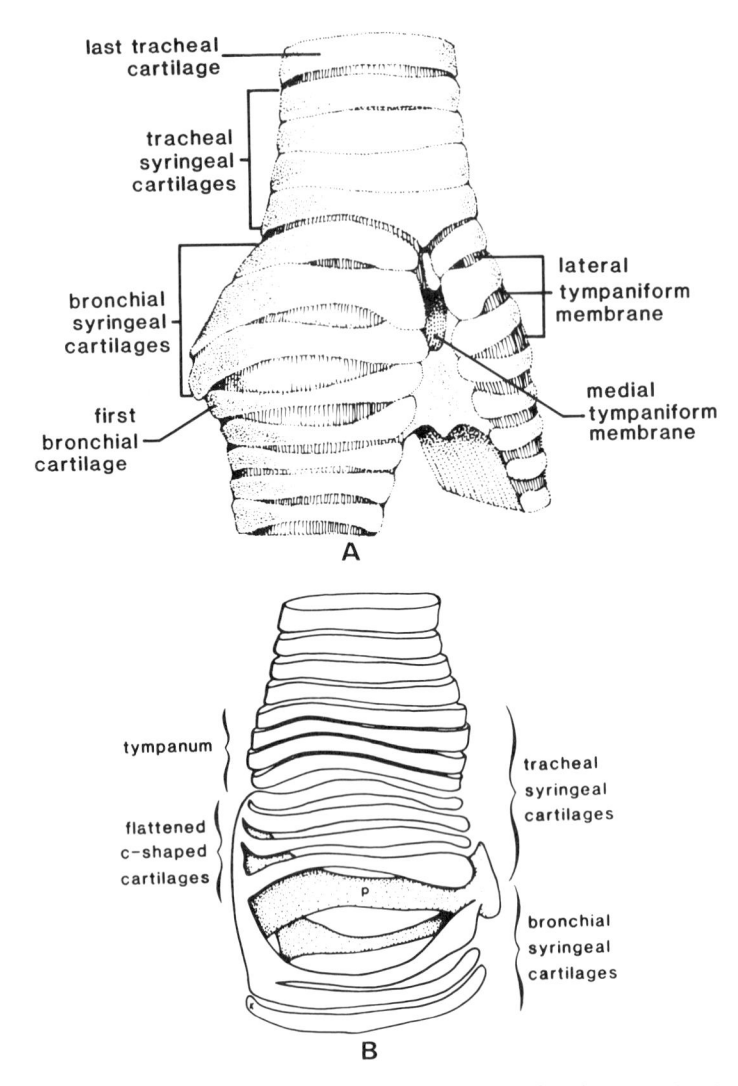

Fig 7-4 **A** Dorsal view of the syrinx of a magpie. The four tracheal syringeal cartilages are fused to form the tympanum. From Haecker (1900). **B** Lateral view of the left side of the syringeal cartilages of *Gallus*. P = pessulus. Based on Myers (1917).

extrapulmonary primary bronchus in being enlarged ventrally. In the bronchial type of syrinx the first cartilages caudal to the trachea are often complete rings, as many as ten such rings occurring in anis. The free ends of the bronchial syringeal cartilages in birds generally support the medial tympaniform membrane.

The vibrating structures of the syrinx

The paired *medial tympaniform membrane* appears to be well developed in birds generally (Fig 7-4A). Each forms most of the medial surface of the

RESPIRATORY SYSTEM

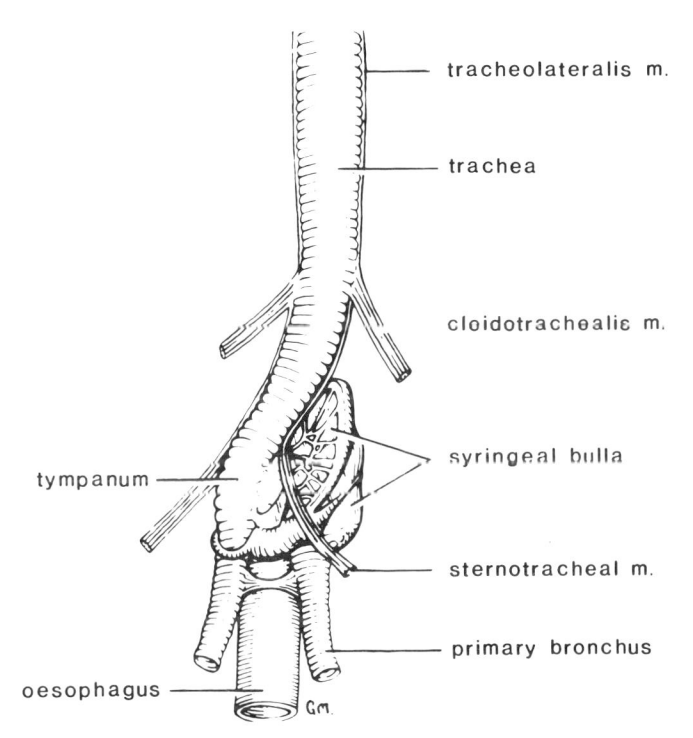

Fig 7–5 Ventral view of the syrinx of a male Tufted Duck. In this species the syringeal bulla is extensively membranous. The membranous parts of the wall resemble thin transparent windows stretched tightly between the slender bars of the bony framework of the bulla. From King (1979), with kind permission of the publisher.

divided part of the syrinx, being held between the free ends of the bronchial syringeal cartilages. It is the primary vibrating membrane. The paired *lateral tympaniform membrane* (Fig 7–4A) is the membranous area which stretches between the cartilages on the lateral aspect of the syrinx. Often it takes the form of indistinct straps between the bronchial syringeal cartilages as in songbirds; more rarely, as in the domestic fowl, it consists of a relatively extensive sheet between the last tracheal syringeal cartilage and the first bronchial syringeal cartilage. In the domestic fowl the inward curve of this membrane gives the syrinx its characteristic constricted appearance (Fig 12–2). The *tracheal membrane* is a modified lateral tympaniform membrane and stretches between the tracheal syringeal cartilages in suboscine passeriforms possessing a tracheal type of syrinx. The *lateral labium* is a pad of elastic tissue, sometimes with a cartilaginous basis, which projects into the lumen of the syrinx from the cartilage of the lateral wall. The *medial labium* projects into the lumen from the pessulus and is probably less constantly present than the lateral labium. The labia occur variably in passerines and some other species including anseriforms, but both labia are absent in the domestic fowl. In many male ducks of the subfamily Anatinae (including the domestic duck) the syrinx is extensively modified to form an asymmetrical dilation on the left side, the *syringeal bulla* (Fig 7–5). The bulla apparently arises from a widening of the bronchial syringeal cartilages, and is generally partly or entirely bony although in some species it is extensively membranous as in Fig 7–5. In the Shelduck the dilation occurs on the right side of the syrinx, while in the Dendrocygnini the swelling is bilaterally symmetrical. A bulla is not present in true swans and geese.

The muscles of the syrinx

The syringeal muscles are very variable. Most are extrinsic muscles which have one attachment outside the syrinx. In songbirds there are five pairs of

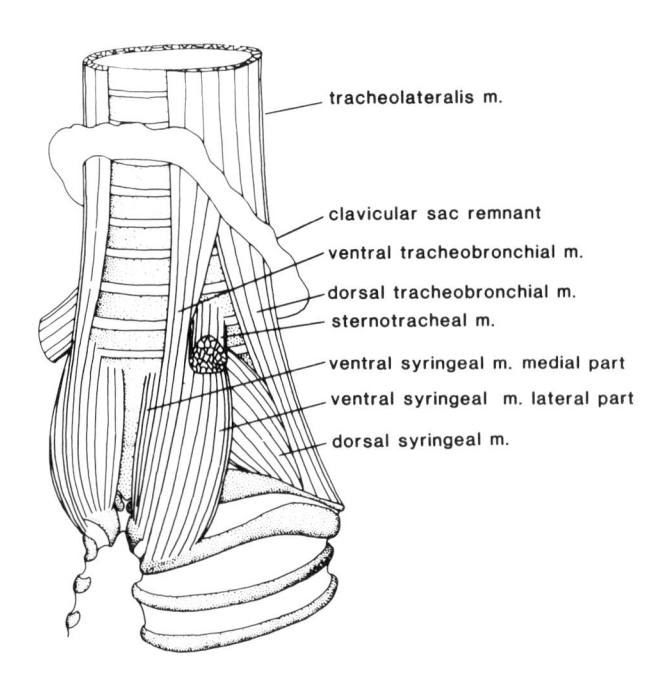

Fig 7–6 Ventrolateral view of the syrinx of the Common Crow. Of the five pairs of syringeal muscles (see text) only four are shown, the tracheobronchialis brevis being absent. From Ames (1971), with kind permission of The Peabody Museum of Natural History.

syringeal muscles (tracheobronchialis dorsalis, tracheobronchialis brevis and tracheobronchialis ventralis; syringealis dorsalis and syringealis ventralis) (Fig 7–6). Suboscine passerines have four pairs of syringeal muscles (vocalis dorsalis and vocalis ventralis; obliquus ventralis and obliquus lateralis). In the great majority of non-passeriform species there is usually only one syringeal muscle, the tracheolateralis (Fig 7–5). However, there are certainly two pairs of short syringeal muscles in some parrots. In a small number of species (including some ratites and galliforms) there are said to be no syringeal muscles whatever, not even tracheolateralis.

Function of the syrinx

The function of the syrinx is to produce the voice. It has been generally assumed that the primary source of the sound is vibration of the tympaniform membranes. A recent alternative view is that when the membranes are in the position for vocalizing they form two slots (where the left and right bronchial elements join the median part of the syrinx) which result in two streams of turbulent air, each producing sound much as a whistle can be produced by human lips. A whistle mechanism would have the advantage of producing loud sounds at the cost of only small volumes of air. Vocalization is apparently produced in the expiratory phase only. It has been suggested that, unlike man who sings by means of a continuous expiration, the songbird may be able to trill or warble by a series of very rapid oscillatory 'mini-breaths', but another hypothesis is that warbles and trills are produced by fast pulses of contraction of the abdominal muscles causing a rapidly intermittent but prolonged expiration. However, it may be necessary to postulate 'mini-breaths' in order to explain how some warblers and especially the Skylark can maintain an unbroken series of pulsed sounds for such prolonged periods. The coordination of breathing and vocalization appears to reside in the intercollicular nucleus of the mesencephalon, and perhaps in parts of the lateral mesencephalic nucleus.

It seems likely that the vibration of the membranes is associated with fluctuating pressure gradients between the clavicular air sac, which surrounds the syrinx, and the lumen of the syrinx. At the onset of expiration the pressure rises in the clavicular sac and momentarily closes the syrinx by forcing the tympaniform membranes into the syringeal lumen. Tension is then applied to the membranes by the intrinsic or extrinisc muscles of the syrinx, thus drawing the membranes partly from the airway. Air then flows past the tautened membranes, which are thus thrown into vibration and sound is produced. The labia may well be of considerable importance in regulating the voice.

THE LUNG

External appearance

The avian lungs are dorsal in position (Figs 5–1, 6–3 and 7–12). The liver, not the lungs as in mammals, lies on each side of the heart (Figs 6–3 and 6–8). Each lung is small, flattened, and in most birds quadrilateral in shape (Fig 7–7); in supposedly primitive birds like penguins, the lung is more triangular, the blunt apex pointing ventrally. In no species of bird is the lung lobed as in many mammals. The vertebral ribs are deeply embedded in costal grooves on the dorsomedial part of the avian lung (Figs 7–7 and 7–12), about one-quarter of the volume of the lung of the domestic fowl being thus enclosed between the ribs. In most species the cranial end of the lung reaches the movable rib which is carried by the last cervical vertebra (Fig 4–3). Usually the caudal end reaches the cranial border of the ilium, but in some species such as storks,

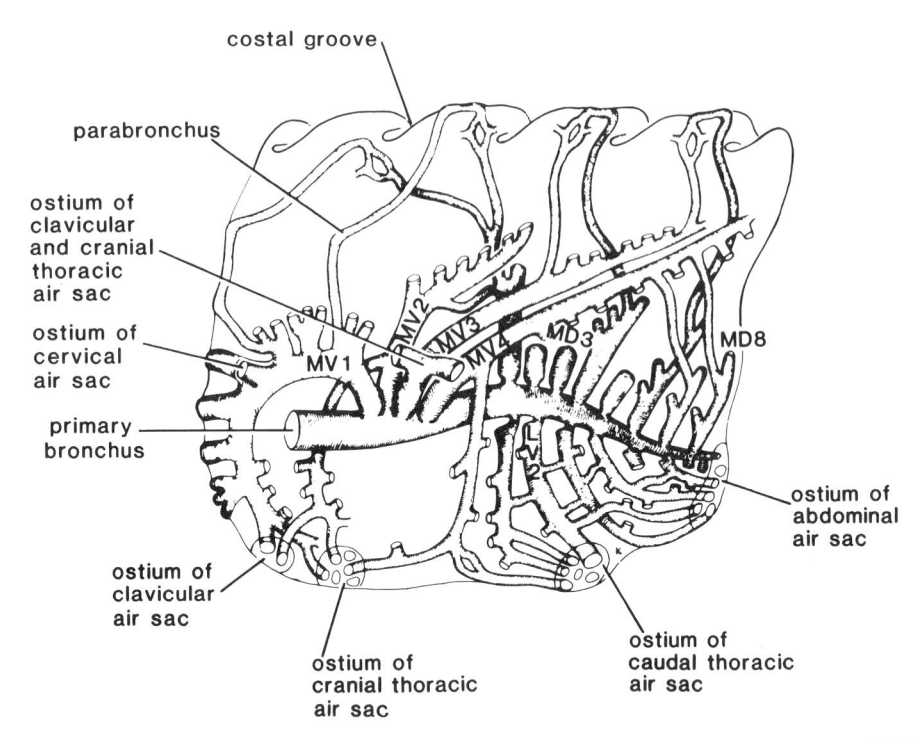

Fig 7–7 The septal (ventromedial) surface of the right lung of the domestic fowl. The lung is drawn as though transparent, to show the bronchi. The primary bronchus is the large tube running right through the lung. The first secondary bronchi to arise from it are the four medioventral secondaries (MV1 to MV4). Eight mediodorsal secondary bronchi (e.g. MD3 and MD8) arise from the dorsal aspect of the caudal half of the primary bronchus, and eight lateroventral secondaries (e.g. LV2) arise from its ventral aspect. Only a few examples of the many parabronchi are shown. The air sacs connect to the lung at six sites, known as ostia. Four of these occur on the costoseptal border of the lung and are indicated by rings. The fifth, for the cervical sac, is in the craniomedial region of the lung. The sixth is somewhat cranial to the centre of the septal surface, and consists of two large direct connections, one to the clavicular and one to the cranial thoracic sac. From King (1966), with kind permission of the publisher.

geese and the Hoatzin, it extends almost to the level of the hip joint. In transverse section (Fig 5–1) the lung is characteristically wedge-shaped and consequently has three surfaces. Its dorsolateral surface is in contact with the ribs and is therefore known as the costal surface. The dorsomedial surface is in contact with the vertebrae and is named the vertebral surface. The ventromedial surface is in contact with the horizontal septum and is termed the septal surface. Several of the larger air passages (i.e. the secondary bronchi) are visible externally at the surface of the lung; in some species, such as the domestic fowl, which have parabronchi of relatively large calibre, these smaller airways are also visible on the lung surface. The hilus, where the primary bronchus and the pulmonary artery and vein enter the lung, is situated on the septal surface (Fig 7–7). Whilst the *weight* of the avian lung is not less in relation to total body weight than in mammals, the *volume* per gram body weight is about 25 per cent smaller.

Primary bronchus

The two extrapulmonary primary bronchi, left and right, are formed by the bifurcation of the airway at the syrinx. Each perforates the septal surface of the lung and then continues as the intrapulmonary primary bronchus to the caudal extremity of the lung (Fig 7–7). The calibre of the primary bronchus is variable. Usually it is widest at the point where it enters the lung, and from there tapers progressively to its caudal end. A spindle-shaped dilation called the vestibule has been described at various points between the entry into the lung and the caudal exit of the primary bronchus, but there is no reliable evidence for its presence in any species.

Histologically the mucosa of the primary bronchus consists of a pseudostratified epithelium with goblet cells, intraepithelial mucous alveoli and projecting ridges carrying cilia. Beneath this is a well-developed layer of smooth muscle. A series of C-shaped cartilages is present in the extrapulmonary portion and in the cranial part of the bronchus within the lung.

Four groups of secondary bronchi arise from the primary bronchus.

The secondary bronchi

The secondary bronchi include all those bronchi of the second order, that is all those which arise from the primary bronchus (Fig 7–7). There are four groups, named according to the regions of the lung which they supply. The first two groups are the medioventral and mediodorsal secondary bronchi; their branches, and the parabronchi which arise from them, form the thick medial part and the whole of the cranial part of the lung (Fig 7–7). The other two groups are the lateroventral and laterodorsal secondary bronchi, and these with their parabronchi form the thin lateral region of the lung. All the larger secondary bronchi tend to be constricted at their point of origin and then immediately expanded.

The medioventral secondary bronchi. The number of medioventral secondary bronchi is usually four but there are species with five (e.g. cassowaries and storks) and even six (e.g. pelicans). In calibre these are the largest secondary bronchi. They originate one after the other from the *dorsomedial wall* of the *cranial third* of the primary bronchus (Fig 7–7), and run medially on the septal (ventromedial) surface of the lung where their main trunks are externally visible. Their branches, and the parabronchi which arise from these branches, spread over about three-quarters of the septal surface of the lung, except for the caudoventral region. In earlier terminologies the medioventrals have been called ventrobronchi, or anterior dorsal or craniomedial secondary bronchi.

All the other secondary bronchi arise along the *caudal two-thirds* of the primary bronchus.

The mediodorsal secondary bronchi. The seven to ten mediodorsal secondary bronchi (eight in the domestic fowl) arise in a linear sequence from the *dorsal* wall of the primary bronchus (Fig 7-7). They become progressively smaller in calibre towards the caudal end of the series. In some species they tend to radiate like a fan from their origins along the primary bronchus, but in the domestic fowl they tend to run vertically and parallel with each other as in Fig 7-7. Their terminal branches form parabronchi which extend over the costovertebral border (between the costal and vertebral surfaces) onto the vertebral surface of the lung. When the laterally situated neopulmonic portion of the lung (p. 126) is absent as in penguins and the Emu or poorly developed as in cranes, cormorants and auks, the main trunks of the mediodorsal secondary bronchi lie on the costal surface of the lung and are therefore visible externally. With increasing development of the neopulmo, however, greater stretches of the mediodorsal bronchi become buried within the lung so that in most species (including the domestic fowl) only their terminal branches can be seen on the costal surface. The mediodorsals were previously known as the dorsobronchi, or the posterior dorsal or caudodorsal secondary bronchi.

The lateroventral secondary bronchi. The lateroventral bronchi also arise in a line from the primary bronchus at the same craniocaudal level as the mediodorsal bronchi, but from its *ventral* wall, that is directly opposite to the openings of the mediodorsal secondary bronchi (Fig 7-7). In birds generally there are two or three large lateroventrals at the cranial end of the series, the rest dwindling away caudally into small irregular openings with a calibre resembling parabronchi. In the domestic fowl there are about eight lateroventral secondary bronchi, the first five or so being quite large (Fig 7-7). The second lateroventral bronchus (LV2 in Fig 7-7) is particularly large in diameter, and in the domestic fowl and many other species it forms the direct connexion to the caudal thoracic air sac. In a few species (e.g. penguins) the openings of the lateroventral bronchi are joined to the openings of the mediodorsal bronchi by crescentic ridges of mucosa. The parabronchi of the lateroventral bronchi form the thin ventrolateral region of the lung. In the Emu, cormorants and penguins the lateroventral secondary bronchi lie on the costal surface of the lung and therefore can be seen externally, but in other birds they are covered to a varying extent by the parabronchi of the neopulmo (p. 126). In older nomenclatures the lateroventrals were called laterobronchi or posterior ventral or caudoventral secondary bronchi.

The laterodorsal secondary bronchi. These arise from the same craniocaudal level of the primary bronchus as the mediodorsal and lateroventral bronchi but from the *lateral* wall of the primary bronchus. Unlike the other groups of secondary bronchi they have a scattered origin instead of arising in a line. The laterodorsals are the most variable group of secondary bronchi. They are entirely absent in penguins and are very reduced in number in cormorants and storks. They are most numerous in the domestic fowl, gulls, pigeons and songbirds. The first two, and sometimes the first five or six as in the Mute Swan, may be quite large, but the others are of small (parabronchial) diameter and erratic in position. In the domestic fowl (but not shown in Fig 7–7) there are three to five large ones at the cranial end of the series, and caudal to these there are about twenty to twenty-five small ones, many of the latter being of parabronchial dimensions. The laterodorsal secondary bronchi extend mainly laterally towards the costal surface of the lung. They were previously named dorsolateral secondary bronchi or caudolateral secondary bronchi, but some authors excluded them from the secondary bronchi because of their small calibre.

The constricted necks of the secondary bronchi, for a length of 1–2 mm, have the same histological structure as the primary bronchus. Subsequently, the trunks of the mediodorsal, lateroventral and laterodorsal secondary bronchi become lined with a simple squamous epithelium and their walls have atria leading to infundibula and air capillaries; thus they resemble parabronchi in their structure.

The parabronchi

After a course of only a few millimetres, the secondary bronchi of all four groups give off a large number of tubes of small but uniform calibre, the parabronchi (Fig 7–7). The course of the parabronchi varies in detail in the four different groups of secondary bronchi. However, all have certain features in common: they are not blind-ending but anastomose freely with other parabronchi, they carry atria and exchange tissue in their walls (Fig 7–8), and they tend to have a fairly constant mean internal diameter within each species (ranging from 0.5 mm in hummingbirds and the Goldcrest to 2 mm in the King Penguin and the domestic fowl). The parabronchi in some previous nomenclatures were referred to as tertiary bronchi.

The cranial and dorsomedial parts of the lung, constituting about two-thirds of the lung, are made up of layer upon layer of long hooplike parabronchi

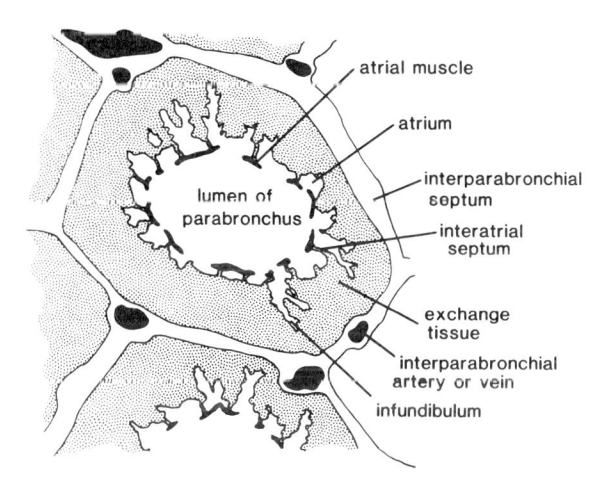

Fig 7–8 A drawing of a histological section showing parabronchi of the domestic fowl in transverse section.

running between the medioventral and mediodorsal secondary bronchi (Fig 7–7). Together these secondary bronchi and their parabronchi form an integrated functional unit, the *medioventral–mediodorsal system* of bronchi. An anastomotic line along the vertebral surface of the lung shows where the parabronchi of the medioventral secondary bronchi met and joined the parabronchi of mediodorsal secondaries during development (Fig 7–7). In this medioventral–mediodorsal system the gas flows constantly in the same direction, that is from the mediodorsal to the medioventral secondary bronchi (p. 139).

The remainder of the lung, that is the thin ventrolateral region of the caudal part of the lung, is made up of anastomosing networks of short parabronchi joining the lateroventral and laterodorsal secondary bronchi both to each other and to the other two groups of secondary bronchi. In fact each of the four groups is connected to all of the other three.

The paleopulmo and neopulmo. The medioventral–mediodorsal system of bronchi and its parabronchial connections to the first two or three lateroventral secondary bronchi constitute the paleopulmo. This name has been given because all birds possess this bronchial complex, and it is therefore presumed to be phylogenetically primitive.

On the other hand the anastomosing bronchial network in the ventrolateral part of the lung, together with the connections of this network to the caudal air sacs, forms the neopulmo. This component of the lung is absent, or virtually absent, in the Emu and penguins, which have been regarded as relatively primitive birds. It is minimally developed in storks, and poorly developed in ducks, gulls, owls, buzzards, cormorants, auks and cranes. However, in most birds, including the galliform species (e.g. the domestic fowl) and the very numerous passeriform species, it is well developed. Consequently, this part of the lung has been regarded as phylogenetically recent, and hence received the name neopulmo. Actually, the phylogenetic implications of the terms paleo- and neo- are questionable at this stage of our knowledge of avian pulmonary anatomy and avian phylogeny, but the recognition of these two components of the lung is functionally convenient because they appear to differ physiologically.

In birds in which the neopulmo is fully developed it forms about 20 to 25 per cent of the lung. As already mentioned above under mediodorsal secondary bronchi, when the neopulmo is well developed it covers over the main trunks of the mediodorsal secondary bronchi. Because the neopulmo is reduced in ducks some of the mediodorsal trunks are visible on the costal surface of the lung. The accessibility of the mediodorsal secondary bronchi via the lateral body wall in the duck for collecting data on flow and partial pressures of pulmonary gas makes the duck the species of choice, rather than the domestic fowl, for much of the research on pulmonary function.

Atrial muscles. The lumen of the parabronchus is lined by a simple squamous epithelium. Beneath this is a network of spiral bands of smooth muscle (Fig 7–8), homologous to the helical bronchial muscle of mammals. These atrial muscles are innervated, and appear to be capable of regulating the diameter of the parabronchi and their atria.

Atria. The atria are numerous pocketlike polygonal cavities, between 100 and 200 μm in diameter, which open into the lumen of the parabronchus between the atrial muscles (Fig 7–8). Their walls are lined by a flat or cuboidal epithelium, containing many osmiophilic bodies which are believed to be the source of surfactant. At the bottom of each atrium are several funnel-shaped openings, the *infundibula*, which lead into the air capillaries. The *interatrial septa*, which separate the atria, contain numerous elastic fibres. The floor of each atrium also carries elastic fibres, but these fibres cease abruptly at the junction between the atrial floor and the exchange tissue. The atria and the atrial muscles tend to be largest in birds which fly poorly or not at all (e.g. coots and the domestic fowl). In other species, especially small songbirds, they tend to be relatively reduced in size.

Interparabronchial septa. In histological sections of the lung of some species, including the domestic fowl, the parabronchi and their exchange tissue are enclosed within roughly hexagonal areas by interparabronchial septa of connective tissue (Fig 7–8). These septa carry the interparabronchial arteries and veins of the pulmonary circulation. In many other species, notably small songbirds, the septa are much reduced or absent.

Air capillaries

The air capillaries (Fig 7–9) are tortuous narrow tubes arising from the infundibula and atria. They branch and anastomose freely with each other, forming an extensive network of air-carrying tubules. Their diameter is greatest in penguins, swans and coots, where it reaches about 10 μm. Smaller air capillaries about 3 μm in diameter occur in songbirds. (The diameter of the alveoli in the smallest mammal is about 35 μm). The diameter of the air capillary probably changes little during the respiratory cycle. The very small diameter of the terminal airway in birds means that the pressure gradient for diffusion of oxygen is much greater in birds than in mammals. *Surfactant* probably arises from the osmiophilic bodies of cells, the granular cells, which line the atria; a thin layer also appears to be present in the air capillaries, where it presumably restricts diffusion of fluid from the blood plasma rather than promotes dilation of the air capillaries, the forces of surface tension being so prohibitively high in tubules of such small radius that anything more than minimal dilation seems improbable.

The air capillaries are intimately entwined with a profuse network of blood capillaries (Fig 7–9), and it is here that the gaseous exchanges occur. The volume of pulmonary capillary blood per gram body weight is about 20 per cent greater in birds than in mammals. As in mammals the *blood–gas barrier* consists of three essential elements, that is the endothelial cell, a common basal lamina and the squamous epithelial cell (Fig 7–9). The barrier is much thinner than in mammals, however, having an arithmetic mean thickness of about 1.06 μm in birds generally compared to 1.50 μm in mammals generally. The harmonic mean thickness, which is a better indicator of the resistance of the barrier to oxygen diffusion, is also much thinner (33 per cent) in birds than in mammals, the mean values being 0.13 μm in birds and 0.53 μm in mammals. The reduction in thickness of the barrier in birds is due to the thinness of the epithelial cell lining the air capillary.

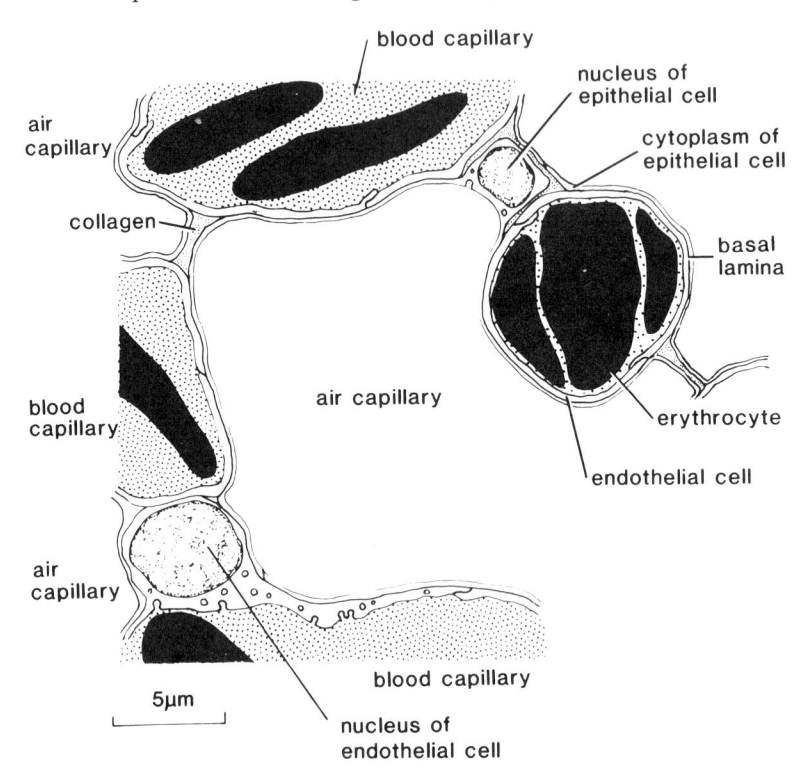

Fig 7–9 A drawing of an electron micrograph of the exchange tissue of the lung of the domestic fowl.

The pulmonary circulation

The pulmonary arteries give rise to numerous *interparabronchial arteries* (Figs 7–8 and 7–10) which travel alongside and obliquely around the parabronchi (in the interparabronchial septa when these are present, as in the domestic fowl). Each of these arteries forms a series of *intraparabronchial arterioles* which penetrate the exchange tissue and terminate in the *blood capillary* networks associated with the air capillaries (Figs 7–9 and 7–10). The blood capillaries drain mainly via *intraparabronchial venules* into large *atrial veins* lying beneath the interatrial septa (Fig 7–10). Some capillaries drain into the atrial veins via small *septal venules* lying below the atrial muscles (Fig 7–10). The atrial veins also receive blood directly from the capillaries (Fig 7–10). The blood flow in the capillaries is therefore essentially from the periphery of the

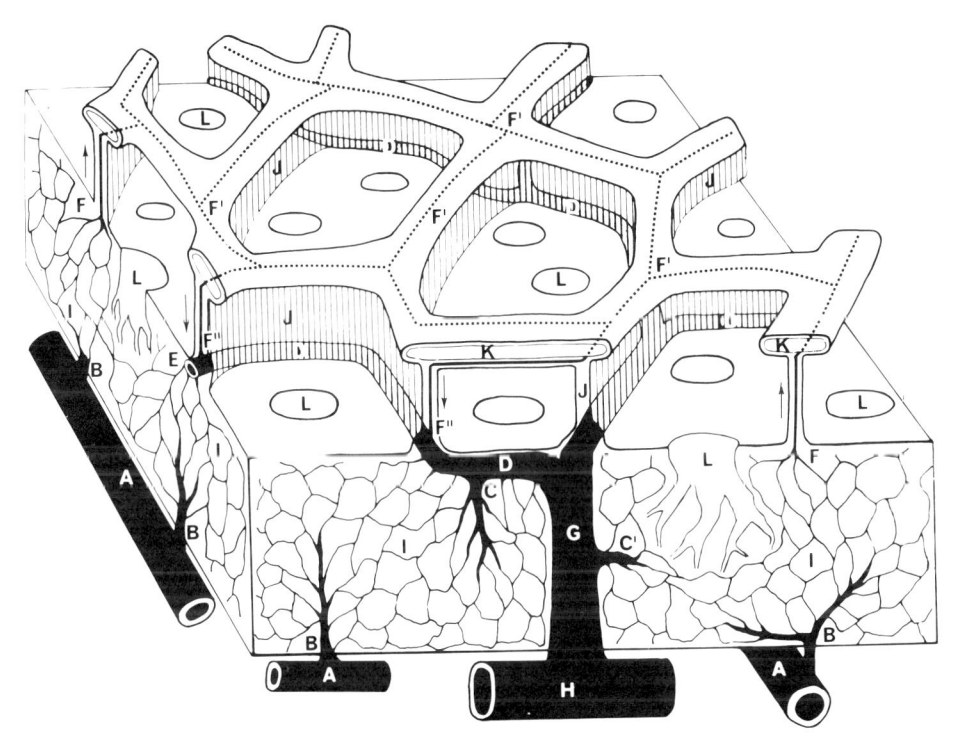

Fig 7–10 A stereogram of the blood supply of the parabronchial wall. The blood in the interparabronchial arteries (A) flows into blood capillaries of the exchange tissue via intraparabronchial arterioles (B) which penetrate the exchange tissue for varying distances. Most of the blood in the capillaries of the exchange tissue is collected by intraparabronchial venules (C) which empty into atrial veins (D) lying below the interatrial septa. The atrial veins also receive blood directly from the capillaries (E). Some capillaries form septal venules which climb the interatrial septa as at F, the direction of blood flow in the vessels being indicated by two arrows pointing upwards. The septal venules form a network (the broken lines at F') below the atrial muscles. The network drains into the atrial veins by septal venules which descend the interatrial septa as at F'', the direction of blood flow being indicated by two downward-pointing arrows. Blood in the atrial veins flows into an intraparabronchial vein (G) which passes through the exchange tissue to drain into an interparabronchial vein (H). On its course through the exchange tissue an intraparabronchial vein sometimes receives an intraparabronchial venule (C'). I = exchange tissue; J = interatrial septum; L = infundibulum. From Abdalla and King (1975), with kind permission of the editor of *Respiration Physiology*.

parabronchus towards the parabronchial lumen. Each blood capillary therefore makes contact with air capillaries of only a short fraction of the total parabronchial length. The atrial veins drain via *intraparabronchial veins* which travel outwards through the exchange tissue and empty into *interparabronchial veins* lying in the interparabronchial septa (Figs 7–8 and 7–10). The intraparabronchial veins also drain some intraparabronchial venules (Fig 7–10). The abundant smooth muscle in the walls of the small interparabronchial arteries has a noradrenergic innervation, and may therefore be capable of regulating blood flow through the exchange tissue. This may be a factor in the matching of blood perfusion against parabronchial ventilation (p. 141).

The arrangement of the pulmonary circulation provides a *cross-current system* in the sense that the blood flow approaches the parabronchus more or less at right angles to the air which is flowing along the parabronchial lumen (Fig 7–11). Deoxygenated blood of uniform composition is transmitted by each heart beat simultaneously to all parts of the parabronchus. In contrast the composition of the air is not uniform; on the contrary, the concentration of carbon dioxide and oxygen in the air changes progressively during its passage along the parabronchus, due to continuous gaseous exchanges with the blood.

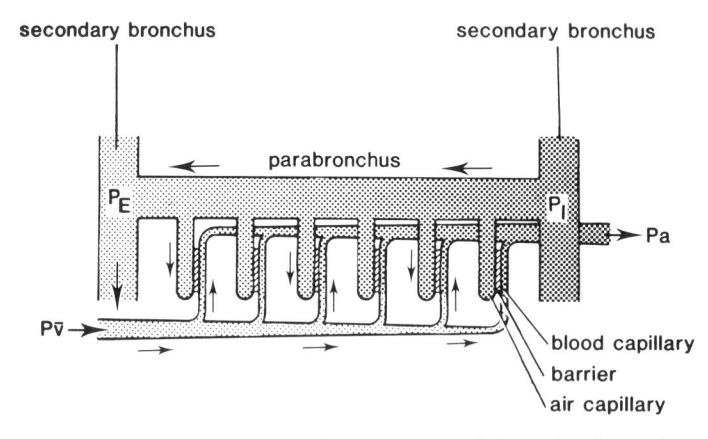

Fig 7–11 Schematic drawing of the exchange tissue of the avian lung. A cross-current arrangement exists between bulk parabronchial gas flow and the flow of blood in the blood capillaries. The blood capillaries are arranged essentially parallel to the air capillaries to form an auxiliary counter-current arrangement. The intensity of shading indicates the relative partial pressures of O_2 in the gas and blood. Thus the partial pressure is relatively high in gas entering the parabronchus (PI) and low in gas leaving the parabronchus (PE); the partial pressure is relatively low in the mixed venous blood entering the exchange tissue ($P\bar{v}$) and high in arterial blood leaving the exchange tissue (Pa). Gas and blood come into equilibrium across the barrier between each air capillary and blood capillary. Arrows indicate the direction of flow. From Scheid (1979), with kind permission of the editor of *Review of Physiology, Biochemistry and Pharmacology*.

Therefore, the blood in the pulmonary capillaries becomes arterialized to different degrees depending on how far along the parabronchus the blood capillary is located. Arterial blood is a mixture of the blood leaving all of these blood capillaries. The pulmonary circulation also provides an auxiliary *counter-current system* within the exchange tissue, the centripetal direction of flow of the deoxygenated blood arriving in the intraparabronchial arterioles and blood capillaries being opposite to the centrifugal direction of diffusion of gas from the parabronchial lumen into the air capillaries (Fig 7–11). Strictly, this is not a true counter-current system, since there is convective flow (i.e. a bulk flow of fluid) in the blood capillary but not in the air capillary (where there is diffusion only); nevertheless this relationship enhances gas exchange.

The air sacs

In the embryo there are six pairs of air sacs (Fig 7–13). In the great majority of birds two pairs have fused at or soon after hatching to form a single median sac (the clavicular sac). In the domestic fowl and a number of other species another pair fuses at about this time to form a further single median sac (the cervical sac), the other three pairs (cranial thoracic, caudal thoracic and abdominal sacs) remaining paired; in the adult there are therefore eight air sacs altogether in such species, i.e. one cervical and one clavicular sac, and two

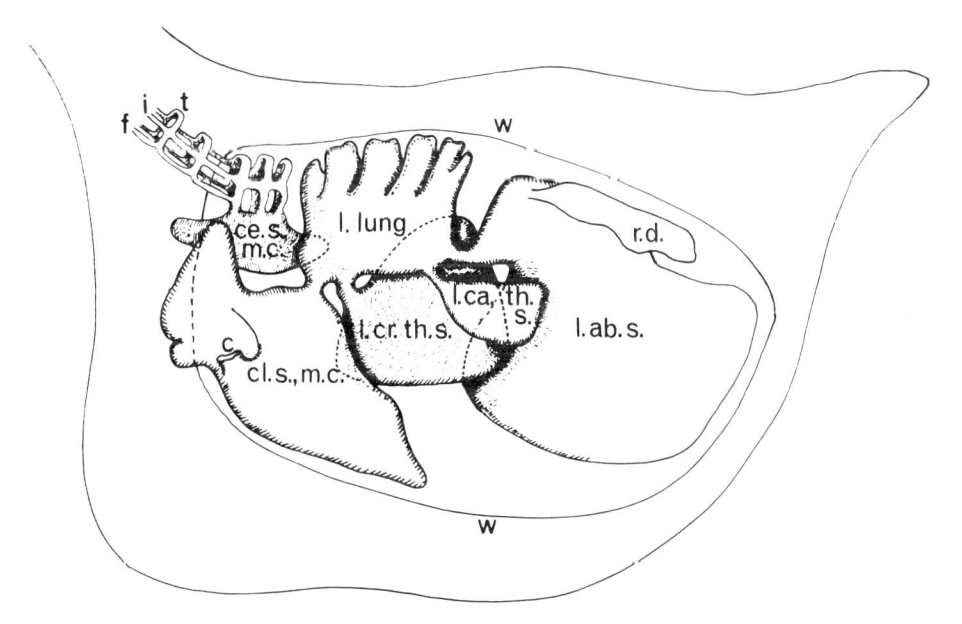

Fig 7–12 Diagrammatic view of the left lung and air sacs of an adult domestic fowl. The lateral part of the clavicular sac is not shown. c = canal (cut) connecting the median part of the clavicular sac and the left lateral part; ce.s., m.c. = main chamber of cervical sac; cl.s., m.c. = main chamber of clavicular sac; f = vertebral diverticulum of cervical sac inside transverse foramen; i = vertebral diverticulum of cervical sac inside neural canal; l.ab.s. = left abdominal sac; l.ca, th.s. = left caudal thoracic sac; l.cr.th.s. = left cranial thoracic sac; l = left lung; r.d. = perirenal diverticulum of the left abdominal sac; t = transverse intervertebral connection between the diverticula of the cervical sac; w = body wall. From King (1966), with kind permission of the publisher.

cranial thoracic, two caudal thoracic and two abdominal sacs. The following account applies to the domestic fowl.

Cervical sac. The cervical sac (Fig 7–12) consists of a median chamber, lying between the lungs and dorsal to the oesophagus, which leads to a pair of tubular *vertebral diverticula* on each side of the vertebral column, one inside the neural canal and one outside. The latter diverticulum goes through the transverse foramina of the cervical vertebrae as far cranially as the axis. The diverticula of the cervical sac also invade the vertebrae.

Clavicular sac. The clavicular sac originates from four primordial sacs, a medial pair and a lateral pair, which fuse to form one sac on each side and then fuse along the midline to form a single unpaired sac (Figs 7–12 and 7–13). In the adult this is a large and complicated sac occupying the thoracic inlet. It consists of *intrathoracic diverticula* which extend around the heart and along the sternum, and *extrathoracic diverticula* which spread between the bones and muscles of the thoracic girdle and shoulder joint, and invade certain of the bones. Numerous blood vessels and nerves, and the oesophagus, trachea, syrinx and associated muscles, are suspended within folds of the clavicular air sac or between the clavicular and cervical sacs.

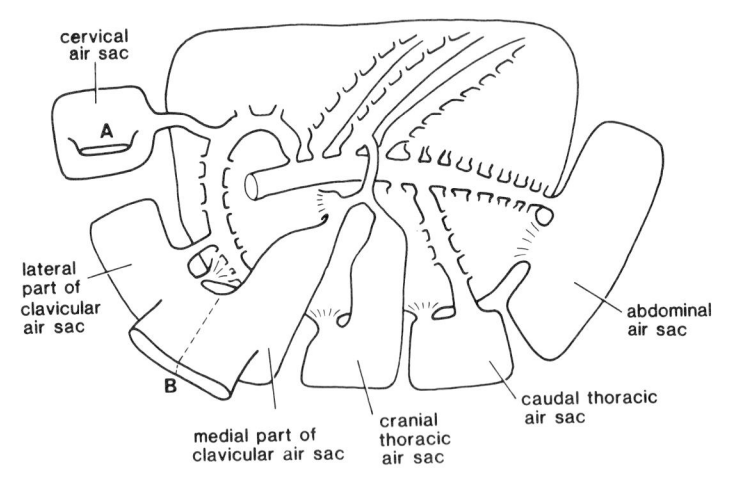

Fig 7–13 Diagram of the right lung of the domestic fowl, showing the developmental origins of the air sacs. There are six primordial pairs of air sacs. The first pair, the cervical sacs, are fused across the midline. The second and third pairs are fused to each other and across the midline, forming a single median sac, the clavicular sac. A and B are the fusions across the midline, transected. From King and Atherton (1970), with kind permission of the editor of *Acta Anatomica*.

Cranial thoracic sac. This sac is paired in birds generally. Each consists of a more or less symmetrical cushionlike cavity, lying between the horizontal and oblique septa, and is therefore essentially dorsolateral in position within the thoracic cage (Fig 5–1). It has no diverticula.

Caudal thoracic sac. This sac again is paired in birds generally. It lies in a dorsolateral position, caudal to the cranial thoracic sac (Fig 7–12). In many species it occupies the caudal part of the space between the horizontal and oblique septa. In the domestic fowl it is very small and is covered medially by the cranial thoracic sac and abdominal sac which press it against the body wall. It possesses no diverticula.

Abdominal sac. The abdominal sac is paired in all birds. At the point where it arises from the lung each abdominal sac pierces the horizontal septum. It then spreads caudally like a thin-walled balloon between the loops of intestine, except where it is attached to the dorsal abdominal wall (Figs 5–1, 5–2 and 7–12). Three paired *perirenal diverticula* (Fig 7–12) spread dorsally between the kidney and the pelvis and synsacrum and invade the adjacent vertebrae and pelvic girdle. In many species extensive *femoral diverticula* invade the bones and muscles of the pelvic limb; in the domestic fowl three small diverticula escape from the pelvic cavity but do not invade bones. When artificially inflated the abdominal sac has an enormous capacity in the domestic fowl, but in life most of it is compressed to a series of irregular potential spaces. Eversion of the abdominal air sac through the primary bronchus as far as the syrinx, thus obstructing the pathways to both lungs, has been described in the domestic fowl.

Air sacculitis. Air sacculitis is a general term covering several diseases of the air sacs in domestic fowl, turkeys and ducks; the identity of these individual diseases cannot be distinguished with certainty at postmortem examination in the processing plant. When trimming a carcase affected with air sacculitis serious consideration must be given to the extensive diverticula of the cervical sac (vertebral diverticula), clavicular sac (extrathoracic diverticula) and abdominal sacs (perirenal and femoral diverticula), since infection of these sacs may require rejection of the parts of the carcase invaded by the diverticula.

The cranial and caudal air sacs. Anatomically and physiologically it is convenient to divide the air sacs into two groups. The *cranial sacs* include the cervical, clavicular and cranial thoracic sacs. All of these arise from the medioventral secondary bronchi (see below) and receive vitiated air from the lung (see section on **Air pathways**). The *caudal sacs* consist of the caudal thoracic and abdominal sacs. These arise from lateroventral secondary bronchi or the primary bronchus, and receive relatively fresh air from the trachea.

Species variations in the air sacs. In nearly all birds the air sacs arise in the embryo from the six primordial pairs of air sacs. This suggests that the maximum possible number of air sacs in a bird should be 12. However, the actual number is apparently always less than this, usually because of the fusion of the four primordial clavicular sacs into a single sac. This composite midline sac, together with the other four pairs, makes nine sacs, and this appears to be the most usual total. Among the few birds with a larger number are storks, in which each caudal thoracic sac is apparently divided into two, making 11 sacs. Also in some species (e.g. loons) the four primordial clavicular sacs may remain separate; however, in loons the cervical sacs seem to be absent, reducing the total to ten sacs. In the domestic fowl there are eight sacs, due to the fusion of the cervical sacs, in addition to the usual fusion of the four primordial clavicular sacs (Fig 7–13). In songbirds as a group the cranial thoracic sacs are apparently fused to the single median clavicular sac, making seven sacs in all (the small paired cervical sacs, a single clavicular-thoracic sac and paired caudal thoracic and abdominal sacs). One of the most curiously modified of all birds is the domestic turkey. In this species, the caudal thoracic sacs are entirely absent, never appearing even in the embryo; two of the four primordial clavicular sacs fuse across the midline and then blend with the fused cervical sacs, forming a single cervicoclavicular sac; the other pair of clavicular sacs remains small but separate, and the cranial thoracic and abdominal sacs are also paired. These modifications give the turkey seven definitive sacs. The abdominal sac is potentially the most capacious in birds generally, but the main chambers of all the other sacs are held open by their attachments to the structures which surround them and in the living bird are often more roomy than the abdominal sac. Usually, the caudal thoracic sac is larger than the cranial thoracic sac, although the opposite occurs in the pigeon and domestic fowl. In a few species such as hummingbirds the caudal thoracic sac is the largest of all the sacs.

The connections between the lungs and the air sacs

Most of these connections lie on the ventrolateral, that is costoseptal, border of the lung (see the lower border of Fig 7–7), but there are some on the septal surface near the entrance of the primary bronchus into the lung (Figs 7–7, 7–12 and 7–13). All of them have to pierce the horizontal septum (see section on **Pleural cavity**). The connections are of two types.

The direct connections. With the exception of the abdominal sac (see below) each air sac is connected *directly* to a secondary bronchus. The cranial air sacs are connected to the medioventral secondary bronchi (the cervical sac to the first, the cranial thoracic sac to the third and the clavicular sac to the third and to the distal end of the first medioventral bronchi). The caudal thoracic sac is connected to one of the lateroventral secondary bronchi (the second, in the domestic fowl and most other species). The abdominal sac opens directly from the end of the primary bronchus itself. The end of the primary bronchus usually becomes very reduced in calibre (less than 1 mm in diameter) before opening into the sac.

The indirect connections. These are possessed by all sacs except the cervical. About three to five of these lesser connections join each sac to the lung. On penetrating into the lung they branch and extensively anastomose with ordinary parabronchi. The indirect connections of the abdominal sac are particularly extensive and are exceptionally large. The parabronchi forming the indirect connections of the abdominal air sac in most birds open into the sac by means of a single large funnel-like bronchus called a *saccobronchus*. In many species including the domestic fowl a saccobronchus also collects the parabronchi connecting the caudal thoracic sac. The individual indirect connections of each sac are smaller in diameter than its direct connection (except for the abdominal sac). However, since there are several indirect connections to each sac, their total cross-sectional area substantially exceeds that of the direct connection. Consequently, the indirect connections offer an important airway between the lung and air sac. In some older terminologies the individual indirect parabronchial connections were referred to as recurrent bronchi. (The term saccobronchus has been used by some authors, both in the distant past and more recently, as a synonym for recurrent bronchus, and this is a source of confusion which should now be avoided).

These connections seem to be remarkably constant in birds generally. The only known major exception occurs in the domestic duck, and possibly in some other Anatidae and penguins. In the duck there is no direct connection of the clavicular sac to the third medioventral secondary bronchus; this connection is replaced by a direct connection to the first medioventral secondary bronchus, close to its origin from the primary bronchus; the other direct connection of this sac is in its typical position at the distal end of the first medioventral secondary bronchus. The general area of attachment of an air sac to the lung is known as an *ostium* (Fig 7–7). Within this zone are the orifices of the direct and indirect connections of the sac. It is visible at dissection as a cluster of small holes

Penetration of the air sacs into the skeleton and subcutaneous tissues

The medullary cavities of some bones in the avian skeleton are occupied by diverticula of the air sacs. The number of these aerated bones varies greatly in different species. It has been claimed that in a few varieties all bones are aerated including the phalanges of fore- and hindlimbs, while in others none is aerated; attempts have been made to relate these variations to power of flight, but there seems to be little or no correlation. Probably in the majority of birds, the sternum, scapula, humerus, femur, pelvis, and the cervical and thoracic vertebrae are aerated. In the domestic fowl the scapula and femur are not aerated, but the other bones in this list are pneumatized, and so also are the coracoid, various ribs and the synsacrum.

In some birds (e.g. the Gannet) extensive diverticula from the air sacs invade the fascial planes under the skin and between the skeletal muscles. No clear relationship has been demonstrated between the occurrence of these diverticula and the mode of life of the species possessing them. In the domestic fowl the extrathoracic diverticula of the clavicular sac are of this subcutaneous and intermuscular type, but compared to species such as the Gannet they are not very extensive.

Histology of the walls of the air sacs

The walls of the air sacs consist mainly of a thin layer of simple squamous epithelium supported by a small amount of connective tissue, but around the connections to the air sacs the epithelium becomes ciliated columnar. Macroscopically the walls are thin, glistening and transparent. The blood supply of the walls is small so that they play no part in the gaseous exchanges. This can be demonstrated by introducing carbon monoxide into an air sac which has had all its connections to the lung blocked; not enough carbon monoxide is absorbed to kill the animal.

Pleural cavity

As in mammals, the parietal pleura is reflected over the lung as the visceral pleura. Unlike that of mammals, however, the parietal pleura of birds is joined to the visceral pleura by fibrous strands (Fig 5–1), and it seems likely that in all species of bird some degree of obliteration of the cavity, sometimes total, takes place in embryonic development. Nevertheless, extensive areas of the pleural cavity persist in the adult of a number of species including the domestic fowl. In the latter the cavity is best preserved on the dorsolateral aspect of the lung (Fig 5–1). In these areas the strands uniting the two pleurae are sparse and delicate, and fail to prevent the lung from collapsing inwards if the pleural cavity is opened; in other regions the strands are so numerous that the pleural cavity is almost obliterated.

The horizontal and oblique septa

The diaphragm as seen in mammals is absent. The term 'diaphragm' is often applied to the horizontal and oblique septa but neither of these membranous septa is physiologically similar or embryologically homologous to the mammalian diaphragm, and the term should not be used in birds.

During embryonic development in both mammals and birds the pleural cavity becomes separated from the peritoneal cavity by the pulmonary fold, a horizontal double-layered sheet lying ventral to the lung; the dorsal layer of this sheet is presumptive parietal pleura and the ventral layer is presumptive parietal peritoneum. The developing lung is thus located dorsally, and in birds (though not in mammals) it remains permanently confined in that position.

Dilations of the developing bronchi soon extend from the lung and penetrate into this double-layered sheet, thus forming the cranial and caudal thoracic air sacs. As these sacs expand, they split the dorsal layer from the ventral layer. The dorsal layer becomes parietal pleura immediately over the ventral surface of the lung, and as such constitutes the *horizontal septum* (Fig 5–1). This area of parietal pleura becomes tough and tendinous and acquires fascicles of skeletal muscle, the *costoseptal muscle*, along its lateral edge where it attaches to the ribs (Fig 5–1). Furthermore, it also becomes fused to the adjacent wall of the thoracic air sac and is thus reinforced by air sac wall. This septum is pierced by the connections of the lung to the air sacs. The horizontal septum has also been called the saccopleural membrane and the pulmonary aponeurosis.

The ventral layer of the embryonic pulmonary fold becomes parietal peritoneum and forms the *oblique septum* (Figs 5–1 and 6–10). It is also reinforced by fusion with the wall of the adjoining thoracic air sac. This septum remains thin and looks like a typical air sac wall. Another name for it is the saccoperitoneal membrane.

The costoseptal muscle contracts during *expiration*. Probably it prevents compression of the lung and narrowing of the ostia, both of which would increase the resistance to air flow.

EXTERNAL RESPIRATION

Respiratory mechanics

The ribs move craniolaterally during inspiration, with a pump-handle action as in mammals, pushing the sternum ventrally and cranially (Fig 7–14b). The ribs also move somewhat laterally (Fig 7–14a). These movements of the thoracic cage simultaneously draw the abdominal wall ventrally and laterally also. Thus the dorsoventral, transverse and craniocaudal diameters of the body cavity are all increased in inspiration (as they are in a mammal, except that

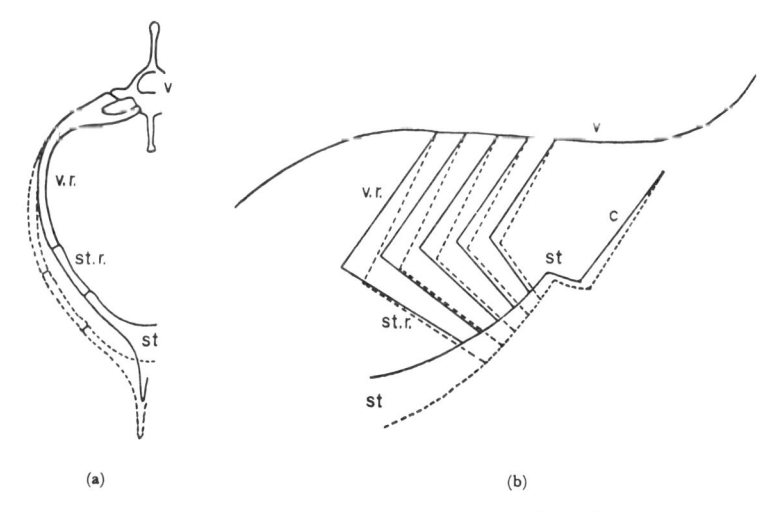

Fig 7–14 The movements of the thoracic cage during breathing. a, cranial view; b, lateral view. Solid lines = position at the end of expiration. Dotted lines = position at the end of inspiration. The inspiratory muscles move the ribs craniolaterally during inspiration, pushing the sternum ventrally and cranially. These inspiratory movements increase the dorsoventral, transverse and craniocaudal diameters of the thoracic cage. v = vertebral column; v.r. = vertebral ribs; st.r. = sternal ribs; st = sternum; c = coracoid. From Baer (1896) *Zeitschrift für Wissenschaftliche Zoologie*, 61, 420; and Zimmer (1935) *Zoologica*, 33, 1.

the craniocaudal increase in mammals is achieved by straightening the diaphragm). This enlargement causes a fall in pressure in the body cavity and hence within the air sacs also, and so air moves through the lungs and into the air sacs. In expiration the return of the thoracic cage to the resting position compresses the sacs thus expelling the air. The air sacs in birds, therefore, have essentially a bellows-like action in moving the air between the lungs and air sacs.

If all muscular forces are eliminated, the thoracic cage (at least in the domestic fowl) comes to rest about midway between full inspiration and full expiration. Therefore passive elastic forces may contribute substantially to the onset of both the inspiratory and the expiratory movement. However, electromyography shows that the inspiratory muscles (external intercostals and triangularis sterni) and expiratory muscles (internal intercostals and abdominal muscles) contribute actively throughout inspiration and expiration, respectively.

Lung design: surface area of exchange tissue

The air sacs act essentially like bellows, pulling the air in through the lung during inspiration and pushing it out through the lung during expiration. Theoretically, gaseous exchanges could occur in inspiration or expiration, or both. The lung itself has no need to expand and contract, and it now seems likely that the lung as a whole undergoes only small changes in volume during the respiratory cycle. One line of reasoning leading to this conclusion is that the surface tension in the narrow air capillaries would be so great that general expansion during inspiration is highly improbable. Thus the exchange area of the lung is now seen as being more or less immobile. Once a lung is no longer obliged to expand and contract substantially with each breath it can enormously increase its surface area for gaseous exchanges. This becomes possible because a relatively non-expanding lung can reduce its smallest airway to the minimum diameter without incurring insuperable problems of surface tension. Thus in the avian lung a very large number of small air capillaries can be packed into the volume which would be occupied by a single mammalian alveolus. Clearly, this large number of small tubules per unit volume of exchange tissue will have a far greater exchange surface area than the single alveolus. In fact the total surface area of the blood–gas barrier available for diffusion of oxygen per gram body weight is about 20 per cent greater in birds than in mammals. Furthermore, the air capillaries of birds are only 3–10 μm in diameter whereas the smallest mammalian alveoli (as in the shrew) are about 35 μm in diameter. This gives birds a greater driving force for diffusion of oxygen than mammals.

Control of airway calibre

Although the air capillaries are believed to remain more or less unchanged in diameter during breathing, the primary bronchus, secondary bronchi and parabronchi may undergo some regulation of their shape and calibre. Certainly they possess the necessary neuromuscular apparatus, i.e. smooth muscle and motor nerves. For example, the parabronchial wall is equipped with its network of innervated spiral atrial smooth muscles. Moreover, the atrial walls are extremely elastic and well endowed with surfactant. Contraction or relaxation of this whole array of bronchial smooth muscle may alter the resistance to air flow and/or the aerodynamic configuration of the bronchial orifices. Such changes may enable the regulation of parabronchial ventilation and may thus contribute to the matching of perfusion and ventilation (p. 141). Very little is known, however, about how such smooth muscle activity is controlled.

Air pathways in the lungs and air sacs

The air pathways in the lower respiratory tract of birds have been intensively investigated during the last 15 years using a variety of physiological techniques. These studies have conclusively established that air flow in the paleopulmo is in the same direction in both inspiration and expiration passing from

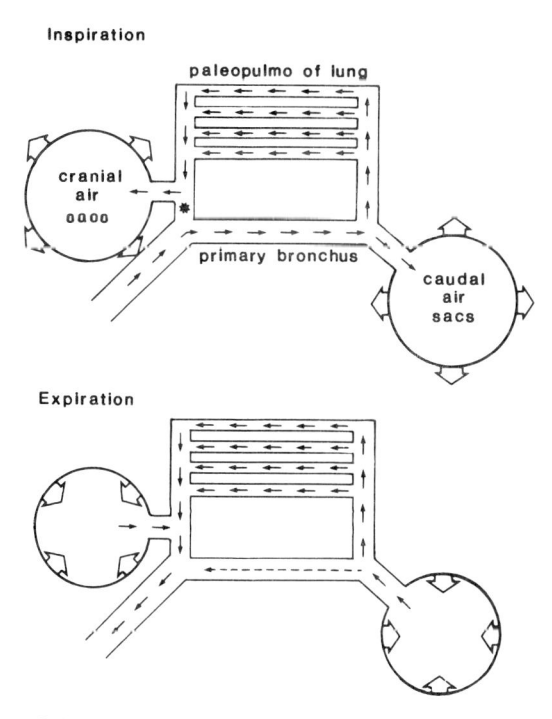

Fig 7–15 Diagram of the air flow in the avian lung and air sacs. In inspiration air in the primary bronchus is drawn into the caudal group of air sacs (the caudal thoracic and abdominal sacs) while air in the paleopulmonic parabronchi is pulled in a cranial direction through the lung and into the cranial group of air sacs (the cervical, clavicular and cranial thoracic sacs); no air passes directly into the cranial sacs from the primary bronchus, because of aerodynamic factors acting at the asterisk. In expiration air in the caudal sacs is expelled cranially into the paleopulmonic parabronchi while air in the cranial sacs is expelled via the primary bronchus and trachea; a small volume of air, indicated by a broken line, passes along the primary bronchus from the caudal sacs and escapes through the trachea. A unidirectional flow of air therefore occurs through the parabronchi of the paleopulmo in both inspiration and expiration. Based on Schmidt-Nielsen (1975), with kind permission of the publisher, and data from Powell (1983).

the mediodorsal secondary bronchi, through the parabronchi, to the medioventral secondary bronchi. In contrast, the flow of air in the neopulmo changes direction with the phase of breathing, travelling towards the caudal group of air sacs in inspiration and away from the sacs in expiration.

The route taken by inhaled air entering the lungs is now agreed to be as follows (Fig 7–15). During *inspiration* a part of the inhaled air, along with a portion of the dead space gas in the trachea and primary bronchus, enters the mediodorsal secondary bronchi and their parabronchi and starts to undergo gaseous exchange. Another part of the inspired air passes via the neopulmonic parabronchi and, by the direct connections (the second lateroventral and the primary bronchus), into the caudal group of air sacs (caudal thoracic and abdominal sacs). The gas in the caudal air sacs therefore is relatively rich in oxygen and relatively poor in carbon dioxide. In the ensuing *expiration* nearly all (probably at least 88 per cent) of the gas in the caudal sacs passes via the neopulmo into the mediodorsal secondary bronchi and their (paleopulmonic) parabronchi; a small amount of gas, probably not more than 12 per cent of the flow from the caudal sacs, may enter the primary bronchus and escape through the trachea. Since the last air which enters the caudal sacs during inspiration leaves in expiration before complete mixing occurs, the gas entering the mediodorsal secondary bronchi has a higher partial pressure of oxygen and a lower partial pressure of carbon dioxide than mixed gas in the sacs. During *inspiration* some of the air passing cranially through the paleopulmonic parabronchi is pulled via the medioventral secondary bronchi into the cranial group of air sacs (cervical, clavicular and cranial thoracic sacs). The gas in the cranial sacs has undergone gaseous exchange and therefore is relatively rich in carbon dioxide and poor in oxygen. During *expiration* air in the cranial sacs is expelled via the openings of the medioventral secondary bronchi into the primary bronchus and trachea. Thus most of the inspired air completes a full circuit of the lungs and air sacs in two breaths.

The mechanism underlying this pattern of air flow almost certainly involves passive aerodynamic factors such as the geometry of the airways and the relatively high velocity of the air flow. While active valving has not been clearly demonstrated the tone of the smooth muscle of the airways and their openings appears to be dependent on the partial pressure of the carbon dioxide in the airways.

The advantage to respiratory efficiency of having a continuous unidirectional flow of gas through the parabronchi of the paleopulmo during both inspiration and expiration, as opposed to the in-and-out tidal flow through the blind-ending bronchial tree of mammals, appears to be two-fold. Firstly, it minimizes the effect of the anatomical dead space by limiting it to the tracheal dead space, and secondly, it makes possible the cross-current relationship to the blood flow. The significance of this is discussed below.

GASEOUS EXCHANGE

As noted already (p. 130) a cross-current arrangement exists in the palaeopulmonic part of the avian lung between bulk parabronchial gas flow and blood flow in the parabronchial blood capillaries (Fig 7–11). The air in its unidirectional course along the parabronchus is continuously changing in composition as it undergoes gaseous exchange with capillary blood, with the result that blood is arterialized less and less along the parabronchus. The blood leaving the lung has a composition which is a mixture of the blood from different parts of the parabronchi. The partial pressure of carbon dioxide in the end-parabronchial gas is higher than that in arterial blood, whereas the partial pressure of oxygen may be lower than that in arterial blood. The gas-exchange efficiency of the cross-current arrangement, therefore, is superior to that of the ventilated pool system of mammalian lungs in which such relationships of $P\text{CO}_2$ and $P\text{O}_2$ are impossible; the best that can be achieved by the mammalian ventilated pool is equilibrium between the $P\text{CO}_2$ and $P\text{O}_2$ of end expired air and the $P\text{CO}_2$ and $P\text{O}_2$ of arterial blood. In other words, birds need less ventilation than mammals to achieve a certain level of arterialization, and with the same ventilation they can achieve greater arterialization than mammals.

In respiration at rest, inspired air and venous blood reach equilibrium a relatively short distance along the parabronchus, and therefore equal perfusion by blood of all levels of the parabronchus would result in an intense arteriovenous shunt in the distal part of the parabronchus. In other words, the pulmonary arterial blood (i.e. mixed venous blood) which is distributed to the *proximal* part of the parabronchus will be fully arterialized and will then pass into the pulmonary vein as arterial blood; in contrast, the pulmonary arterial blood which goes to the *distal* end of the parabronchus will not be arterialized at all but will continue (or be 'shunted', in physiological terms) into the pulmonary vein as mixed venous blood. This situation can also be described as 'uneven matching' of blood (perfusion) and gas (ventilation). Such uneven matching would lead to retention of CO_2 and systemic arterial hypoxaemia.

During exercise the ventilation of the parabronchus will be increased. The gas in the distal part of the parabronchus will then have a relatively high $P\text{O}_2$ and a relatively low $P\text{CO}_2$ compared with the values at rest. This increase must be matched by a corresponding increase in vascular perfusion in the distal part of the parabronchus in order to arterialize the blood.

Presumably, the matching of parabronchial perfusion and ventilation must be achieved by regulation of the smooth muscle in the blood vessels, or in the airways, or both. The most extreme mechanism to attain matching would be by arteriovenous anastomoses (an 'absolute shunt') but these are known not to occur in the avian lung. Regulation of perfusion is more likely to be a function of the small interparabronchial arteries, which have a relatively thick layer of innervated smooth muscle (p. 129). Regulation of ventilation by relaxation and contraction of bronchial and parabronchial smooth muscle (p. 138) is also a possibility. So far, nothing is known about how these intrapulmonary smooth muscles are controlled.

Oxygen and carbon dioxide, respectively, move into and out of the air capillaries from the lumen of the parabronchus by diffusion. The anatomical relationship of the air capillaries to the blood capillaries in the exchange tissue forms a counter-current-like (auxiliary counter-current) arrangement (Fig 7 11). This arrangement is another factor which promotes high gas-exchange efficiency.

The total anatomical diffusing capacity of the lung for oxygen is the maximum amount of oxygen which the lung can diffuse in unit time, assuming perfect ventilation by gas and perfect perfusion by blood. Thus it assumes that the entire exchange surface is engaged in diffusion, which is not in fact something that happens under physiological conditions. The anatomical diffusing capacity is proportional to the surface area of the blood–gas barrier and to the volume of the pulmonary capillary blood, and is inversely proportional to the thickness of the blood–gas barrier. As already noted, the total surface area of the blood–gas barrier per gram body weight is about 20 per cent greater in birds than in mammals, the volume of pulmonary capillary blood per gram body weight is about 20 per cent greater in birds than in mammals, and the mean thickness of the blood–gas barrier is generally only about a third that of mammals. The thinness of the barrier, its increased surface area and the greater volume of pulmonary capillary blood give the bird's lung a total anatomical diffusing capacity for oxygen which is about 20 per cent greater than that of mammals.

The bellows action of the air sacs, the continuous unidirectional air flow through the lung, the cross-current relationship between parabronchial gas and blood, the auxiliary counter-current relationship between blood and air capillaries, the small diameter of the terminal airways, and the total anatomical diffusing capacity of the lung for oxygen, combine to make the avian lung the most efficient gas exchange system among all the air-breathing vertebrates. These pulmonary adaptations are a major factor in the remarkable energetic capacity of birds, especially at high altitude (p. 6).

Further reading

Abdalla, M.A. & King, A.S. (1975) The functional anatomy of the pulmonary circulation of the domestic fowl. *Respir. Physiol.*, 23, 267–290.

Abdalla, M.A., Maina, J.N., King, A.S., King, D.Z. & Henry, J. (1982) Morphometrics of the avian lung. 1. The domestic fowl (*Gallus gallus* variant *domesticus*). *Respir. Physiol.*, 47, 267–278.

Ames, P.L. (1971) The morphology of the syrinx in passerine birds. *Bull. Peabody Mus. nat. Hist.* (New Haven, CT), 37, 1–194.

Bang, B.G. (1971) Functional anatomy of the olfactory system in 23 orders of birds. *Acta anat.*, 79 (Supplement 58), 1–76.

Bouverot, P. (1978) Control of breathing in birds compared with mammals. *Physiol. Rev.*, 58, 604–655.

Brackenbury, J. (1980) Respiration and production of sounds by birds. *Biol. Rev.*, 55, 363–378.

Brackenbury, J.M. (1981) Airflow and respired gases within the lung-air-sac system of birds. *Comp. Biochem Physiol.*, 68A, 1–8.

Burger, R. (1980) Respiratory gas exchange and control in the chicken. *Poult. Sci.*, 50, 2654–2665.

Duncker, H.-R. (1971) The lung air sac system of birds. A contribution to the functional anatomy of the respiratory apparatus. *Ergebn. Anat. EntwGesch.*, 45, 1–171.

Duncker, H.-R. (1972) Structure of avian lungs. *Respir. Physiol.*, 14, 44–63.

Duncker, H.-R. (1974) Structure of the avian respiratory tract. *Respir. Physiol.*, 22, 1–19.

Fedde, M.R. (1976) Respiration. In *Avian Physiology* (Ed.) Sturkie, P.D. New York: Springer-Verlag.

Fedde, M.R. (1980) Structure and gas-flow pattern in the avian respiratory system. *Poult. Sci.*, 59, 2642–2653.

Gaunt, A.S., Gaunt, S.L.L. & Casey, R.M. (1982) Syringeal mechanisms reassessed: evidence from *Streptopelia*. *Auk*, 99, 474–494.

Goodchild, W.M. (1970) Differentiation of the body cavities and air sacs of *Gallus domesticus* post mortem and their location *in vivo*. *Br. Poult. Sci.*, 11, 209–215.

Greenewalt, C.H. (1969) How birds sing. *Scient. Amer.*, 221, 126–139.

Haecker, V. (1900). *Der Gesang der Vögel seine anatomischen und biologischen Grundlagen*. Jena: Fischer.

King, A.S. (1966) Structural and functional aspects of the avian lung and air sacs. In *International Review of General and Experimental Zoology* (Ed.) Felts, W.S.L. & Harrison, R.J. Vol. 2. New York and London: Academic Press.

King, A.S. (1975) Aves respiratory system. In *Sisson and Grossman's The Anatomy of the Domestic Animals* (Ed.) Getty, R. Vol. 2, 5th Edn. Philadelphia: Saunders.

King, A.S. (1979) Systema respiratorium. In *Nomina Anatomica Avium* (Ed.) Baumel, J.J., King, A.S., Lucas, A.M., Breazile, J.E. & Evans, H.E. London and New York: Academic Press.

King, A.S. & Atherton, J.D. (1970) The identity of the air sacs of the turkey (*Meleagris gallopavo*). *Acta anat.*, 77, 78–91.

King, A.S. & Molony, V. (1971) The anatomy of respiration. In *Physiology and Biochemistry of the Domestic Fowl* (Ed.) Bell, D.J. & Freeman, B.M. Vol. 1. London and New York: Academic Press.

Lasiewski, R.C. (1972) Respiratory function in birds. In *Avian Biology* (Ed.) Farner, D.S. & King, J.R. Vol. II. New York and London: Academic Press.

McLelland, J. (1965) The anatomy of the rings and muscles of the trachea of *Gallus domesticus*. *J. Anat.*, 99, 651–656.

McLelland, J. & Molony, V. (1983) Respiration. In *Physiology and Biochemistry of the Domestic Fowl* (Ed.) Freeman, B.M. Vol. 4. London and New York: Academic Press.

McLelland, J., Moorhouse, P.D.S. & Pickering, E.C. (1968) An anatomical and histochemical study of the nasal gland of *Gallus gallus domesticus*. *Acta anat.*, 71, 122–133.

Maina, J.N., Abdalla, M.A. & King, A.S. (1982) Light microscopic morphometry of the lung of 19 avian species. *Acta anat.*, 112, 264–270.

Maina, J.N. & King, A.S. (1982) The thickness of the avian blood–gas barrier: qualitative and quantitative observations. *J. Anat.*, 134, 553–562.

Marples, B.J. (1932) The structure and development of the nasal gland of birds. *Proc. zool. Soc. Lond.*, 2, 829–844.

Morejohn, G.V. (1966) Variations of the syrinx of the fowl. *Poult. Sci.*, 45, 33–39.

Myers, J.A. (1917) Studies of the syrinx of *Gallus domesticus*. *J. Morph.*, 29, 165–215.

Pattle, R.E. (1978) Lung surfactant and lung lining in birds. In *Respiratory Function in Birds, Adult and Embryonic* (Ed.) Piiper, J. Berlin: Springer.

Peaker, M. (1975) Recent advances in the physiology of salt glands. In *Symp. zool. Soc. Lond.*, No. 35 (Ed.) Peaker, M. London and New York: Academic Press.

Piiper, J. & Scheid, P. (1972) Maximum gas transfer efficacy of models for fish gills, avian lungs and mammalian lungs. *Respir. Physiol.*, 14, 115–124.

Powell, F.L. (1983) Respiration. In *Physiology and Behaviour of the Pigeon* (Ed.) Abs, M. New York and London: Academic Press.

Scheid, P. (1978) Analysis of gas exchange between air capillaries and blood capillaries in avian lungs. *Respir. Physiol.*, 32, 27–49.

Scheid, P. (1979) Mechanisms of gas exchange in bird lungs. *Rev. Physiol. Biochem. Pharmacol.*, 86, 137–186.

Scheid, P. (1982) Respiration and control of breathing. In *Avian Biology* (Ed.) Farner, D.S. & King, J.R. Vol. VI. New York and London: Academic Press.

Scheid, P. & Piiper, J. (1970) Analysis of gas exchange in the avian lung. Theory and experiments in the domestic fowl. *Respir. Physiol.*, 9, 246–262.

Scheid, P. & Piiper, J. (1971) Direct measurement of the pathway of respired gas in duck lungs. *Respir. Physiol.*, 11, 308–314.

Schmidt-Nielsen, K. (1975) Recent advances in avian respiration. In *Symp. Zool. Soc. London.*, No. 35 (Ed.) Peaker, M. London and New York: Academic Press.

Warner, R.W. (1971) The structural basis of the organ of voice in the genera *Anas* and *Aythya* (Aves). *J. Zool.* (Lond.), 164, 197–207.

Warner, R.W. (1972) The syrinx in the family Columbidae. *J. Zool.* (Lond.), 166, 385–390.

Warner, R.W. (1972) The anatomy of the syrinx in passerine birds. *J. Zool.* (Lond.), 168, 381–393.

White, S.S. (1975) Aves the larynx. In *Sisson and Grossman's The Anatomy of the Domestic Animals* (Ed.) Getty, R. Vol. 2, 5th Edn. Philadelphia: Saunders.

Chapter 8

FEMALE REPRODUCTIVE SYSTEM

Two bilaterally symmetrical gonads and oviducts develop in the avian embryo. In birds generally, however, and in all the domestic species, the left ovary and oviduct soon exceed the right in their development. In the great majority of species in adult life only the female organs on the *left* side are functional, although rudiments of the right gonad and oviduct sometimes persist. Among the species in which two fully developed ovaries are frequently seen are many birds of prey and the Brown Kiwi, although right and left ovaries have also been observed in birds belonging to at least 16 orders which are usually assumed to have only one ovary. Two oviducts occur much less frequently, but appear to predominate in birds of prey. In the Brown Kiwi which has two functional ovaries, the single left oviduct is reported to be specially positioned to receive oocytes from both the right and left ovaries.

The general form and function of the left ovary and oviduct seem to be remarkably constant in the majority of birds.

THE LEFT OVARY

Growth and form

At an early stage of embryonic growth in the genetic female of the domestic fowl and of several other species, the gonadal region is colonized unequally by the incoming germ cells, more of which arrive at the left gonad than at the right. This initial asymmetry is further augmented by migration of many germ cells from the right to the left gonad. The left gonad therefore becomes larger than the right gonad, even before hatching. The primordial germ cells become incorporated in the so-called germinal epithelium, the other cells of this epithelium being of mesenchymal (peritoneal) origin. During the 6th and 7th days of incubation in the domestic fowl, the germinal epithelium buds off the *primary sex cords* into the depth of the gonad; these give rise to the ovarian medulla in the genetic female, and to the seminiferous tubules in the genetic male. In the genetic female the germinal epithelium multiplies into a thickened peripheral zone of cells, which is separated from the primary sex cords by a layer of connective tissue called the *primary tunica albuginea*.

During the 8th to 11th days of incubation this thickened peripheral zone of cells proliferates a second wave of down-growing cells, the *secondary sex cords*; it is these that form the oogonia. An *oogonium* is a germ cell which is actively multiplying. When the oogonia stop multiplying and begin to enlarge they become *primary oocytes*. This occurs at the time of hatching. At this stage they reach the prophase of their first maturation (reduction) division. Most of the yolk material now accumulates and the primary oocyte gradually grows to full size. The avian primary oocyte is the largest cell in the animal kingdom. In the domestic fowl its final weight is about 20 g. One of the largest cells to occur on this planet was the oocyte of the Madagascan Elephant-bird, which achieved a diameter of about 37 cm and a volume of much the same capacity as a bucket; the enormous eggs of this recently extinct bird were probably the basis for the gigantic Roc of ancient mythology. At the end of their period of growth the oocytes can complete their two maturation divisions, the first of which forms the *secondary oocyte* and the second forms the *ovum*. Thus there are three phases of oogenesis, a period of multiplication, a period of growth and a period of maturation.

From hatching until about four months in the domestic fowl the left ovary grows slowly, reaching about 1.5 cm in length and weighing about 0.5 g. During this phase the ovary consists of a cortex and medulla. The cortex encloses the medulla except where the ovary is in contact with the dorsal body wall. The outer surface of the cortex is lined by a tall cuboidal or flattened peritoneal epithelium which persists into maturity. Beneath the epithelium is a layer of dense connective tissue, the *definitive tunica albuginea.* Between 18 and 24 weeks the rate of growth rapidly accelerates, with many oocytes reaching their final size and the distinction between cortex and medulla becoming virtually lost. The cortex is now represented by ill-defined *parenchymatous zones* containing many immature follicles; the medulla is represented by irregular *vascular zones* containing blood vessels, nerves, smooth muscle and interstitial cells. In the final phase of rapid growth the diameter of the maturing primary oocyte increases from about 6 mm to a maximum of about 40 mm in about six days. The ovary now weighs about 60 g and, suspended by the mesovarium, occupies an extensive area on the roof of the coelom, overlapping the kidneys and lungs.

The light breeds of domestic fowl come into lay at five months and the heavy breeds at six to seven months, but modern hybrids can be persuaded to lay at four-and-a-half months. Most seasonal birds breed in the first spring or, in some species, in the second spring after hatching. Some birds, however, become sexually active much later than that. Fulmars, for example, begin breeding when about eight years old. On the other hand, the Australian Zebra Finch can breed when only a few months old, thus taking advantage of the restricted periods of rainfall, and captive Japanese Quail are sexually active at six weeks of age, a characteristic which enhances their value as laboratory animals.

During sexual activity the left ovary resembles a bunch of grapes because of the many large follicles which hang from it (Fig 8–1). In an actively laying hen about four or five very large follicles, reaching 40 mm in diameter, may be present as well as thousands of smaller ones. During the resting phase the left

FEMALE REPRODUCTIVE SYSTEM

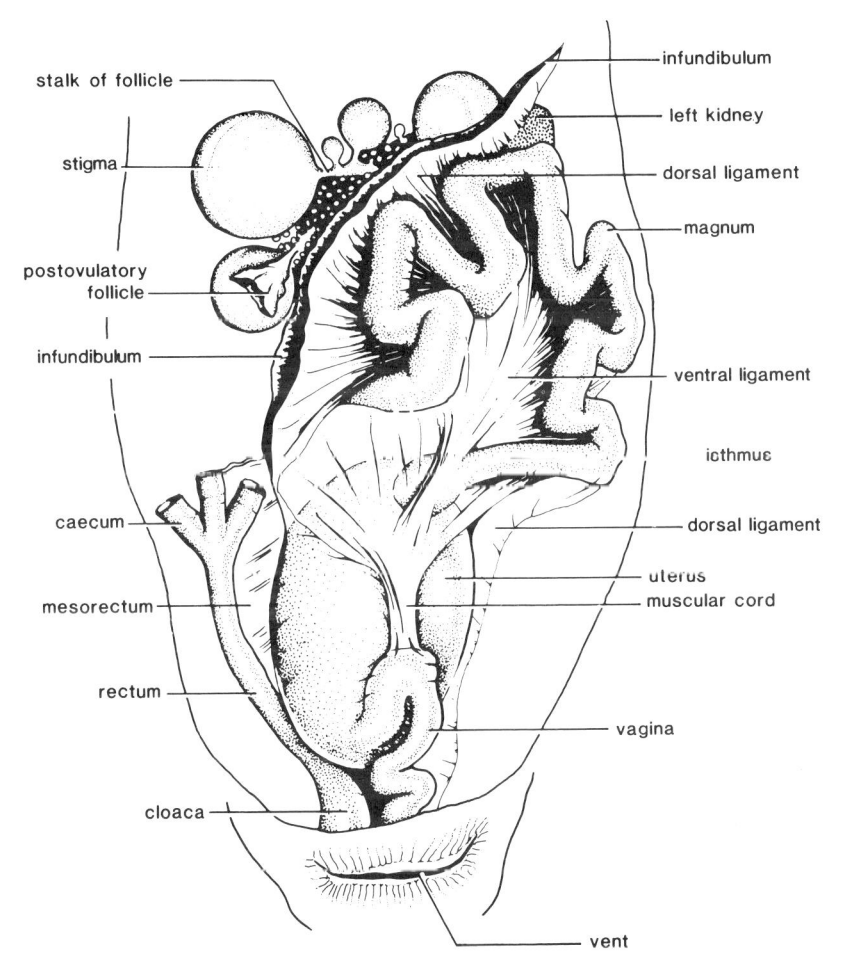

Fig 8–1 Ventral view of the left ovary and oviduct of the laying domestic fowl. The uterus contains an egg. The mesovarium and the dorsal ligament of the oviduct suspend the left ovary and left oviduct, lying on the left side of the dorsal mesentery of the gut.
Based on King (1975), with kind permission of the publisher.

ovary diminishes in size and in the domestic fowl weighs only 2–6 g. In seasonal birds three phases in the ovary can generally be distinguished, a phase of prenuptial acceleration in which it enlarges, a culmination phase when ovulation and laying occur, and possibly a refractory period when the ovary is much reduced in size. These phases are more or less synchronous with comparable phases in the testes, and, as in the male, are under the control of the neuroendocrine system.

In the domestic fowl the ovary is usually supplied with arterial blood via the ovario-oviductal branch of the left cranial renal artery (Fig 13–5). The ovary is drained via two ovarian veins directly into the caudal vena cava. These blood vessels and many nerves enter the broad dorsal surface of the ovary, the ovarian hilus, which is planted on the roof of the coelom.

The follicle

A large follicle is suspended by a stalk (Fig 8–1) which possesses smooth muscle and is abundantly vascularized and innervated. The follicle contains the large primary oocyte. This is enclosed by the *wall of the follicle*, which consists from inside to outside of the following six layers. (a) A *fine inner layer* (Fig 8–2), comprising an inner and outer component. The inner one is the *zona radiata*. This consists of the fine radial processes of the cytolemma of the oocyte, and the secretions and delicate radial processes of the granulosa. It is a transient structure, disappearing shortly before ovulation. The outer component is the *perivitelline lamina*; this consists of electron dense rods which

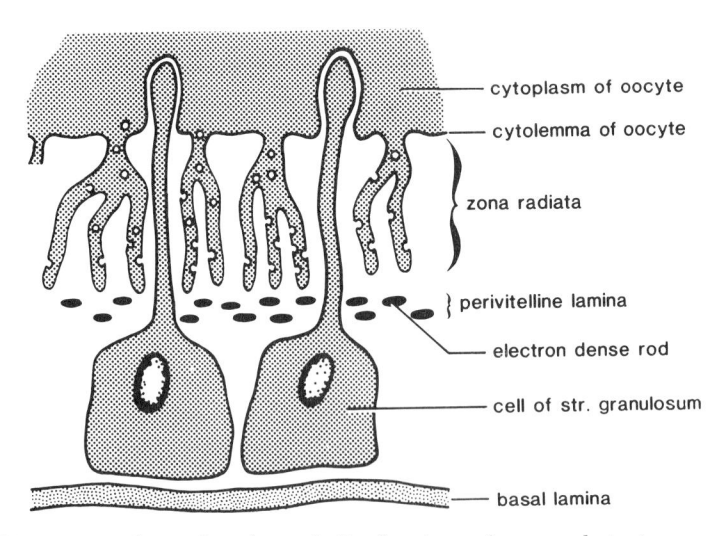

Fig 8–2 Diagrammatic section through the fine inner layer and stratum granulosum of the wall of a mature follicle. This is in the region of the small square in Fig 8–3. The fine inner layer consists of the zona radiata and the perivitelline lamina. The zona radiata is formed mainly by the radial processes of the oocyte, which show micropinocytosis; it disappears shortly before ovulation. The perivitelline lamina consists chiefly of electron dense rods, secreted by the cells of the stratum granulosum, which gradually develop into a fibrous meshwork. Based on King (1975), with kind permission of the publisher.

eventually develop into a meshwork of long fibres. (b) The *stratum granulosum* (Figs 8–2 and 8–3), a layer of cells with an unusually prominent basal lamina. (c) The *theca interna* (Fig 8–3), a compact narrow layer of spindle-shaped fibroblast-like cells and collagen fibres. (d) The *theca externa* (Fig 8–3), a looser, wider layer of flattened elongated fibroblast-like cells and numerous collagen fibres. Groups of interstitial cells usually occur at the boundary between the two thecae. (e) A *superficial tunic* of connective tissue (Fig 8–3). (f) The *superficial epithelium* (Fig 8–3) formed by peritoneal mesothelial cells (the so-called germinal epithelium). The wall of the follicle is highly vascular and quite profusely innervated by cholinergic and adrenergic fibres. A white meridional band, the *stigma*, occurs on all large follicles of most but not all species (Fig 8–1) (an exception is the White-crowned Sparrow). At the stigma the wall of the follicle has no superficial tunic of connective tissue and is less vascular than elsewhere. Contrary to general opinion the stigma seems to be devoid of smooth muscle.

Maturation of the oocyte, ovulation and fertilization

At the end of their period of growth the oocytes continue the process of maturation. As in most mammals, the *first maturation division* (reduction or meiotic division from the diploid to the haploid number of chromosomes), forming the secondary oocyte and first polar body, is completed while the primary oocyte is still in the follicle (about two hours before ovulation). In

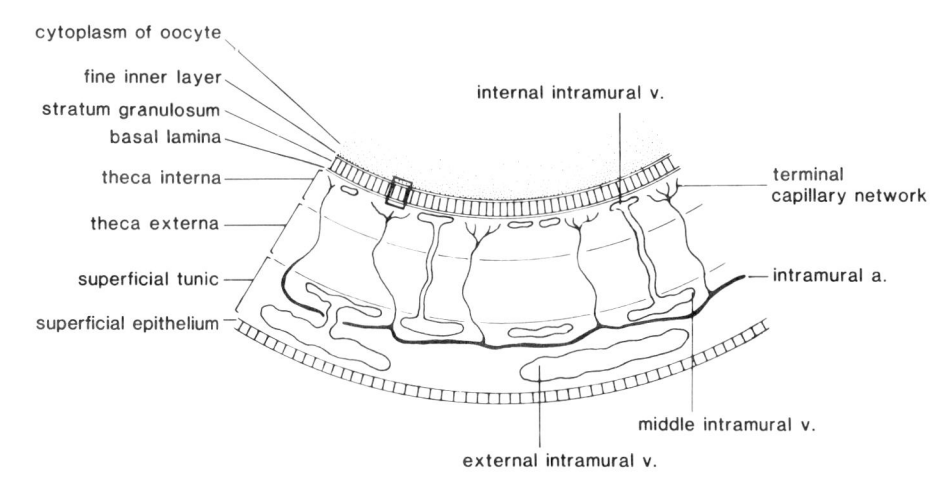

Fig 8–3 Diagrammatic section through the wall of a mature follicle. The region of the small square is shown enlarged in Fig 8–2. The theca interna is a compact cellular capsule, while the theca externa is a wider and looser layer. The superficial epithelium consists of peritoneal mesothelium. Based on King (1975), with kind permission of the publisher.

contrast to mammals and most other vertebrates, however, it is the female bird which is heterogametous and carries the XY combination, and therefore determines the sex of the progeny.

Ovulation occurs next. Luteinizing hormone (LH) released by the adenohypophysis is clearly involved in this process, but the exact mechanism is not known. LH may induce contraction of the smooth muscle of the stalk of the follicle, causing the stigma to split. Or it may induce ischaemia and hence necrosis of the stigma, but such necrosis has not been demonstrated. The number of ovulations in a reproductive cycle varies greatly with the species. Some birds, such as albatrosses and petrels, lay only one egg. Galliform birds, on the other hand, have clutches containing 8 to 12 eggs. A few birds such as the Budgerigar and crows are *determinate layers* and lay a fixed number of eggs. Many other birds, however, including the Red Jungle Fowl, are *indeterminate layers* and can quickly replace eggs which are lost from their clutch. This characteristic of the Red Jungle Fowl has been artificially selected by man with the result that the domestic fowl lays almost continuously for a large part of the year. Domestic fowl have laid up to 352 eggs in a year, and a Japanese Quail has managed 365 eggs thereby losing not a single working day. Most wild birds lay only one clutch of eggs each year, while others (e.g. redpolls) lay two or more clutches. A few species which take more than a year to rear their young, such as albatrosses and penguins, do not lay every year.

Ovulation in the domestic fowl usually follows about half an hour after an egg is laid, the integrating mechanism being hormonal or nervous. The newly released secondary oocyte is grasped and finally swallowed by the infundibulum. This catching process is made easier by the left abdominal air sac, which tightly encloses the ovary in the *ovarian pocket* except caudally where the infundibulum opens. Nevertheless, not all ovulated oocytes are successfully captured by the infundibulum. 'Internal laying' occurs quite often, especially when the bird is just going in or out of 'lay'; at these times the oviduct and ovary tend to get out of phase. Some at least of the oocytes which are thus lost in the coelom are believed to be harmlessly absorbed within 24 hours; others become involved in egg peritonitis, although it is not known why this happens. Egg peritonitis is one of the most common conditions seen in end of lay hens, turkeys and ducks. In the Brown Kiwi, which normally has a functional ovary on the right side as well as the left, the infundibulum of the left oviduct spreads right across the body and can therefore receive ovulated oocytes from both ovaries.

The *second maturation division*, forming the ovum and second polar body, occurs in the oviduct. Probably penetration by the spermatozoon is needed before this division can be completed, as in vertebrates generally. *Penetration by spermatozoa* occurs about 15 minutes after ovulation. Since penetration must occur before the secondary oocyte becomes covered by albumen, it presumably happens in the funnel of the infundibulum where there are no glandular cells. *Fertilization* is the actual fusion of the male and female pronuclei. Polyspermy is quite common. When present the supernumerary male pronuclei divide and initiate local centres of division, but these nuclei soon stop dividing and then degenerate.

The postovulatory follicle

Immediately after ovulation the follicle shrinks to a thin-walled sac, devoid of blood clots, which then undergoes quite rapid regression and absorption (Fig 8–1). Within two or three days the lumen of the sac has become filled with hypertrophied granulosa cells containing large amounts of lipid. In a few species thecal luteal cells invade the lumen. This luminal mass in the later stages is penetrated by capillaries. By the 10th day after ovulation, regression is virtually complete in the domestic fowl and in most of the other species that have been examined. Thus there is certainly no persistent postovulatory corpus luteum in birds. The likelihood remains, however, that the postovulatory follicle secretes hormones (perhaps progesterone), although this only happens during the first 24 hours of its life (in the domestic fowl). There is good evidence that these secretions affect oviposition and nesting behaviour.

The ovary in the period after laying

After the period of egg laying the bird enters a non-breeding state during which it is involved in incubation and caring for the chicks. In the period of *follicular atresia* any developing oocyte which may be present undergoes resorption. Several different forms of atresia have been observed. These include 'bursting atresia' which is characterized by rupture of the follicle wall and escape of the yolk into the ovary or body cavity, and 'invasion atresia' which is typified by invasion of the oocyte by cells from the granulosa layer or theca and reabsorption of the yolk in situ. Both of these types of atresia occur in the domestic fowl and in many other species. The disappearance of the follicle may or may not leave a scarlike area in the ovary. Following atresia of the follicle the resting ovary enters the *interbreeding period* in which its shrunken appearance is similar to that of the juvenile ovary (Fig 6–10).

The endocrine secretions of the left ovary

There is evidence for secretion of (a) oestrogens, by thecal interstitial cells; (b) androgens, by thecal interstitial cells and by interstitial cells which lie in the ovarian tissue outside the follicles; and (c) progestagens by granulosa cells of the preovulatory follicle and perhaps also from the postovulatory follicle. The term *interstitial cell* refers to the large vacuolated cells of the ovary, all of which are histologically similar whether they lie in the medulla, cortex or the theca of the follicular wall. It is possible that all these interstitial cells are merely the same endocrine cell at different stages of development. On the other hand, some of them may arise from the embryonic medulla, that is from the primary (medullary) sex cords, and are then the so-called 'medullary interstitial cells'; some of these medullary cells may then migrate into the cortex and become the so-called 'cortical interstitial cells'. Alternatively, the cortical cells may come from the secondary (cortical) sex cords. Finally, some of the cortical interstitial cells may embed themselves in the thecae and become the 'thecal interstitial cells'; another possible source of the thecal interstitial cells is the mesenchyme of the developing gonad, rather than the epithelium of the cortical sex cords. The true origins of the interstitial cells of the ovary remain controversial.

Sex determination by laparoscopy

Laparoscopy is a technique successfully used to determine the sex of wild birds in captivity which do not show sexual dimorphism. The procedure involves incising the body wall between the last two ribs on the left side and introducing the laparoscope into the left caudal thoracic air sac. However, in all but the smallest of birds the left gonad can only be visualized if the double-layered membrane separating the caudal thoracic and abdominal sacs is also broken down.

THE LEFT OVIDUCT

In the laying bird the left oviduct fills most of the dorsal and caudal part of the left side of the coelom. In the domestic fowl it weighs about 75 g and is about 65 cm in length. When the bird goes out of lay the size of the oviduct becomes much reduced (to about 5 g weight and 15 cm length, in the domestic fowl). The juvenile oviduct is shown in Fig 6–10. Seasonal growth and differentiation of the oviduct is under the control of the ovarian hormones, oestrogens, progesterone and androgens.

The left oviduct has five parts: infundibulum, magnum, isthmus, uterus (shell gland) and vagina.

Infundibulum

This part has two components, a funnel followed by a tubular region (Fig 8–1). In the domestic fowl their combined length is about 7 cm. The *funnel* has a thin wall and low mucosal folds. The opening of the funnel into the coelom faces into the 'ovarian pocket'. It is an elongated slit (about 9 cm long in the domestic fowl). The funnel tapers rapidly into the tubular part. The *tubular part*, also called the chalaziferous region, has a slightly thicker wall and taller mucosal folds with more secondary folds than in the funnel. In the Brown Kiwi the infundibulum extends the whole width of the coelomic cavity and is therefore able to receive oocytes from the left and right ovaries both of which are functional in this species.

Plates of glandular cells occur at the bottom of the grooves in the wall of the funnel region. The tubular part has some convoluted branched tubular glands, but these are confined to the region adjoining the magnum. The cells of these glands differ from those of the magnum; their secretory granules are smaller and do not compress the nuclei so strongly into a flattened form and basal position. Penetration by the spermatozoon presumably occurs in the funnel, before the secondary oocyte is covered by the first coat of albumen.

Magnum

The transition into the magnum is abrupt being marked by a sudden great enlargement of the mucosal folds. This is by far the longest and most coiled part of the oviduct (about 34 cm long in the domestic fowl) (Fig 8–1). The great thickness of the wall is caused by the presence of numerous tubular glands which are packed into the massive mucosal folds (Fig 8–4). These folds are taller and thicker than those of any other region of the oviduct, increasing the secretory area of the mucosa by a factor of about three. There are about 22 primary folds, which are seen to be devoid of true secondary folds when examined histologically; the few indentations in the epithelium are due to ducts of the glands.

The branched convoluted tubular glands in the lamina propria reach their greatest development in the magnum. Their cells contain large eosinophilic granules and have small rounded basal nuclei. The ducts open anywhere on the luminal surface, but are difficult to see in ordinary histological sections except after their secretion has been released. Immediately before ovulation the cells of these glands are so bulging with their secretion (which forms the bulk of the egg-white protein) that the lumen of the glands and the interglandular connective tissue are almost invisible. Because of the presence of

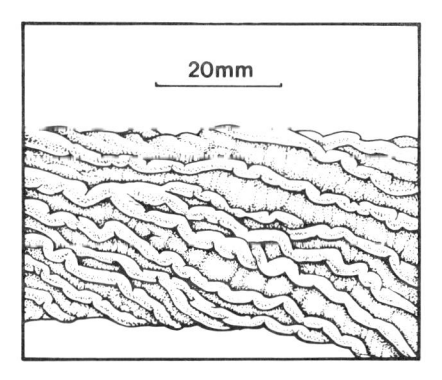

Fig 8–4 The mucosal surface of the magnum of the domestic fowl showing the thick spiral mucosal folds. The long axis of the oviduct is longitudinal in the illustration.

this secretory material the magnum has a milky-white colour. In hens that lay 'watery-white' eggs the secretory cells are abnormal. After discharge of the secretion the lumen and the outlines of the individual glands are more easily distinguished histologically. The stimulus to discharge may be mechanical, arising from the passage of the egg along the magnum. However, some glands seem to remain full even though many others are emptied, and this may indicate that discharge is controlled by more complex factors than simple mechanical influences.

The last few centimetres of the magnum are modified to form the so-called mucous region of the magnum. The folds and tubular glands here are much reduced, and the glandular cells contain relatively abundant mucus.

Isthmus

This region is short (about 8 cm long in the domestic fowl) and reduced in calibre (Fig 8–1). A narrow and sharply distinguished translucent band of tissue (1–3 mm wide in the domestic fowl) marks the exact junction of magnum and isthmus. The folds of the isthmus are less prominent than those of the magnum. Unlike those of the magnum, histologically they carry secondary folds. The translucent region is unusual in having no tubular glands, but the rest of the isthmus does have tubular glands histologically resembling those of the magnum. However, only the cells of the isthmus glands possess sulphur-containing proteins, which is consistent with the production by the isthmus of shell membranes of a keratinous nature.

Uterus

There is no distinct anatomical boundary between the isthmus and uterus (shell gland). The uterus is a relatively short region (about 8 cm long in the domestic fowl) (Fig 8–1). The cranial part is short and relatively narrow (the so-called 'red region') through which the egg passes rapidly. The major part is pouchlike and holds the egg during shell formation; it is here that the egg remains during the greater part of its journey through the oviduct. The longitudinal folds of the uterus are intersected by transverse furrows, thus forming numerous leaflike lamellae (about 4 mm tall). When an egg is inside the uterus these lamellae flatten themselves against the shell.

The tubular glands differ from those of the magnum, their cells being less granular and more vacuolated and possessing larger basally or centrally placed nuclei.

Vagina

The junction of the uterus with the vagina is marked by the vaginal sphincter which belongs to the beginning of the vagina. The vagina is fixed in a permanent S-shape by smooth muscle and connective tissue (Fig 8–1). When dissected free the vagina is about 8 cm long in the domestic fowl. The powerful muscle of the vaginal wall is thicker than that of any other part of the oviduct. The mucosal folds are relatively thin and low, and possess secondary folds. The mucosa has no secretory tubular glands. In the region of the sphincter the folds carry the tubular *spermatic fossulae* (sperm-host glands) which are the main site for *storage of spermatozoa*. The fossulae are lined by tall columnar cells possessing apical microvilli. In the juvenile bird the entrance to the cloaca from the vagina is closed by a membrane, at least in the domestic fowl and in wild anseriform birds; this occluding plate breaks down at sexual maturity, its removal apparently being under hormonal control.

The general structure of the wall of the oviduct

The wall of the adult oviduct possesses an epithelial lining, multicellular glands and smooth muscle. The *epithelium* consists of ciliated cells alternating with unicellular glands which resemble goblet cells, the proportions of the two cell types varying between different regions of the oviduct. This arrangement occurs in all parts of the oviduct in spite of the distinctive contribution made by each region. The *unicellular glands* are particularly large and numerous in the magnum, where they discharge their contribution to the albumen as each egg goes by, and refill between successive eggs. While the *cilia* in all regions mainly beat in the direction of the cloaca, the cilia in the magnum in a tract adjacent to the attachment of the dorsal ligament beat towards the infundibulum. The functions of the oviductal cilia have not been agreed, but they probably provide a protective mucociliary carpet, as in the upper respiratory tract. The *multicellular glands* are of two sorts, tubular glands and glandular grooves. The tubular glands open on the mucosal folds of the tubular part of the infundibulum, the magnum, the isthmus (except in the translucent region) and the uterus. Glandular grooves are found only in the infundibulum.

The *mucosal folds* (Fig 8–4) are more or less continuous throughout the oviduct, though varying in height and thickness. They are slightly spiral and thus rotate the egg as it goes down the oviduct.

The *smooth muscle* layers (inner circular and outer longitudinal) are thickest in the vagina and uterus and thinnest in the infundibulum. One function is to transport spermatozoa rapidly up the oviduct by oviductal retroperistalsis. Another is to drive the egg down the oviduct by peristaltic waves.

Ligaments of the oviduct

The oviduct is suspended from the roof of the coelomic cavity by a double-layered sheet of peritoneum forming the dorsal and ventral ligaments of the oviduct (Fig 8–1). The dorsal ligament extends from the roof of the coelomic cavity to the oviduct. The ventral ligament extends ventrally from the oviduct and has a free border. Both the dorsal and ventral ligaments contain much smooth muscle which is continuous with the outer longitudinal layer of muscle in the wall of the oviduct. Caudally, the smooth muscle of the ventral ligament condenses into a massive muscular cord about 5 mm in diameter in the domestic fowl, which fuses with the ventral surface of the uterus and vagina (Fig 8–1). The smooth muscle of the oviductal ligaments may help to move the egg along the oviduct, especially in the magnum where the intrinsic musculature appears too feeble to perform this function unaided.

The blood and nerve supply of the oviduct

In the domestic fowl the arterial supply to the oviduct (Fig 13–5) is generally derived from three oviductal arteries, the cranial, middle and caudal oviductal arteries, which are branches of the left cranial renal artery, the left ischiadic artery, and the left pudendal artery respectively. Sometimes in the domestic fowl there is a fourth oviductal artery, the cranial accessory oviductal artery, arising from the left external iliac artery; in some species, such as the domestic duck, this vessel is constantly present. The oviductal arteries form a series of longitudinal anastomosing arterial arcades along the dorsal and ventral surfaces of the oviduct in the dorsal and ventral ligaments. Within the wall of the oviduct the arteries form plexuses between the muscle layers and in the lamina propria, arterioles from the plexuses ending in a capillary network around the tubular glands. The uterus is especially vascular, as would be expected from the fact that in the domestic fowl calcium is deposited in the shell during the last 15 hours of shell formation at a rate of 100–150 mg/h. The veins draining the cranial part of the oviduct empty into the caudal vena cava, while those draining the caudal part empty into the hepatic portal or renal portal systems. The oviduct is innervated by both sympathetic and parasympathetic nerves, although the precise origin of these nerves does not appear to have been established. The most densely innervated region of the oviduct is the uterus and its junction with the vagina.

Storage of spermatozoa in the oviduct

Spermatozoa must be stored somewhere in the oviduct. This is evident from the fact that, in the domestic fowl and several other species, the capacity for fertilization is retained for several weeks after a single insemination; this characteristic is likely to be common to birds in general. Within minutes of insemination sperm reach the top of the oviduct, but within 24 hours they disappear, only to reappear in the lumen in small numbers at the time of each subsequent laying or ovulation. In the domestic fowl the *spermatic fossulae* in the vagina (the sperm-host glands) are the main sites of residence, but it is not known how the sperm enter the fossulae, how they survive and what causes them to be released at the time of laying or ovulation; mechanical, nervous and vascular factors have been considered. Some sperm may also be stored in the glandular grooves and tubular glands of the infundibulum.

FORMATION OF THE EGG

The egg traverses the oviduct in about 25 hours. The raw materials of the yolk (protein and lipids) are synthesized in the liver and travel in the blood plasma to the granulosa cells, which then pass them to the oocyte (vitellogenesis). The oocyte reorganizes them into yolk spheres and fluid. There is no biochemical synthesis of yolk in the oocyte itself.

The egg passes through the *infundibulum* in about 15 minutes. The chalaziferous layer of the albumen (Fig 8–5) is laid down by the tubular glands there. It consists of only a thin layer of dense albumen, immediately surrounding the yolk. The tubular part of the infundibulum forms the *chalaza* (Fig 8–5) at each end of the egg. The chalaza at the sharp end is a double strand, while that at the blunt end is a single strand. The chalazae appear to suspend the yolk between the two ends of the egg. They become twisted, probably from the rotation of the egg as it travels down the oviduct.

The egg takes about three hours to traverse the *magnum*. During this time it acquires albumen (Fig 8–5) secreted by the tubular glands, with a contribution also from the unicellular glands. Sodium, magnesium and calcium are added mainly in the magnum.

Movement through the *isthmus* is slow, taking about 75 minutes in all. The inner and outer shell membranes (Fig 8–5), lining the shell, are the main components which are formed here (from the tubular glands). Before these are laid down, however, a small amount of protein (about 10 per cent of the total) is added to the albumen. Calcification of the shell appears to be initiated in the isthmus.

The egg occupies the *uterus* for about 20 hours, during which time the shell (i.e. the testa, cuticle and pigment) is formed (Fig 8–5). *Plumping* occurs here. This consists of the rapid addition of watery solutions (probably from the glands of the relatively short and narrow cranial part of the uterus) into the egg, mainly during the first eight hours, doubling the weight of the albumen. During plumping, calcification of the shell is slow, but during the last 15 hours it is rapid. Every 15 minutes the uterus withdraws from the blood a weight of calcium equalling the total amount circulating at any one moment; the extreme vascularity of the uterus presumably contributes to this remarkable activity. Potassium is added mainly in the uterus. Usually, in the domestic fowl the egg lies in the uterus with its sharp end pointing caudally, and the egg is laid with this orientation. In some species (e.g. gulls and ducks) the egg in the uterus turns round just before oviposition, so that the blunt end comes out first. The biological significance of this rotation of the egg, and the way the muscles of the oviduct and its ligaments achieve it, are not known.

The egg travels through the *vagina* in a matter of seconds.

Obstruction of the oviduct is relatively common in both domestic and wild birds. The most usual causes include the presence in the oviduct of necrotic egg material, a broken-shelled egg, a normal soft-shelled egg or localized infection. Amongst wild species a common factor associated with obstruction of the oviduct is senility.

STRUCTURE OF THE EGG

The egg consists of a germinal disc, yolk, the membranes surrounding the yolk, albumen and a shell (Fig 8–5).

The germinal disc

The germinal disc (blastoderm if fertilized, blastodisc if unfertilized) is a small disc of cytoplasm (Fig 8–5) containing the remnant of the nucleus. It can be seen on the surface of the yolk of a fresh egg as a circular, opaque white spot, which in the domestic fowl is 3–4 mm in diameter.

The yolk

The yolk is a thick viscous material containing about 50 per cent solids of which 99 per cent are proteins (30 per cent of yolk is lipo- and phosphoproteins). As in reptiles it forms the main source of nutrition for the embryo. There are two kinds, white and yellow yolk (Fig 8–5). The *white yolk* is about two-thirds protein and one-third fat. It consists of a small spherical mass (the centre of the latebra), joined by a slender column (the neck of the latebra) to a conical disc (the disc of the latebra) which lies beneath the germinal disc. The *yellow yolk*, which is about two-thirds fat and one-third protein, is often organized in the domestic fowl into alternating yellow and white strata. However, this stratification is an artefact depending on the diet, the pale strata being deficient in carotenoid pigments. When the diet is well balanced these strata disappear. Another kind of stratification of the yellow yolk can be discerned, although only by means of special techniques. These strata do reflect the real form of the yolk, but little is known about their structural basis.

The 'yolk membranes'

These membranes form a barrier (in the domestic fowl probably about 24 μm thick in the fresh state) between the yolk and albumen, which has great mechanical strength but is permeable to water and salts (Fig 8–5). The electron microscope shows that the membranes consist of four layers. The two

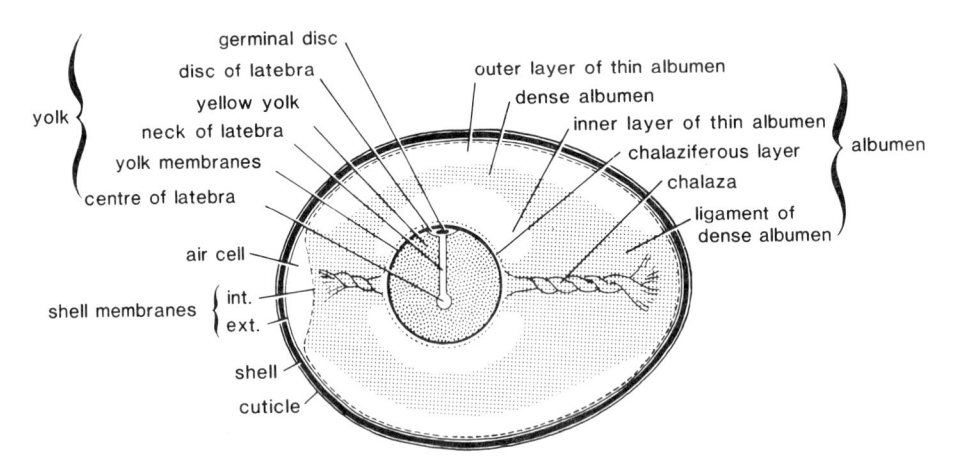

Fig 8–5 An egg. The white yolk is unshaded and the yellow yolk is stippled. Based on King (1975), and Gilbert (1979), with kind permission of the publisher.

next to the yolk, i.e. (1) the *oocyte cytolemma* and (2) the meshwork of coarse fibres known as the *perivitelline lamina*, are derived from the follicle of the ovary; they are the same structures as those which are shown in Fig 8–2, and are described on page 148 under layer (a) of the wall of the follicle. While the fertilized ovum is in the infundibulum the gaps in the meshwork of the perivitelline lamina are wide enough to admit the head of a spermatozoon, but further down the oviduct the gaps become obliterated. The two outer layers of the yolk membrane are laid down by the oviduct soon after the egg has entered. One of them, (3) the *continuous lamina*, is a very thin granular layer; the outermost, (4) the *extravitelline lamina*, consists of delicate lattice-works of fine fibres in concentric layers and forms about two-thirds of the total thickness of all these four membranes together.

The albumen

The albumen is much less viscous than the yolk. The solid component is composed almost entirely of protein (ovomucin). Dense albumen contains a relatively large amount of ovomucin and possibly some mucin fibres; thin albumen is more watery and contains less ovomucin and almost no mucin fibres. The *chalaziferous layer* is a thin layer of dense albumen which encloses the yolk membranes (Fig 8–5). The two *chalazae* are twisted strands (absent in reptilian eggs) of fine ovomucin fibres, which are formed from the chalaziferous layer (Fig 8–5). The chalaza at the sharp end is a double strand, while that at the blunt end is a single strand. The chalazae are continuous at one end with the chalaziferous layer, and at the other end with the middle layer of dense albumen which is connected to the shell membranes at each end of the egg by the so-called albumen ligament (or polar albumen). Thus the chalazae suspend the yolk more or less in the centre of the egg. They become twisted by the rotation of the egg as it passes down the oviduct. The *inner layer* is a layer of thin albumen which is almost devoid of mucin fibres (Fig 8–5). The *middle layer* is a layer of dense albumen, which at each end of the egg attaches by fibres to the inner shell membrane forming the two *albumen ligaments* (Fig 8–5). The *outer layer* of thin albumen surrounds the dense layer except at the albumen ligaments (Fig 8–5). The albumen contributes to the aqueous environment of the embryo, has anti-bacterial properties, and in many if not all birds is a source of nutrition for the embryo.

The shell

The shell consists of the shell membranes, the testa and the cuticle.

The outer and inner *shell membranes* (together about 70 µm thick in the domestic fowl) are each composed of several layers of fibres (Fig 8–5). At the blunt end of the egg the outer and inner membranes separate from each other as the egg cools immediately after hatching, forming the air cell (Fig 8–5). The head of the embryo comes to lie close beneath this space, which is absent in reptiles. The outer shell membrane is firmly attached to the testa. At each end of the egg the inner shell membrane is fused to the albumen ligament of the dense layer of albumen.

The major part of the shell is formed by the *testa* (270–370 µm thick in the domestic fowl) which contributes most of the thickness of the shell. Shell thickness varies greatly between species but the eggs of large birds in general have proportionately thicker shells, being about 2 mm thick in the Ostrich but only paper thin in hummingbirds. The testa consists of an organic matrix of fine fibres, and a far more bulky (98 per cent of the total) solid inorganic component consisting mainly of calcite (a crystalline form of calcium carbonate).

The fibrous *organic matrix* of the testa is arranged into a relatively thin inner mamillary layer (about one-third of the thickness of the testa) and a thicker outer spongy layer (the remaining two-thirds of the testa). The mamillary layer consists of conical-shaped knobs, the apices of which are embedded in the outer shell membrane. The fibres of the spongy layer are arranged parallel to the surface of the egg. The crystalline *inorganic component* of the testa is arranged into a thin inner layer (the layer of cones and basal caps) corresponding to the mamillary layer of the organic matrix, and a thicker outer layer (the palisade layer) corresponding to the spongy layer of the organic matrix. During the calcification of the testa, the crystals appear first in the mamillary layer (thus forming the layer of cones and basal caps), and then accumulate in columns in the spongy layer gradually growing towards the surface of the shell (thus forming the palisade layer). In most species of birds thousands of fine pores open on the surface of the shell, and extend between the crystals right through to the shell membranes. In the domestic fowl the pores are frequently more closely grouped at the blunt end of the egg near the air cell, and hence near the head of the chick. The pores are covered by the cuticle, but because of radial cracks in the cuticle here the pores still remain permeable to gases.

Overlying the testa and pores is a continuous organic layer, the *cuticle*, 90 per cent of which is peptide (Fig 8–5). Although extremely thin (10 μm thick in the domestic fowl) the cuticle is water repellant, reduces water loss and acts as a barrier to bacteria.

Pigmentation is frequently present in the shell, in contrast to the eggs of reptiles which are almost always white. Most of the pigmentation occurs in the inorganic palisade layer and sometimes in the cuticle. Rarely is it present in the mammillary layer.

Near the end of incubation the beak of the embryo penetrates the inner shell membrane where it forms the inner wall of the air cell, and at about this time the lungs usually become functional. The embryo breathes into the air cell for several hours and then cracks or 'pips' the shell, in the process of which the outer shell membrane is usually ruptured. In the domestic fowl a period of about 20 hours elapses between pipping and hatching.

OVIPOSITION

At the onset of oviposition the muscle of the uterus drives the egg through the relaxed vaginal sphincter. The presence of the egg in the vagina then initiates the 'bearing down' reflex which forces the egg through the cloaca (the vagina is not capable by itself of expelling the egg from the body). The duration of oviposition varies greatly. In the Cuckoo it lasts only a few seconds since speed is essential to avoid upsetting the host, but in the domestic turkey and goose it takes several hours. The time of oviposition varies greatly between species. Songbirds tend to lay in the early morning. The domestic fowl usually lays in the morning, but may do it in the afternoon. Pigeons utilize the afternoon and pheasants the evening, while the American Coot operates very privately at about midnight. The rate of oviposition in the domestic fowl, and probably in birds generally, usually reflects the rate of ovulation. In most birds the average laying interval ranges from 24 hours (as in most passerines and the domestic fowl) to four or five days (as in the Andean Condor), but it may be much longer as in the Brown Kiwi where the interval may extend to 44 days.

PHYSICAL CHARACTERISTICS OF THE EGG

Eggs of different species show an enormous range of external appearance.

Size. Egg size in general is related to the size of the parent, although the relative weight of the egg usually diminishes as body weight increases. Also the eggs of precocial species, such as the domestic fowl of which the young are very active immediately after hatching, are relatively larger than those of altricial birds such as passerines whose young are helpless after hatching. Among living birds the largest egg (1400 g) belongs to the Ostrich, whilst the smallest (0.5 g) is that of the Vervain Hummingbird. The small size of the egg of the Cuckoo relative to body weight (3 per cent of the body weight compared to 7 to 11 per cent in a large number of birds) is possibly due to the fact that the parent bird parasitizes the nests of much smaller species. The largest egg relative to body weight (25 per cent) occurs in the Brown Kiwi.

Shape. The shape of birds' eggs varies greatly—conical, spherical, oval, and cylindrical forms being common. Many adaptive reasons have been suggested for these variations, for example that conical eggs are laid by birds which nest on cliff ledges since this prevents the eggs from rolling off, but few of these explanations survive close examination. On the other hand, the shape of the egg does seem to be related to the shape of the bony pelvis, spherical eggs occurring when the pelvis is deep dorsoventrally and elongated eggs when the pelvis is narrow dorsoventrally.

Colour. The coloration of egg shells ranges over a broad spectrum from pure white, through blues, greens, browns and brick-red, to a near black colour. This range of colour is produced by two main pigments, porphyrins (red-brown) and biliverdin (blue-green), which are deposited as crystals throughout the calcified part of the testa though more densely near the surface. Commonly superimposed on the ground colour are markings such as scribbling, streaking or spotting, these markings being caused by the presence of the pigments on the surface of the cuticle. In some plovers and falcons the pigment is so superficial that it can be easily rubbed off. The egg of the Cuckoo is remarkable in mimicking the egg-colour of the host species. Despite much speculation, the functional significance of the different colours and markings of birds' eggs does not appear to be known.

Surface texture. The outer surface of the shell in most species is smooth with a slight sheen. However, in some birds the shell may be highly glossy (woodpeckers), greasy (ducks and geese), chalky (cormorants), powdery (flamingos), ridged (the Emu) and pitted (the Ostrich). The significance of these variations is not known.

THE REPRODUCTIVE ORGANS ON THE RIGHT SIDE OF THE GENETIC FEMALE

The right gonad of the genetic female

In the genetic female in birds generally, the growth of the right gonad is normally arrested at a testis-like stage of development, when the primary sex cords have been formed giving rise to a medulla with testicular potential, and only a few secondary sex cords have arisen (which would tend to form a cortex and give rise to oogonia). Although very small, the right gonad subsequently persists in the adult as wrinkled strands of tissue along the ventral side of the caudal vena cava (about 5 mm long in the domestic fowl). If the left ovary is destroyed surgically or by natural disease, the right gonad enlarges to a length of about 2 cm. In the domestic fowl about 90 per cent of such enlarged right gonads are testis-like, but areas of active spermatogenesis are scarce and occur only in birds in which the left gonad was removed within a month of hatching (the primordial germ cells disappear from the primary sex cords about three weeks after hatching). Of the remaining enlarged right gonads, some not only develop their primary sex cords but also form scattered, or more rarely, quite extensive areas of secondary sex cords. These gonads are therefore bipotential and develop into an ovotestis and even occasionally into a more or less active ovary.

In only one species, the Brown Kiwi, is it established that paired functional ovaries normally occur. While both ovaries in falconiforms are often reported to be fully formed it seems to be relatively rare for the right ovary in these birds to be functional. As stated at the beginning of this chapter, paired ovaries have been reported in at least 16 orders. Despite the obvious suggestion that the characteristic reduction of the right ovary and oviduct is an adaptation to flight because it saves weight, no really convincing adaptational advantage from this avian modification has been established.

The right mesonephros and right mesonephric duct

These persist in normal adult genetic females. After removal of the left ovary they enlarge and become associated with the right gonad to form an epididymis and ductus deferens. Thus the right side of the adult genetic female can produce a fully functional male system.

Two natural cases are known of a hen which laid eggs and then turned into a cock and sired two chicks, but this has never been accomplished after surgical removal of the left ovary in spite of many attempts. Natural events of this kind have long been a source of superstitious awe, and at least one such unfortunate 'cock' was tried and duly burnt at the stake as a creature possessed of Satan. 'A whistling maid and a crowing hen are fit for neither Gods nor men.'

The right oviduct

After a normal start, the development of the right oviduct of the genetic female is totally arrested very early in embryonic life. However, in the domestic fowl, a small remnant of the right oviduct can nearly always be found joining the cloaca. Large cystic vestiges (up to 10 cm in diameter), containing fluid derived from the blood plasma, are not uncommon (Fig 8–6). Very large cysts (up to 20 cm in diameter) cause difficulty in breathing and in circulation. Histologically the tissue forming the wall of these cysts is equivalent to that of the magnum or infundibulum, but not the uterus or vagina. Rarely, a fully developed oviduct occurs on both the left and right sides of the domestic fowl, complete with the left ovary or even with functional ovaries on both sides. In

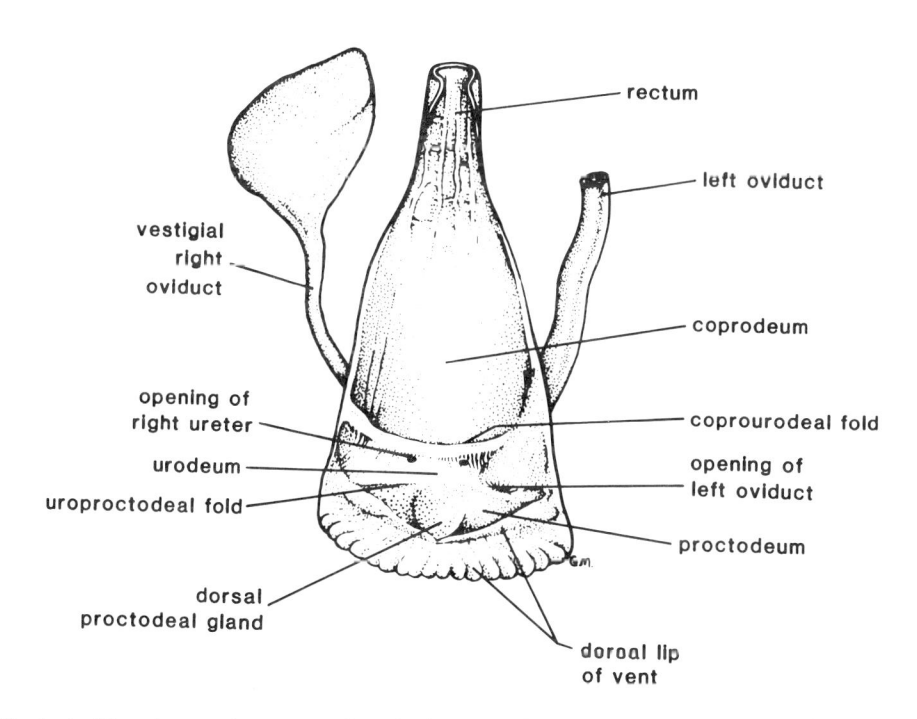

Fig 8–6 The cloaca of a mature female domestic fowl opened along the ventral midline and laid flat. A cystic remnant of the right oviduct is present. From King (1981), with kind permission of the publisher.

such birds, the longitudinal suspension of the rectum from the roof of the abdominal cavity (Fig 8–1) makes it very difficult for ovulated oocytes to cross from the right ovary to the left oviduct (or vice versa), but isolated reports have demonstrated that this really can happen.

Vestiges of the right oviduct have been reported, although somewhat infrequently, in more than 30 other species from many different orders. A fully developed right oviduct is rarely found, except in the domestic fowl. In hawks in which both gonads become fully functional, the right oviduct still remains vestigial. This applies also to the Brown Kiwi, even though both ovaries are regularly present and fully functional in this species.

Further reading

Aitken, R.N.C. (1971) The oviduct. In *The Physiology and Biochemistry of the Domestic Fowl* (Ed.) Bell, D.J. & Freeman, B.M. Vol. 3. London and New York: Academic Press.

Bakst, M.R. & Howarth, B. (1975) SEM preparation and observation of the hen's oviduct, *Anat. Rec.*, 181, 211–226.

Bakst, M.R. & Howarth, B. (1977) The fine structure of the hen's ovum at ovulation. *Biol. Reprod.*, 17, 361–369.

Balch, D.A. & Cooke, R.A. (1970) A study of the composition of hen's egg shell membranes. *Annls Biol. anim. Biochim. Biophys.*, 10, 13–25.

Becking, J.H. (1975) The ultrastructure of the avian egg shell. *Ibis*, 117, 143–151.

Bellairs, R. & Boyde, A. (1969) Scanning electron microscopy of the shell membranes of the hen's egg. *Z. Zellforsch. mikrosk. Anat.*, 96, 237–249.

Blom, L. (1973) Ridge pattern and surface ultrastructure of the oviducal mucosa of the hen (*Gallus domesticus*). *K. danske Vidensk. Selsk. Skr.*, 20, 1–15.

Board, R.G. (1974) Microstructure, water resistance and water repellancy of the pigeon egg shell. *Br. Poult. Sci.*, 15, 415–419.

Board, R.G. & Halls, N.A. (1973) The cuticle: a barrier to liquid and particle penetration of the shell of the hen's egg. *Br. Poult. Sci.*, 14, 69–97.

Burke, W.H., Ogasawara, F.X. & Fuqua, C.L. (1972) A study of the ultrastructure of the utero-vaginal sperm-storage glands of the hen, *Gallus domesticus*, in relation to a mechanism for the release of spermatozoa. *J. Reprod. Fert.*, 29, 29–36.

Creger, C.R., Phillips, H. & Scott, J.T. (1976) Formation of an egg shell. *Poult. Sci.*, 55, 1717–1723.

Freedman, S.L. & Sturkie, P.D. (1963) Extrinsic nerves of the chicken uterus (shell gland). *Anat. Rec.*, 147, 431–437.

Fujii, S. (1974) Further morphological studies on the formation and structure of hen's eggshell by scanning electron microscopy. *J. Fac. Fish. Anim. Husb. Hiroshima Univ.*, 13, 29–56.

Fujii, S. (1975) Scanning electron microscopical observation on the mucosal epithelium of hen's oviduct with special reference to the transport mechanism of spermatozoa through the oviduct. *J. Fac. Fish. Anim. Husb. Hiroshima Univ.*, 14, 1–13.

Fujii, S. (1976) Scanning electron microscopical observations on the penetration mechanisms of fowl spermatozoa into the ovum in the process of fertilization. *J. Fac. Fish. Anim. Husb. Hiroshima Univ.*, 15, 85–92.

Gilbert, A.B. (1967) The formation of the egg in the domestic chicken. In *Advances in Reproductive Physiology* (Ed.) McLaren, A. Vol. 1. London: Logos Press.

Gilbert, A.B. (1971) The Ovary. In *Physiology and Biochemistry of the Domestic Fowl* (Ed.) Bell, D. J. & Freeman, B.M. Vol. 3. London and New York: Academic Press.

Gilbert, A.B. (1971) Egg albumen and its formation. In *Physiology and Biochemistry of the Domestic Fowl* (Ed.) Bell, D.J. & Freeman, B.M. Vol. 3. London and New York: Academic Press.

Gilbert, A.B. (1971) Transport of the egg through the oviduct and oviposition. In *Physiology and Biochemistry of the Domestic Fowl* (Ed.) Bell, D.J. & Freeman, B.M. Vol. 3. London and New York: Academic Press.

Gilbert, A.B. (1971) The egg: its physical and chemical aspects. In *The Physiology and Biochemistry of the Domestic Fowl* (Ed.) Bell, D.J. & Freeman, B.M. Vol. 3. London and New York: Academic Press.

Gilbert, A.B. (1971) The egg in reproduction. In *Physiology and Biochemistry of the Domestic Fowl* (Ed.) Bell, D.J. & Freeman, B.M. Vol. 3. London and New York: Academic Press.

Gilbert, A.B. (1979) Female genital organs. In *Form and Function in Birds* (Ed.) King, A.S. & McLelland, J. Vol. 1. London and New York: Academic Press.

Hodges, R.D. (1965) The blood supply to the avian oviduct with special reference to the shell gland. *J. Anat.*, 99, 485–506.

Hodges, R.D. (1974) *The Histology of the Fowl*. London and New York: Academic Press.

Hoffer, A.P. (1971) The ultrastructure and cytochemistry of the shell membrane-secreting region of the Japanese Quail oviduct. *Am. J. Anat.*, 131, 253–288.

Howarth, B. (1971) Transport of spermatozoa in the reproductive tract of turkey hens. *Poult. Sci.*, 50, 84–89.

King, A.S. (1975) Aves urogenital system. In *Sisson and Grossman's The Anatomy of the Domestic Animals* (Ed.) Getty, R. Vol. 2, 5th Edn. Philadelphia: Saunders.

King, A.S. (1979) Systema urogenitalia. In *Nomina Anatomica Avium* (Ed.) Baumel, J.J., King, A.S., Lucas, A.M., Breazile, J.E. & Evans, H.E. London and New York: Academic Press.

King, A.S. (1981) Cloaca. In *Form and Function in Birds* (Ed.) King, A.S. & McLelland, J. Vol. 2. London and New York: Academic Press.

Kinsky, F.C. (1971) The consistent presence of paired ovaries in the kiwi (*Apteryx*) with some discussion of this condition in other birds. *J. Orn., Lpz.*, 112, 334–357.

Lillie, F.R. (1952) *Lillie's Development of the Chick* (Revised by Hamilton, H.L.) 3rd Edn. New York: Holt, Rinehart and Winston.

Lofts, B. & Murton, R.K. (1973) Reproduction in birds. In *Avian Biology* (Ed.) Farner, D.S. & King, J.R. Vol. III. New York and London: Academic Press.

McIndoe, W.M. (1971) Yolk synthesis. In *Physiology and Biochemistry of the Domestic Fowl* (Ed.) Bell, D.J. & Freeman, B.M. Vol. 3. London and New York: Academic Press.

Murton, R.K. & Westwood, N.J. (1977) *Avian Breeding Cycles*. Oxford: Clarendon Press.

Perry, M.M., Gilbert, A.B. & Evans, A.J. (1978) Electron microscope observations on the ovarian follicle of the domestic fowl during the rapid growth phase. *J. Anat.*, 125, 481–497.

Romanoff, A.L. (1960) *The Avian Embryo*. New York: Macmillan.

Romanoff, A.L. & Romanoff, A.J. (1949) *The Avian Egg*. New York: Wiley.

Rothwell, B. & Solomon, S.E. (1977) The ultrastructure of the follicle wall of the domestic fowl during the rapid growth phase. *Br. Poult. Sci.*, 18, 605–610.

Simkiss, K. & Taylor, T.G. (1971) Shell formation. In *Physiology and Biochemistry of the Domestic Fowl* (Ed.) Bell, D.J. & Freeman, B.M. Vol. 3. London and New York: Academic Press.

Solomon, S.E. (1983) The oviduct. In *Physiology and Biochemistry of the Domestic Fowl* (Ed.) Freeman, B.M. Vol. 4. London and New York: Academic Press.

Talbot, C.J. & Tyler, C. (1974) A study of the progressive deposition of shell in the shell gland of the domestic hen. *Br. Poult. Sci.*, 15, 217–224.

Tingari, M.D. & Lake, P.E. (1973) Ultrastructural studies on the uterovaginal sperm-host glands of the domestic hen, *Gallus domesticus*. *J. Reprod. Fert.*, 34, 423–431.

Witschi, E. (1935) Origin of asymmetry in the reproductive system of birds. *Am. J. Anat.*, 56, 119–141.

Wyburn, G.M., Johnston, H.S., Draper, M.H. & Davidson, M.F. (1970) The fine structure of the infundibulum and magnum of the oviduct of *Gallus domesticus*. *Q. Jl exp. Physiol.*, 55, 213–232.

Wyburn, G.M., Johnston, H.S., Draper, M.H. & Davidson, M.F. (1973) The ultrastructure of the shell-forming region of the oviduct and the development of the shell of *Gallus domesticus*. *Q. Jl exp. Physiol.*, 58, 143–151.

Chapter 9

MALE REPRODUCTIVE SYSTEM

Testis

The bean-shaped left and right testes are symmetrically arranged on either side of the midline in the dorsal coelom, near the caudal end of the lung and the cranial end of the kidney (Fig 9–2). In the domestic fowl and several passeriforms, and probably in most other birds (but not in several species of hawk), the left gonad of the genetic male, like the left gonad of the genetic female, acquires many of the primordial germ cells from the right gonad early in embryonic growth. Consequently, in the immature bird the left testis tends to be larger than the right. In the domestic cock this relationship persists until about six months of age, but after that the right testis tends to become heavier than the left. Each testis is suspended by a short mesentery, the mesorchium. It is surrounded by the abdominal air sac, but is not cooled by the air sac, contrary to earlier suggestions. When cut the testis exudes a milky fluid containing spermatozoa.

The *dimensions* of the testis increase greatly with sexual activity. In heavy breeds of domestic fowl the weight of both testes together ranges from about 0.05 g at one month to about 30 g at 18 months, and the length varies from about 2 cm shortly before sexual maturity to about 5.5 cm during sexual activity. This great enlargement at sexual maturity is due mainly to the much increased length and diameter of the seminiferous tubules and to greater numbers of interstitial cells, and is therefore directly related to an enhanced capacity to produce semen. In seasonal birds the testes enlarge to a remarkable degree at each culmination (nuptial) phase, the increase in weight of the organ being 300 to 500 times in some passerines (Fig 9–2). The *colour* also varies with sexual activity. In most breeds of domestic fowl and in many seasonal birds the immature or inactive testis is yellow from the accumulation of lipids in the interstitial cells. When the testis enlarges with sexual activity, it turns white through the dispersal of the interstitial cells by the expanding seminiferous tubules. In some species, e.g. the Common Starling and the Black Leghorn breed of domestic fowl, the resting testis is black because of melanocytes, but again it turns from grey to white when it enlarges with sexual activity. The surface of the testis is covered by a very thin *tunica albuginea*. In contrast to mammals, septa, and hence lobulation, are absent and there is no

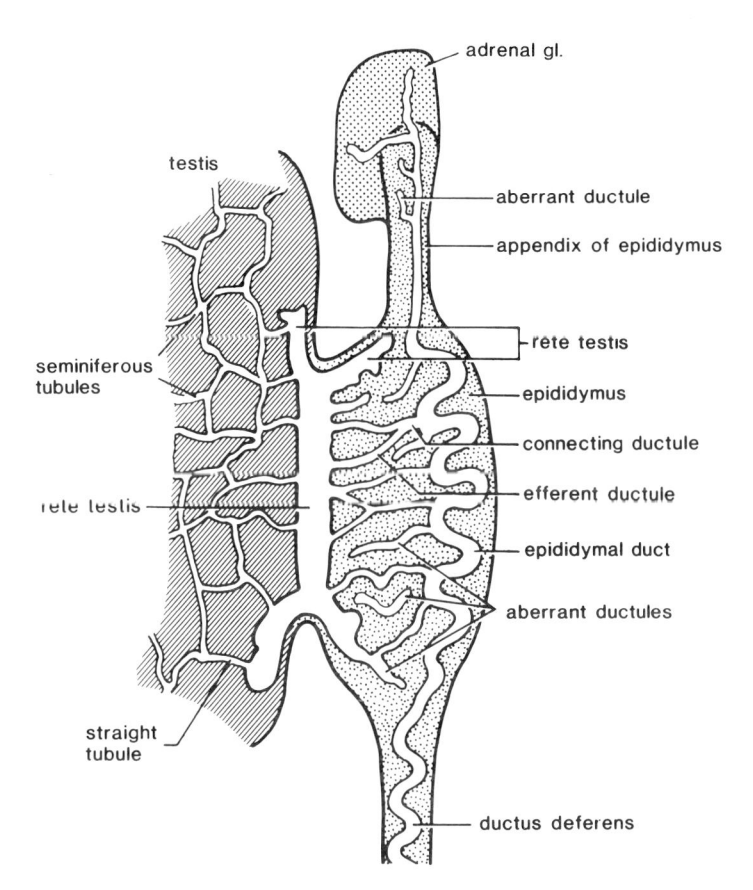

Fig 9–1 Diagram of the testis and epididymis of the domestic fowl. The seminiferous tubules anastomose with each other. The straight tubules are very short and indistinct in histological sections. The rete testis is a network of cavernous channels embedded in the dorsomedial surface of the testis adjoining the epididymis. The numerous coiled and anastomosing efferent ductules arise as wide tubes all along the epididymis and open via the connecting ductules into the epididymal duct. There are numbers of blind-ending aberrant ductules in the epididymis and in the appendix of the epididymis. When the testis is removed the epididymal appendix remains in the body attached to the adrenal gland.

mediastinum testis. In the domestic fowl most of the arterial blood reaches the testis in the testicular artery (Fig 13–4), a branch of the cranial renal artery. A pampiniform plexus, which in mammals acts to keep the temperature of the testis below that of the body, is absent. The testis is drained by several short testicular veins into the caudal vena cava.

The bulk of the testis consists of thousands of convoluted *seminiferous tubules*. The individual tubules essentially resemble those of mammals but anastomoses (Fig 9–1) are much more numerous in birds. The seminiferous tubule is lined by a spermatogenetic epithelium consisting of germ cells and sustentacular (Sertoli) cells. Like oogenesis, *spermatogenesis* proceeds in three phases: the first is a period of multiplication in which the cells are known as

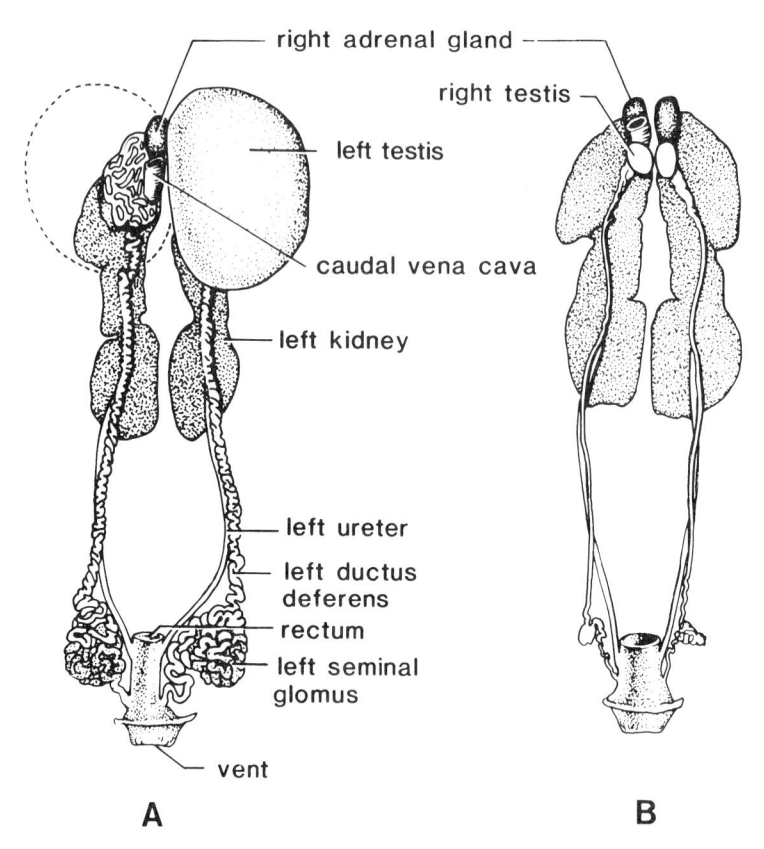

Fig 9–2 Ventral views of the male genital organs of the House Sparrow in **A**, the breeding condition and **B**, the non-breeding condition, showing the great change which occurs in the size and appearance of the testis and ductus deferens during the breeding season.

spermatogonia; the second is a period of growth in which the primary spermatocytes enlarge; and finally there is a period of maturation in which the first maturation division forms the secondary spermatocytes and the second maturation division forms the spermatids. Each spermatid develops into a spermatozoon. The spermatozoa develop in clusters with their heads attached to the sustentacular cells and their tails projecting into the lumen. Spermatogenesis is of a much shorter duration in birds than in mammals. Each sustentacular cell extends the whole depth of the epithelium and therefore provides mechanical support for the germ cells. Sustentacular cells are also believed to be involved in steroid hormone production and may have a phagocytic role.

When the spermatozoa are mature they detach themselves and pass towards the very short outgoing *straight tubules* which end by opening into the rete testis at about six points (Fig 9–1). The straight tubules lose the capacity for spermatogenesis and are lined by a low epithelium which seems to consist only of sustentacular cells. In the domestic fowl the *rete testis* consists of thin-walled irregular channels, embedded in fibrous connective tissue on the dorsomedial surface of the testis adjoining the epididymis; it also extends along the surface of the cranial and caudal ends of the epididymis (Fig 9–1). The rete testis is present in some passerines (e.g. finches) but absent in others (e.g. corvids) and in terns; possibly it is not a general feature of birds.

Interstitial cells (cells of Leydig) occupy the interstitial spaces between the tubules. Mature interstitial cells are polygonal and are characterized by smooth endoplasmic reticulum, mitochondria with tubular cristae, and cholesterol-rich lipid droplets. The lipid droplets are believed to be the precursor material involved in the production of androgenic steroids. As in mammals, testosterone seems to be the principal testicular androgen in birds, the interstitial cells being the main source. Androgens have a major effect on the growth of the deferent ducts, on the development of secondary sexual characteristics including plumage and appendages such as wattles and comb, and on song and courtship behaviour. Other cells in the interstitium include the melanocytes which cause the dark colour of the inactive testis in some birds.

Epididymis

In the domestic fowl and birds generally the epididymis lies on the dorsomedial surface of the testis (Fig 10–1) and is therefore almost invisible in a ventral dissection, even during the breeding season. It is relatively small compared to the mammalian epididymis, being only about 3 mm thick in the sexually mature domestic cock. In seasonal birds it enlarges greatly during the culmination (breeding) phase, and then becomes somewhat more prominent although it is still concealed in a ventral view. Head, body and tail cannot be distinguished, since the *efferent ductules* (Fig 9–1) are not confined to the cranial or head end of the epididymis but usually arise along its whole length (in the Jackdaw the efferent ductules are concentrated at the caudal extremity of the testis). The numerous coiled and anastomosing efferent ductules are initially of wide calibre but then taper and finally open via short *connecting ductules* into the epididymal duct (Fig 9–1). The efferent ductules are lined with a ciliated secretory pseudostratified columnar epithelium but in the distal parts of the connecting ductules the ciliated cells disappear. In the domestic fowl and the wild species that have been examined the *epididymal duct* is relatively straight and therefore much shorter than in mammals. Whilst it usually spans the entire length of the epididymal border of the testis, in some specimens of Jackdaw it lies as a tangled knot at the caudal extremity of the testis. The duct is lined with a non-ciliated secretory pseudostratified columnar epithelium. In sexually active domestic cocks the efferent and connecting ductules contain less and less spermatidis and more and more spermatozoa as they approach the epididymal duct, which is uniformly packed with spermatozoa. In contrast, in seasonal birds spermatozoa are seldom numerous in the epididymal duct, even during the culmination phase, being stored instead in the seminal glomus (see below).

In the domestic fowl the *epididymal appendix*, which is about 3–10 mm long, extends cranially from the epididymis, being firmly attached to the ventral border of the adrenal gland by connective tissue (Fig 9–1). It consists of many blind-ending *aberrant ductules* which connect with a cranial extension of the epididymal duct within the appendix (Fig 9–1); some of these aberrant ductules of the appendix invade the adrenal gland (Fig 9–1). A few other aberrant ductules open into the main part of the epididymal duct, and into the rete testis (Fig 9–1). When the testis is removed surgically, the main part of the epididymis comes away with the testis, but the appendix remains attached to the adrenal gland. In the surgically castrated male the aberrant ductules of the appendix of the epididymis sprout nodules which secrete androgens. These nodules lie in or deep to the capsule of the adrenal gland.

Seasonal reconstruction of the testis and epididymis

In birds with distinct breeding cycles, but not in the domestic fowl, three well defined phases can be recognized in the testis and epididymis. In the *regeneration phase* the seminiferous tubules, interstitial cells, tunica albuginea, and efferent and connecting ductules of the epididymis, largely disintegrate and are then reconstructed, thus restoring reproductive and endocrine function. In the *acceleration phase* (in autumn or spring) gonadotrophins induce a renewal of gametogenetic activity leading to the production of spermatozoa. In this phase the interstitial cells are large and their lipid content is greatly increased. The epithelial cells lining the epididymal ductules begin merocrine secretion. In the *culmination phase* gametogenesis reaches its peak as reproduction takes place. The interstitial cells become depleted of lipids. The epithelial cells lining the epididymal ductules undergo intermittent apocrine secretion.

Ductus deferens

The anatomy of the deferent duct of the domestic fowl resembles that of the wild species which have been investigated. It forms a tight zigzag parallel with the ureter and near the midline (Fig 10–1). Its diameter is increased progressively by smooth muscle (attaining a maximum of about 3.5 mm at its entry into the cloaca in the domestic fowl). It penetrates into the dorsal wall of the urodeum, ending in a spindle-shaped dilation, the *receptacle of the ductus deferens*. The receptacle is embedded in cloacal muscle except for the last 2–3 mm which project as the short *papilla of the ductus deferens* (Fig 11–1). Compared to that of the epididymal duct the epithelial lining is taller, tends to be pseudostratified, and is non-ciliated and less secretory. There are no glands, even in the most caudal part of the duct where the lumen is widest and the wall is thickest. In the sexually active domestic cock the ductus deferens is about as densely packed with spermatozoa as is the ductus epididymis; the spermatozoa take about one to four days to travel from the rete testis to the end of the ductus deferens.

In seasonal birds it is unusual for spermatozoa to be numerous in the ductus deferens. During the resting phase the main part of the ductus deferens is a relatively straight tube, its cranial part being embedded in the ventral surface of the kidney (Fig 9–2B). As the culmination phase approaches the duct increases greatly in length, lifts away from the kidney and becomes highly convoluted (Fig. 9–2A).

In passerines the caudal end of the ductus deferens forms a mass of convolutions called the *seminal glomus* (Fig 9–2A). At the onset of the culmination phase this structure becomes more convoluted and enlarged, so that the left and right seminal glomera together force out the cloacal wall into a prominent projection, the *cloacal promontory*. Essentially, the promontory consists of a rounded or conical bulge, about 10 mm in diameter in finches, which carries the vent about 5–7 mm caudal to its usual position. The promonotory can be used for sexing passerines. In these birds the seminal glomera are the main storage site for spermatozoa, the temperature there being about 4°C below the deep rectal temperature.

Accessory genital glands

No organs homologous to the mammalian seminal vesicle, prostate, bulbourethral glands and ampulla occur in birds. Contrary to earlier suggestions the seminal glomus is not homologous to the seminal vesicle of mammals; furthermore, the receptacle of the ductus deferens has no glands in its wall and therefore is not homologous to the glandular ampulla of mammals.

Semen

Studies of semen are based mainly on the domestic fowl. In this species the spermatozoa undergo maturation in the male tract, and must reach the ductus deferens before becoming fully fertile. In sexually active birds this journey takes from one to four days. The seminal plasma comes from sustentacular cells and the epithelial cells lining the efferent and connecting ductules, epididymal duct and ductus deferens. During artificial ejaculation in the domestic fowl (and to a lesser extent in the domestic turkey and some seasonal birds), a lymphlike fluid (the 'transparent fluid') arises from the tumescent lymphatic folds which lie on the floor of the proctodeal compartment of the cloaca (p. 193). A similar fluid arises from the cloacal floor in the domestic duck, again during the artificial collection of semen. Although these fluids must mix with the semen to a greater or lesser extent, they are harmful to spermatozoa probably because they contain blood clotting agents (which agglutinate the spermatozoa) and high concentrations of calcium and chloride ions. The volume of ejaculate is about 0.5–1.0 ml. The total number of sperm in an ejaculate ranges from about 1 to 2×10^9 million. For artificial insemination, ejaculation can be induced by abdominal massage; 100 million sperm are required for optimal fertility. In the absence of the seminal glomus the main site of storage of sperm in the domestic fowl is the ductus deferens, which has a capacity of a little less than 1 ml.

The *spermatozoon* has three components, the acrosome, the head (containing the nucleus) and the tail (Fig 9–3). The tail is long and threadlike, and can be divided into four regions: the relatively short neck (enclosing the proximal centriole), the relatively short middle piece (enclosing the distal centriole and the helically-arranged mitochondria), the very long principal piece (enclosing the axial filament complex), and the very short end piece (enclosing the remainder of the axial filament complex). In general, the slender form of the avian spermatozoon gives it a smaller volume than that of the mammalian spermatozoon.

Structurally, there are two kinds of avian spermatozoon, a complex type characterizing the passeriform birds and a simple form typifying all the remaining orders. The *simple type* is exemplified by the spermatozoon of the domestic fowl (Fig 9–3a). It is about 100 μm long, about one-third longer than

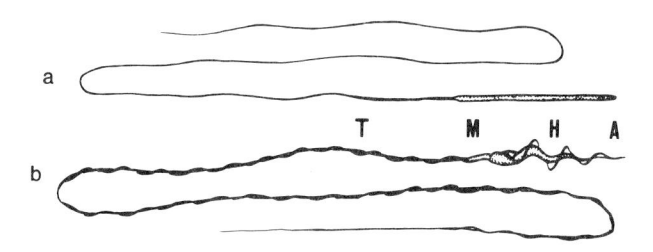

Fig 9–3 Diagrams of the two kinds of avian spermatozoa, a, the simple type exemplified by that of the domestic fowl and b, the complex type characteristic of passeriform birds. A = acrosome; H = head; M = neck and middle piece of tail; T = principal piece of tail.

that of man. The homogeneous conical acrosome attaches only at the extremity of the head (in contrast to that of mammals which spreads over the membrane of the head); it has been shown to contain a proteolytic enzyme which assists the spermatozoon to penetrate the oocyte. The head is a long, slightly curved and very slender cylinder (about 0.5 μm in diameter, compared to 2.5–3.5 μm in man). On ejaculation the tail become actively motile, creating an undulating movement. The *complex type* is typified by one outstanding feature, a predominantly spiral structure (Fig 9–3b). The acrosome, the head and the mitochondria of the tail all have a spiral form, with the same pitch. The corkscrew turns are most conspicuous in the tail, but nevertheless extend throughout the entire length of the cell. It is not known whether its spiral form affects the motion of the passeriform spermatozoon, but instead of undulating it does rotate rapidly around its longitudinal axis.

Phallus

There are two main varieties of phallus in male birds, the truly intromittent organ and the non-intromittent (non-protruding) type of phallus. The intromittent form occurs in the ratites (large flightless birds), tinamous, anseriforms (ducks, geese, swans, etc.), and probably in the galliform Cracidae. The non-intromittent type is known with certainty to occur in some of the domestic galliforms (domestic fowl and turkey) where it has been thoroughly investigated, and perhaps is also present in a group of passeriform birds (buntings, juncos and sparrows, mostly belonging to the family Emberizidae). These structures and their function in coitus are described in Chapter 11.

Caponization

In the chemical sterilization of cockerels for sale as *capons* (now illegal under EEC regulations) an oestrogen compound is subcutaneously implanted at about eight weeks of age in the form of a slow-release pellet. Since the head is discarded at slaughter, the pellet is introduced just behind the comb. An interval of at least six weeks is usually recommended between implantation and slaughter in order to avoid contamination of the meat. Surgical castration of cockerels is similarly illegal in Britain.

Further reading

Aire, T.A. (1979) The epididymal region of the Japanese Quail (*Coturnix coturnix japonica*). *Acta anat.*, 103, 305–312.
Bailey, R.E. (1953) Accessory reproductive organs of male fringillid birds: seasonal variations and response to various sex hormones. *Anat. Rec.*, 115, 1–20.
Budras, K.D. & Sauer, T. (1975) Morphology of the epididymis of the cock (*Gallus domesticus*) and its effect upon the steroid sex hormone synthesis. I. Ontogenesis, morphology and distribution of the epididymis. *Anat. Embryol.*, 148, 175–196.
Budras, K.D. & Sauer, T. (1975) Morphology of the epididymis of the cock (*Gallus domesticus*) and its effect upon the steroid sex hormone synthesis. II. Steroid sex hormone synthesis in the tubuli epididymidis and the transformation of the ductuli aberrantes into hormone producing noduli epididymidis in the capsule of the adrenal gland of the capon. *Anat. Embryol.*, 148, 197–213.
Burke, W.H. (1973) Testicular asymmetry in the turkey. *Poult. Sci.*, 52, 1652–1654.
Cooksey, E.J. (1975) Ultrastructural studies on the testis of the domestic fowl (*Gallus domesticus*). PhD Thesis, University of Strathclyde.
Eroschenko, V.P. & Wilson, W.O. (1974) Histological changes in the regressing reproductive organs of sexually mature male and female Japanese Quail. *Biol. Reprod.*, 11, 168–179.
Erpino, K.H. (1969) Seasonal cycle of reproductive physiology in the Black-billed Magpie. *Condor*, 71, 267–279.
Gasc, J.M. (1978) Growth and sexual differentiation in the gonads of the chick and duck embryos. *J. Embryol. exp. Morph.*, 44, 1–13.
Gunawardana, V.K. (1977) Stages of spermatids in the domestic fowl: a light microscope study using Araldite sections. *J. Anat.*, 123, 351–360.
Gunawardana, V.K. & Scott, M.G.A.D. (1977) Ultrastructural studies on the differentiation of spermatids in the domestic fowl. *J. Anat.*, 124, 741–755.
Hess R.A. & Thurston, R.J. (1977) Ultrastructure of the epithelial cells in the epididymal region of the turkey (*Meleagris gallopavo*). *J. Anat.*, 124, 765–778.
Hess, R.A., Thurston, R.J. & Biellier, H.V. (1976) Morphology of the epididymal region and ductus deferens of the turkey (*Meleagris gallopavo*). *J. Anat.*, 122, 241–252.
Humphreys, P.N. (1972) Brief observations on the semen and spermatozoa of certain passerine and non-passerine birds. *J. Reprod. Fert.*, 29, 327–336.
Humphreys, P.N. (1975) Ultrastructure of the Budgerigar testis during a photoperiodically induced cycle. *Cell Tissue Res.*, 159, 541–550.
King, A.S. (1975) Aves urogenital system. In *Sisson and Grossman's The Anatomy of the Domestic Animals* (Ed.) Getty, R. Vol. 2, 5th Edn. Philadelphia: Saunders.
King, A.S. (1979) Systema urogenitale. In *Nomina Anatomica Avium* (Ed.) Baumel,

J.J., King, A.S., Lucas, A.M., Breazile, J.E. & Evans, H.E. London and New York: Academic Press.

Lake, P.E. (1957) The male reproductive tract of the fowl. *J. Anat.*, 91, 116–129.

Lake, P.E. (1971) The male in reproduction. In *Physiology and Biochemistry of the Domestic Fowl* (Ed.) Bell, D.J. & Freeman, B.M. Vol. 3. London and New York: Academic Press.

Lake, P.E. (1980) Male genital organs. In *Form and Function in Birds* (Ed.) King, A.S. & McLelland, J. Vol. 2. London and New York: Academic Press.

Lake, P.E. & El Jack, M.H. (1966) The origin and composition of fowl semen. In *Physiology of the Domestic Fowl* (Ed.) Horton Smith, C. & Amoroso, E.C. British Egg Marketing Board Symposium No. 1, 44–51. Edinburgh: Longman.

Lake, P.E., Smith, W. & Young, D. (1968) The ultrastructure of the ejaculated fowl spermatozoon. *Q. Jl exp. Physiol.* 53, 356–366.

Lofts, B. & Murton, R.K. (1973) Reproduction in birds. In *Avian Biology* (Ed.) Farner, D.S. & King, J.R. Vol. 3. New York and London: Academic Press.

Maretta, M. (1975) The ultrastructure of the spermatozoon of the drake. I. Head. *Acta vet. hung.*, 25, 47–52.

Maretta, M. (1975) The ultrastructure of the spermatozoon of the drake. II. Tail. *Acta vet. hung.*, 25, 53–60.

Mehrotra, P.N. (1962) Cyclical changes in the epididymis of the goose, *Anser melanotus*. *Q. Jl microsc. Sci.*, 103, 377–383.

Mehrotra, P.N. (1964) On the microscopic anatomy of the epididymis of *Anser melanotus* L. *Trans. Amer. microsc. Sci.*, 83, 456–460.

Nishida, T. (1964) Comparative and topographical anatomy of the fowl. XLII. Blood vascular system of the male reproductive organs. *Jap J. vet. Sci.*, 26, 211–221.

Riddle, O. (1925) On the sexuality of the right ovary of birds. *Anat. Rec.*, 30, 365–383.

Romanoff, A.L. (1960) *The Avian Embryo*. New York: Macmillan.

Rothwell B. (1973) The ultrastructure of Leydig cells in the testis of the domestic fowl. *J. Anat.*, 116, 245–253.

Rothwell, B. & Tingari, M.D. (1973) The ultrastructure of the boundary tissue of the seminiferous tubule in the testis of the domestic fowl *(Gallus domesticus)*. *J. Anat.*, 114, 321–328.

Scott, D.M. & Middleton, A.L.A. (1968) The annual testicular cycle of the Brown-headed Cowbird *(Molothrus ater)*. *Can. J. Zool.*, 46, 77–87.

Serventy, D.L. & Marshall, A.J. (1956) Factors influencing testis coloration in birds. *Emu*, 56, 219–222.

Stanley, A.J. & Witschi, E. (1940) Germ cell migration in relation to asymmetry in the sex glands of hawks. *Anat. Rec.*, 76, 329–342.

Test, F.H. (1944) Testicular asymmetry in the woodpecker *Colaptes*. *Pap. Mich. Acad. Sci.*, 30, 347–353.

Tingari, M.D. (1971) On the structure of the epididymal region and ductus deferens of the domestic fowl *(Gallus domesticus)*. *J. Anat.*, 109, 423–435.

Tingari, M.D. (1972) The fine structure of the epithelial lining of the excurrent duct system of the testis of the domestic fowl *(Gallus domesticus)*. *Q. Jl exp. Physiol.*, 57, 271–295.

Tingari, M.D. (1973) Observation on the fine structure of spermatozoa in the testis and excurrent ducts of the male fowl, *Gallus domesticus*. *J. Reprod. Fert.*, 34, 255–265.

Traciuc, E. (1967) L'anatomie microscopique de l'épidydyme chez *Sterna hirundo* L. *Anat. Anz.*, 121, 381–386.

Traciuc, E. (1969) La structure de l'epididyme de *Coloeus monedula* (Aves, Corvidae). *Anat. Anz.*, 125, 49–67.

Venzke, W.G. (1954) The morphogenesis of the indifferent gonads of chicken embryos. *Am. J. vet. Res.*, 15, 300–308.

Witschi, E. (1935) The origin of asymmetry in the reproductive system of birds. *Am. J. Anat.*, 56, 119–141.

Chapter 10

URINARY SYSTEM

External form of the kidney

In birds generally the kidneys lie symmetrically in bony depressions of the synsacrum, the renal fossae (Fig 4–10). They reach the lungs cranially and the end of the synsacrum caudally (Fig 6–10). In the adult domestic fowl they are about 7 cm long, and about 2 cm wide at their broadest point. The kidneys generally form over 1 per cent of the body weight in small species, and less than 1 per cent in large species. Extending between the kidneys and the pelvis are diverticula of the abdominal air sacs; because of this relationship the kidneys are removed during the postmortem processing of poultry in the slaughter plant if the abdominal air sacs are infected.

Each kidney is divided into *cranial, middle* and *caudal divisions* (Fig 10–1). These divisions are in no way homologous to the lobes of mammalian kidney. The boundaries between them are not always distinct, being formed by the external iliac artery and ischiadic artery as in Fig 10–1. In birds of most non-passerine orders, including the domestic fowl, the divisions are distinct, but in passerines the middle division is not well defined, being largely fused to the other two divisions, particularly the caudal one. Also, the relative sizes of the divisions vary greatly. Nevertheless the presence of the three divisions appears to be characteristic of birds in general. In many birds, including herons, puffins and penguins, but not the domestic fowl, the caudal divisions of the left and right kidneys are fused in the midline. Spinal nerves of the lumbar plexus and sacral plexus pass through the substance of the kidney.

The surface of each division is covered with small rounded projections, about 1–2 mm in diameter in the domestic fowl. These are renal lobules reaching the surface of the kidney (see Fig 10–1, left kidney).

The architecture of the avian kidney is founded on renal lobules, the rough outlines of which can be identified without too much difficulty. It is also possible to recognize the presence of larger units, the renal lobes, but these are much less obvious. The lobes and lobules form areas of cortex and medulla. The general organization of all these components appears to be broadly similar in the domestic fowl and the other species which have been examined, the only differences being those of detail.

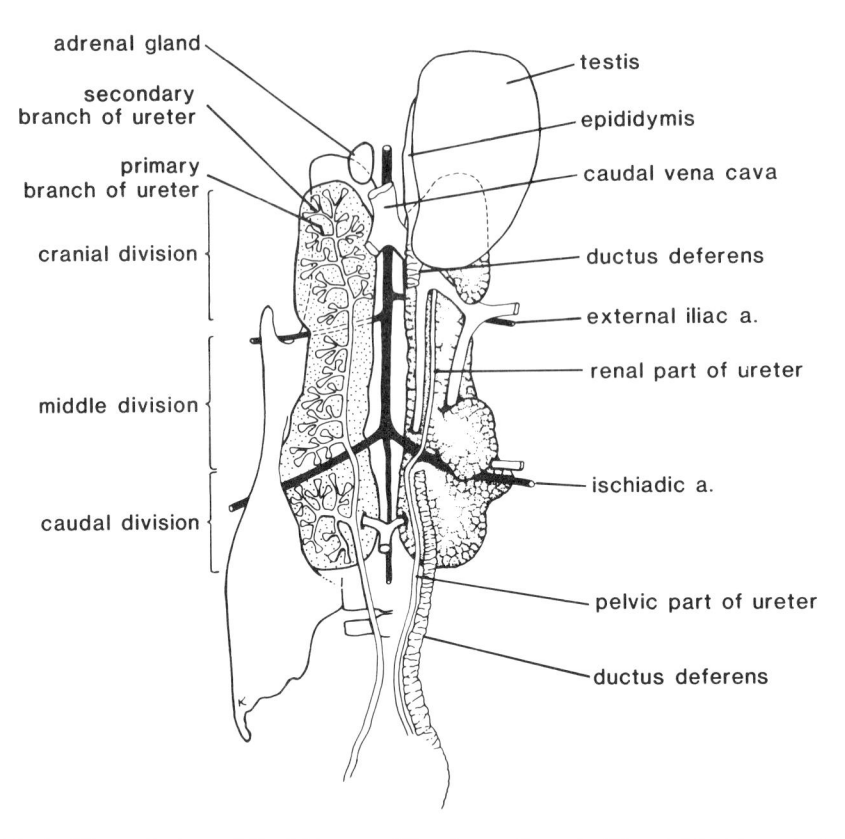

Fig 10–1 Ventral view of the kidneys of the domestic cock. The right kidney is drawn as though transparent to show the primary branches of the ureter and some of its secondary branches. At the end of each secondary branch of the ureter the diagram shows a conical expansion. Each such conical expansion represents the encapsulated cone-shaped assembly of medullary collecting tubules which forms the medullary part of a lobe. The proximal part of the left ductus deferens has been removed. From King (1975), with kind permission of the publisher.

The renal lobule

In histological sections, which of course are only two-dimensional, an avian renal lobule is seen as a pear-shaped area of tissue wedged between the *inter*lobular veins of the renal portal system (Fig 10–2). There are many such lobules in one kidney, some being superficial and many others buried deep below the surface. Each lobule is enclosed by its perilobular collecting tubules, as the staves enclose a barrel; the perilobular collecting tubules are therefore *inter*lobular in position. The efferent vein which drains the lobule lies in the centre of the lobule, and is therefore the *intra*lobular vein. The artery which supplies the lobule is also in the centre of the lobule, thus constituting the *intra*lobular artery. (These relationships are therefore essentially the reverse of those in mammals, where the collecting tubules, as medullary rays, are *intra*lobular and the arteries are *inter*lobular.)

URINARY SYSTEM

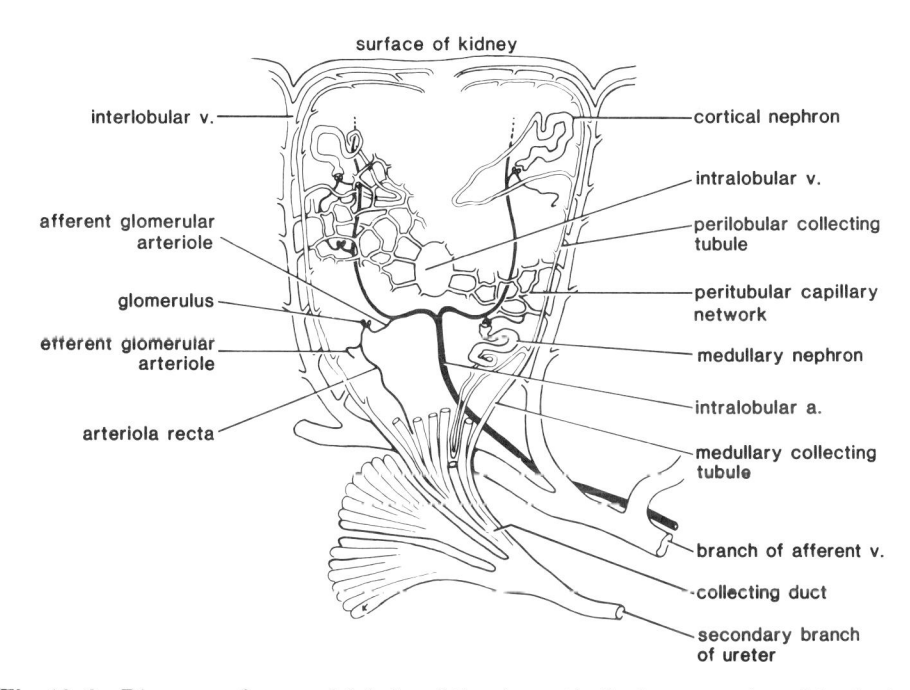

Fig 10–2 Diagram of a renal lobule of the domestic fowl as seen in a histological section. The lobule appears pear-shaped. The cortical region of the lobule is shown in the upper two-thirds of the diagram. In addition to the afferent and efferent blood supply of the lobule, the diagram shows three nephrons in the cortical region of the lobule; two of these nephrons are of the cortical or reptilian type with a short convoluted intermediate segment, and the third is of the medullary or mammalian type with a nephronal loop. Five glomeruli are shown, as double loops on the very short afferent glomerular artery. At the lower centre of the diagram, the collecting tubules and nephronal loops converge into a conical bundle of medullary collecting tubules, forming the stalklike medullary region of the lobule. At the bottom left side of the diagram there are two other conical bundles, representing the medullary regions of two other lobules. The three medullary regions of these three lobules then combine at the bottom centre of the diagram to form a single cone-shaped assembly of medullary collecting tubules, representing the medullary region of a single renal lobe. This empties into a secondary branch of the ureter. Redrawn from King (1975), with kind permission of the publisher.

At the tapering stalk of the pear-shaped lobule, the collecting tubules converge to form a conical bundle of medullary collecting tubules. This conical stalklike part of the lobule is the *medullary region* or *medullary cone* of the lobule; in addition to collecting tubules it contains the nephronal loops (of Henle) belonging to the medullary types of nephron (Fig 10–2). The wide part of the pear is the *cortical region* of the lobule; it contains the nephrons of both the cortical and medullary types, except for the nephronal loops of the latter.

Three-dimensional preparations, as in cleared kidneys, show that the lobule is not really a simple pear-shaped object. It is more elongated, rather like a loaf of bread. The cortical part of the loaf actually drains not into only one medullary region but several. Furthermore, any one medullary region may receive contributions from the cortical regions of several lobules (Fig 10–3). Thus the renal lobule turns out to be much more complex than it appears to be at first sight. Nevertheless the lobule is still a convenient structural concept.

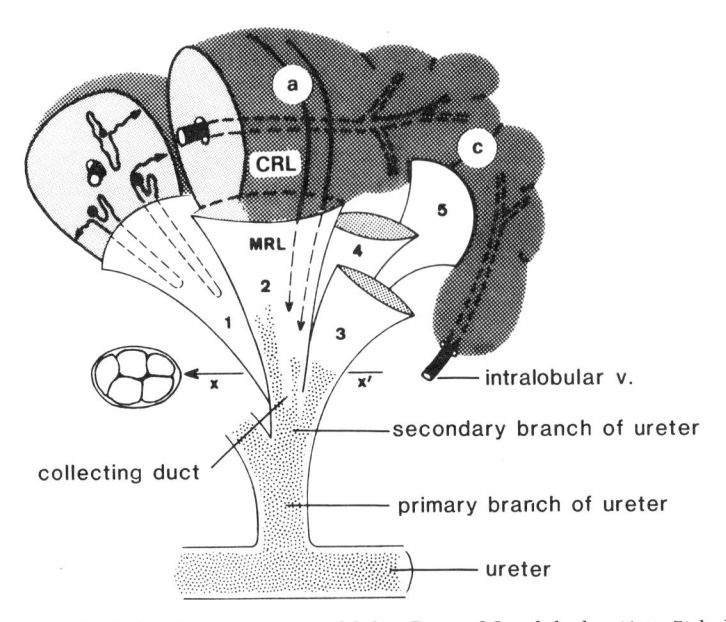

Fig 10–3 Renal lobules forming a renal lobe. Part of five lobules (1 to 5) belonging to one lobe are shown. Each lobule consists of a cortical region (CRL) and a medullary region (MRL), but the cortical regions of lobules 3 and 4 are omitted. The medullary regions of the five lobules converge into a single encapsulated cone-shaped assembly of medullary collecting tubules, cut in transverse section at X–X′, which constitutes the medullary region of the renal lobe. This medullary region, and the lobules associated with it, form a complete renal lobe. The arrangement of the lobule is somewhat more complicated than the pear-shaped structure found in histological sections as in Fig 10–2. When seen in three-dimensional preparations the cortical region of the lobule is elongated. The elongated cortical region drains into several medullary regions, although this is not shown in the diagram. Furthermore, any one medullary region of a lobule may drain more than one cortical region; thus medullary region 5 drains the two independent cortical regions which adjoin each other at 'c'. Three nephrons are shown in lobule 1, the upper one being a cortical nephron and the lower two being medullary nephrons with nephronal loops contributing to the medullary region of the lobule. The two arrows show the direction taken by perilobular collecting tubules at 'a' as they enter the medullary region of lobule 2, where they combine to form increasingly larger ducts. From Johnson (1974), with kind permission of the editor of the *Journal of Morphology*.

The renal lobe

As in Fig 10–2, the conical bundle of medullary collecting tubules at the stalk of a lobule joins several similar conical bundles from several other adjacent lobules. This group of several conical bundles of medullary collecting tubules eventually converges into a single cone-shaped assembly of collecting tubules encapsulated within a connective tissue sheath (cut transversely at X–X′ in

Fig 10–3). One such cone-shaped assembly therefore drains a group of several lobules; consequently it represents the medullary region of a renal lobe, and is probably homologous to the medullary pyramid of a mammalian multilobular kidney as in the ox. Together, the encapsulated cone-shaped assembly of collecting tubules and the cortical regions which drain into it form a renal lobe. It drains into a single large collecting duct (Fig 10–2). Several collecting ducts combine to form a secondary branch of the ureter (Figs 10–2 and 10–3).

The renal cortex and medulla

The cortex is formed by the wide cortical regions of the lobules, and the medulla by the tapering stalklike medullary regions of the lobules. However, the lobules, and therefore also the lobes, are embedded at varying depths in the kidney, and consequently the cortex and medulla do not form the continuous outer and inner strata respectively which typify the mammalian kidney. Instead, large areas of cortex enclose relatively small cone-shaped islands of medulla.

The number of these cone-shaped medullary regions per unit volume of kidney tends to be considerably greater in species which are particularly successful at conserving water (e.g. the salt marsh Savannah Sparrow). In such species each medullary region must drain a relatively small volume of cortex; this presumably implies a relatively high proportion of mammalian type nephrons (which have nephronal loops) and therefore suggests relatively good counter-current concentration.

The nephron

The avian kidney has two types of nephron. The cortical type of nephron, which forms the majority, is devoid of a nephronal loop (of Henle) and is confined to the cortical region of the lobule (Figs 10–2 and 10–3). These nephrons are reptilian in form. The medullary type has a nephronal loop which penetrates the conical medullary region of the lobule (Figs 10–2 and 10–3). These are mammalian in form.

Both types of nephron always begin with a *renal corpuscle*, which consists of the glomerular capsule (Bowman's capsule), deeply indented by the *glomerulus*, which is a tuft of capillaries. The renal corpuscles lie about midway between the inter- and intralobular veins (Figs 10–2 and 10–3). The *cortical nephron* has a convoluted *proximal part* (Fig 10–4) which tends to be localized at the periphery of the lobule and forms about half of the total length of the nephron. There is an ill-defined, short convoluted *intermediate segment*, followed by the *distal part* of the nephron which forms compact convolutions near the intralobular vein (Fig 10–4). Two zones can be distinguished in histological sections of the avian cortex, i.e. an outer zone which contains the convolutions of the proximal parts of the nephrons, and an inner zone which contains the convolutions of the distal parts of the nephrons surrounding the intralobular vein (Fig 10–4).

The *medullary nephron* has convoluted proximal and distal parts, like those of the cortical nephron, but the intermediate segment forms a nephronal loop (nephronal ansa or loop of Henle) which descends usually within the medullary region of the lobule (Figs 10–2 and 10–3) but is quite often outside it. The avian nephronal loop is always like the short variety of loop in the mammal, the calibre enlarging *before* the bend (Fig 10–4). In passerines the medullary region of the lobule is relatively highly organized, with a ring of medullary collecting tubules which encloses the thin descending limbs of the nephronal loops and is in turn surrounded by layers of ascending thick limbs. In the

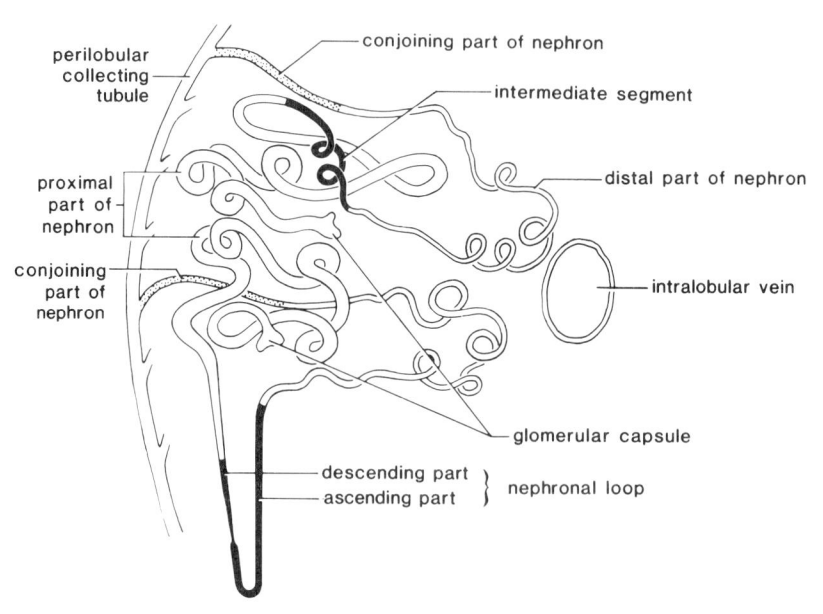

Fig 10–4 Diagram showing the disposition of the various parts of a cortical (reptilian) type of nephron (above), and a medullary (mammalian) type of nephron below.

domestic fowl and many other non-passerines there is a general intermingling of these medullary elements.

The *collecting tubules* (Fig 10–2) which lie superficially on the surface of the cortical region of the lobule are known as perilobular collecting tubules; those that pass within the medullary region of the lobule are named medullary collecting tubules. The medullary collecting tubules from one lobule typically combine into a single large vessel, the *collecting duct* (Fig 10–2). Several collecting ducts in turn combine to form a *secondary branch of the ureter* (Fig 10–2).

A complete *juxtaglomerular complex* is present in birds. It includes the macula densa which is a thickening of the epithelium of the beginning of the distal part of the nephron, where the latter comes into contact with the afferent arteriole. It also incorporates the secretory juxtaglomerular cells (modified myoepithelial cells) of the adjacent afferent arteriole. At the vascular pole of the glomerulus there is a third component, the juxtavascular insula (or extraglomerular mesangium) with its juxtavascular cells (or Polkissen cells). A vasopressor renin-like substance is present in the kidney of the domestic

URINARY SYSTEM 181

fowl; it may be formed in the juxtaglomerular cells and stored in the other components of the juxtaglomerular complex.

The arteries of the kidney

The cranial, middle and caudal renal arteries supply the cranial, middle and caudal divisions of the kidney respectively (Fig 10–5). Their branches eventually form the *intralobular arteries* (Fig 10–2), which pass towards the surface

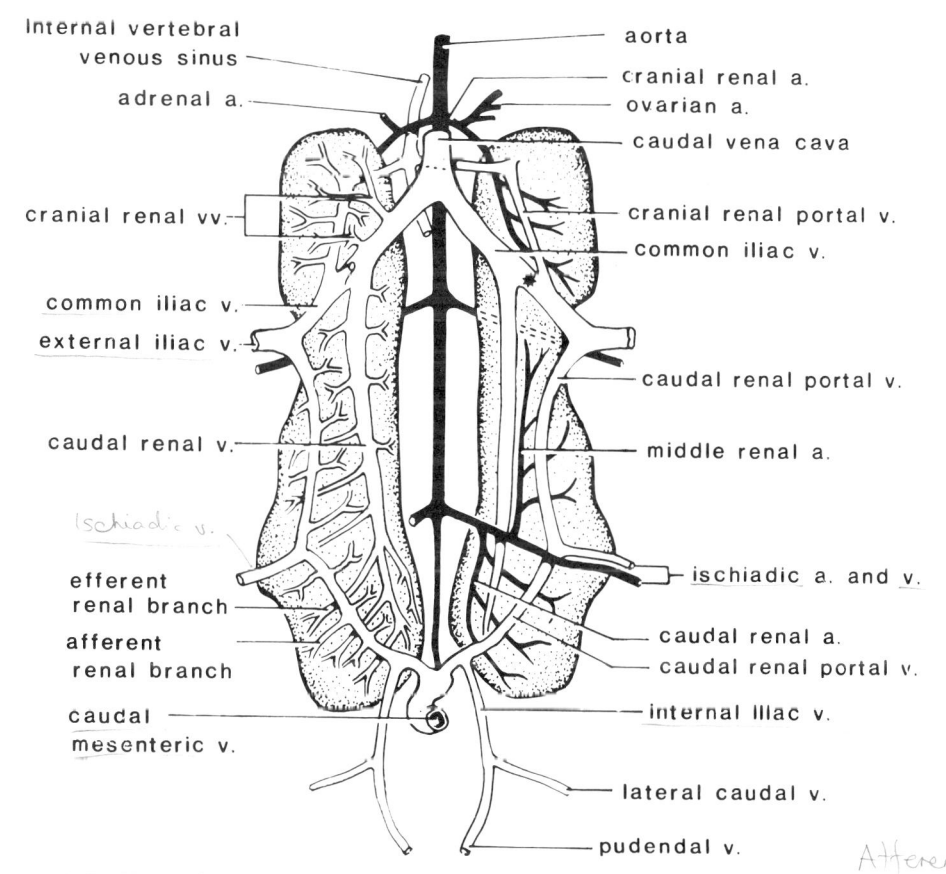

Fig 10–5 Ventral view of the kidneys of the domestic fowl showing the blood supply. The kidneys are drawn as though transparent, in order to reveal the vessels within. The left side of the diagram shows the renal portal veins and the efferent veins; the right side shows the arteries. The asterisk marks the site of the renal portal valve in the left kidney. In the right kidney the conical valve is shown diagrammatically. From King (1975), with kind permission of the publisher.

of the lobule lying about half-way between the inter- and intralobular veins. The intralobular arteries give rise to the very short *afferent glomerular arterioles*, each of which almost immediately forms its *glomerulus* (Fig 10–2). The capillary tufts of the glomerulus, which are much simpler and smaller than in mammals, consist of only two or three capillaries, and continue as the *efferent glomerular arterioles* (Fig 10–2). The latter empty into the cortical *peritubular capillary plexus*, their entry point being near the periphery of the lobule (Fig 10–2). The peritubular plexus finally discharges into the intralobular vein in the centre of the lobule (Fig 10–2). In the more cortical region of the lobule, the capillaries of the peritubular network are closely applied to the epithelium of the adjacent nephrons. The medullary region of the lobule is supplied by *vasa recta* resembling those of mammals in their looped arrangement; they consist of *arteriolar rectae* which arise from the efferent glomerular arterioles lying nearest to the medulla (Fig 10–2), and *venulae rectae* which form the drainage pathway.

The afferent or renal portal veins

The cranial and caudal renal portal veins form a venous ring which encompasses both kidneys (Fig 10–5). The ring is completed at the cranial end by anastomosis of the left and right cranial renal portal veins with the internal vertebral venous sinus, and at the caudal end by anastomosis of the left and right caudal renal portal veins with the caudal mesenteric (coccygeomesenteric) vein. Portal blood can enter the ring, and hence the kidneys, via the external iliac veins, the ischiadic veins, the internal iliac veins and the caudal mesenteric vein. Many smaller veins, known as the *afferent renal branches*, arise from the ring (Fig 10–5) and penetrate the parenchyma of the kidney. Muscular sphincters at the roots of these branches control the volume of blood entering the renal parenchyma from the venous ring. Within the parenchyma of the kidney the smaller afferent (portal) branches become the *interlobular veins*, and these finally form the peritubular capillary plexus at the periphery of the lobule (Fig 10–2).

A conical valve, the *renal portal valve*, is so sited within the lumen of the common iliac vein (Fig 10–5) that when it is open it can divert the portal flow away from the tissues of the kidney and directly into the caudal vena cava. The smooth muscle of the valve is richly innervated by adrenergic and cholinergic fibres, and is inhibited by adrenaline and stimulated by histamine or acetylcholine. Nevertheless, the physiological mechanisms which control the valve in the live animal are not yet known. However, cineradiographic studies in the domestic fowl have shown that the portal blood can be shunted (a) through the portal valve into the caudal vena cava, (b) into the caudal mesenteric vein towards the liver, or (c) into the internal vertebral venous sinus within the neural canal. These shunts sometimes bypass the kidney completely, but usually only a part of the portal flow is diverted from the kidney and the rest enters it, eventually traversing the peritubular capillary network. The direction of flow in the caudal mesenteric vein is usually towards the kidney; reversal of this flow sends the blood into the hepatic portal circulation. The shunt through the portal valve directly into the caudal vena cava may be a part of a general physiological mechanism for increasing the venous return to the heart as in exercise. Or it may be part of a more specialized mechanism to allow an increase in the blood flow through the legs, to a higher level than that required by the kidneys, when the bird is running.

The caudal renal vein and the cranial renal veins

The caudal renal vein drains the middle and caudal divisions of the kidney (Fig 10–5). The initial tributaries are the intralobular veins (at the centres of the lobules), which drain the peritubular capillary network (Fig 10–2). The intralobular veins drain into efferent renal veins (efferent renal branches in Fig 10–5), which eventually empty into the caudal renal vein. This ends by opening into the common iliac vein, immediately on the heart side of the renal portal valve.

The intralobular veins of the cranial division of the kidney drain into efferent renal veins which in turn form several cranial renal veins; the latter empty into either the common iliac vein (Fig 10–5) or the caudal vena cava directly.

The ureter

The renal part of the ureter starts within the depth of the cranial division of the kidney but continues caudally in a groove on the ventral surface of the other two divisions (Figs 6–10, 9–2, 10–1). It receives a series of tributaries (about 17 in the domestic fowl), which are the primary branches of the ureter (Fig 10–1). Each primary branch receives about five or six secondary branches; each secondary branch in turn drains the medullary component of a renal lobe (Figs 10–2 and 10–3). In histological sections the lumen of the pelvic part of the ureter has a stellate shape and is lined mainly by a mucus-secreting, pseudostratified, columnar epithelium.

The pelvic part of the ureter (Fig 10–1) opens into the urodeum (Fig 11–1). A ventral (allantoic) urinary bladder is absent in all birds. The Ostrich is sometimes said to have one, but the 'urinary bladder' of this species opens dorsally and is merely the cloacal bursa which has a very wide opening in the large flightless birds (ratites).

Excretion

Birds, like mammals, can produce a concentrated urine which is hypertonic to blood plasma. The concentrating power of the kidneys lies in the countercurrent medullary nephronal multiplier systems of the medullary nephronal loops. These systems are better developed in mammals than birds, and consequently birds have less ability than mammals to concentrate their urine.

Birds resemble reptiles in being uricotelic, that is they excrete the end product of nitrogen metabolism as uric acid. The uric acid is synthesized in the liver, and the kidney excretes it from the blood plasma partly by means of glomerular filtration and mainly by tubular secretion. The uric acid and the salts (urates) arising from it can form colloidal solutions with concentrations of up to 2 per cent, and this allows transport through the tubules and collecting ducts without precipitation. Precipitation could obstruct the collecting ducts. In the ureter the urine has been found to be viscous and stringy, and mucus may be needed to lubricate the movement of precipitated urate down the ureter.

It is widely believed that the excretion of nitrogen as urate enables birds to economise on water by excreting a semi-solid urine instead of the much more watery solution of mammals. However, it has been calculated that the excretion of uric acid by an adult domestic fowl could entail the use of 200 ml of water per gram of nitrogen: to excrete urea a man could use 150 ml of water per gram of nitrogen. The advantages of uric acid probably lie more in overcoming the difficulties of the cleidoic (closed-box) egg, where the supply of water is strictly limited. The embryo bird stores its nitrogenous wastes in its allantois. Uric acid and its salts are less soluble than urea. In fact the urate is held in the avian allantois as a crystalline anhydrous deposit, which allows the transporting water to be reabsorbed and, because of its low solubility, is not toxic to the embryo. Moreover, if urea were the waste product it would hold water by osmotic forces, and would thereby reduce the availability to the growing embryo of the limited water supply inside the egg.

The renal portal system takes part in the secretion of urates. It achieves this by supplying venous blood to the peritubular capillary plexus at the periphery of the lobule (Fig 10–2), where the convoluted proximal parts of the nephrons are mainly situated (Fig 10–4). The proximal part of the nephron is responsible for the tubular secretion of urates. Tubular secretion can be demonstrated in the domestic fowl by injecting a dye into one leg; the dye appears first in the ureter on the same side of the body as the injection, and appears in the other ureter only after a further lapse of time.

Glomerular filtration also contributes to the elimination of urates, but is insufficient alone to account for urate clearance. It has been estimated that about two-thirds of the blood to the kidney is supplied by the renal portal veins. All of this blood goes to the peritubular capillaries and none to the glomeruli. It therefore seems likely that the renal portal system makes a major contribution to the elimination of urates.

The urine which leaves the ureter enters the urodeum and is then moved by retroperistalsis into the coprodeum and rectum, a small amount reaching as far cranially as the caeca. It is then stored in these regions until defaecation, there being an opportunity during storage for the recovery of water or solute. In the dehydrated or salt-depleted domestic fowl about 14 per cent of the urates and 66 per cent of the NaCl are reabsorbed. Therefore under these conditions the cloaca is important for water and salt conservation. However, in the normal domestic fowl, the evidence appears to be against any essential role for the cloaca in the maintenance of water balance. On the other hand, in the Galah, a xerophilic Australian parrot, the extrarenal conservation of water has been shown to be much greater than in the domestic fowl; possibly, the resorptive mechanisms in the rectum and cloaca are in general more highly developed in xerophilic species than in other birds.

Further reading

Akester, A.R. (1964) Radiographic studies of the renal portal system in the domestic fowl (*Gallus domesticus*). *J. Anat.*, 98, 365–376.

Akester, A.R. (1967) Renal portal shunts in the kidney of the domestic fowl. *J. Anat.*, 101, 569–594.
Akester, A.R., Anderson, R.S., Hill, K.J. & Osbaldiston, G.W. (1967) A radiographic study of urine flow in the domestic fowl. *Br. Poult. Sci.*, 8, 209–212.
Akester, A.R. & Mann, S.P. (1969) Adrenergic and cholinergic innervation of the renal portal valve in the domestic fowl. *J. Anat.*, 104, 241–252.
Baumel, J.J. (1975) Aves heart and blood vessels. In *Sisson and Grossman's The Anatomy of the Domestic Animals* (Ed.) Getty, R. Vol. 2, 5th Edn. Philadelphia: Saunders.
Braun, E. (1976) Intrarenal blood flow distribution in the desert quail following salt loading. *Am. J. Physiol.*, 231, 1111–1118.
Braun, E.J. & Dantzler, W.H. (1972) Function of mammalian-type and reptilian-type nephrons in kidney of desert quail. *Am. J. Physiol.*, 222, 617–629.
Das, B.K. (1924) On the intra-renal course of the so-called 'renal portal' veins in some common Indian birds. *Proc. zool. Soc. Lond.*, 50, 757–773.
Emery, N., Poulson, T.L. & Kintner, W.B. (1972) Production of concentrated urine by avian kidneys. *Am. J. Physiol.*, 223, 180–187.
Goodchild, W.M. (1956) Biological aspects of the urinary system of *Gallus domesticus* with particular reference to the anatomy of the ureter. MSc Thesis, University of Bristol.
Johnson, O.W. (1968) Some morphological features of avian kidneys. *Auk*, 85, 216–228.
Johnson, O.W. (1974) Relative thickness of the renal medulla in birds. *J. Morph.*, 142, 277–284.
Johnson, O.W. (1979) Urinary organs. In *Form and Function in Birds* (Ed.) King, A.S. & McLelland, J. Vol. 1. London and New York: Academic Press.
Johnson, O.W. & Mugaas, J.N. (1970) Some histological features of avian kidneys. *Am. J. Anat.*, 127, 423–435.
Johnson, O.W. & Mugaas, J.N. (1970) Quantitative and organizational features of the avian renal medulla. *Condor*, 72, 288–292.
Johnson, O.W. & Ohmart, R.D. (1973) The renal medulla and water economy in Vesper Sparrows (*Pooecetes gramineus*). *Comp. Biochem. Physiol.*, 44A, 655–661.
Johnson, O.W. & Ohmart, R.D. (1973) Some features of water economy and kidney microstructure in the Large-billed Savannah Sparrow (*Passerculus sandwichensis rostratus*). *Physiol. Zool.*, 46, 276–284.
Johnson, O.W., Phipps, G.L. & Mugaas, J.N. (1972) Injection studies of cortical and medullary organization in the avian kidney. *J. Morph.*, 136, 181–190.
King, A.S. (1975) Aves urogenital system. In *Sisson and Grossman's The Anatomy of the Domestic Animals* (Ed.) Getty, R. Vol. 2, 5th Edn. Philadelphia: Saunders.
King, A.S. (1979) Systema urogenitalia. In *Nomina Anatomica Avium* (Ed.) Baumel, J.J., King, A.S., Lucas, A.M., Breazile, J.E. & Evans, H.E. London and New York: Academic Press.
Kurihara, S. & Yasuda, M. (1973) Comparative and topographical anatomy of the fowl. LXXIII. Size and distribution of corpuscula renis. *Jap. J. vet. Sci.*, 35, 311–318.
Kurihara, S. & Yasuda, M. (1975) Morphological study of the kidney in the fowl. I. Arterial system. *Jap. J. vet. Sci.*, 37, 29–47.
Kurihara, S. & Yasuda, M. (1975) Morphological study of the kidney in the fowl. II. Renal portal and venous system. *Jap. J. vet. Sci.*, 37, 363–377.
McNabb, R.A. & McNabb, F.M.A. (1975) Urate excretion by the avian kidney. *Comp. Biochem. Physiol.*, 51A, 252–258.
Michel, G. & Junge, D. (1972) Zur mikroskopischen Anatomie der Niere bei Huhn und Ente. *Anat. Anz.*, 131, 124–134.
Oelofsen, B.W. (1973) Renal function in the penguin (*Spheniscus demersus*) with special reference to the role of the renal portal system and renal portal valves. *Zool. afric.*, 8, 41–62.
Oelofsen, B.W. (1977) The renal portal valves of the Ostrich (*Struthio camelus*). *S. Afr. J. Sci.*, 73, 57–58.

Poulson, T.L. (1965) Countercurrent multipliers in avian kidneys. *Science, NY*, 148, 389–391.
Radu, C. (1975) Les fosses renales des oiseaux domestiques (*Gallus domesticus, Meleagris gallopavo, Anser domesticus* et *Anas platyrynchos*). *Anat., Histol., Embryol.*, 4, 10–23.
Shoemaker, V.H. (1972) Osmoregulation and excretion in birds. In *Avian Biology* (Ed.) Farner, D.S. & King, J.R. Vol. II. New York and London: Academic Press.
Siller, W.G. (1971) Structure of the kidney. In *Physiology and Biochemistry of the Domestic Fowl* (Ed.) Bell, D.J. & Freeman, B.M. Vol. 1. London and New York: Academic Press.
Siller, W.G. (1983) Structure of the kidney. In *Physiology and Biochemistry of the Domestic Fowl* (Ed.) Freeman, B.M. Vol. 4. London and New York: Academic Press.
Siller, W.G. & Hindle, R.M. (1969) The arterial blood supply to the kidney of the fowl. *J. Anat.*, 104, 117–135.
Skadhauge, E. (1973) Renal and cloacal salt and water transport in the fowl (*Gallus domesticus*). *Dan. med. Bull.*, 20 (Supplement 1), 1–82.
Skadhauge, E. (1974) Renal concentrating ability in selected West Australian birds. *J. exp. Biol.*, 61, 269–276.
Sturkie, P.D. (1976) Kidneys, extrarenal salt excretion, and urine. In *Avian Physiology* (Ed.) Sturkie, P.D. New York: Springer-Verlag.
Sykes, A.H. (1971) Formation and composition of urine. In *Physiology and Biochemistry of the Domestic Fowl* (Ed.) Bell, D.J. & Freeman, B.M. Vol. 1. London and New York: Academic Press.

Chapter 11

CLOACA AND VENT

The cloaca: external form

The cloaca is the chamber which receives the terminal parts of the digestive and urogenital systems and opens to the outside at the vent. Its basic organization seems to be fairly uniform throughout birds generally, the main variations being associated with the phallic region of the proctodeum. Externally the cloaca appears as a bell-shaped dilation of the end of the rectum (Figs 6–3, 6–10 and 11–1). In the adult domestic fowl it is about 2.5 cm long and 2.0–2.5 cm wide, although the size varies with the volume of faeces inside it. In the mature male the cloaca is in the midline, but in the mature female the enlarged left oviduct displaces it somewhat to the right; in immature birds the dorsally situated cloacal bursa is larger than the cloaca itself and compresses

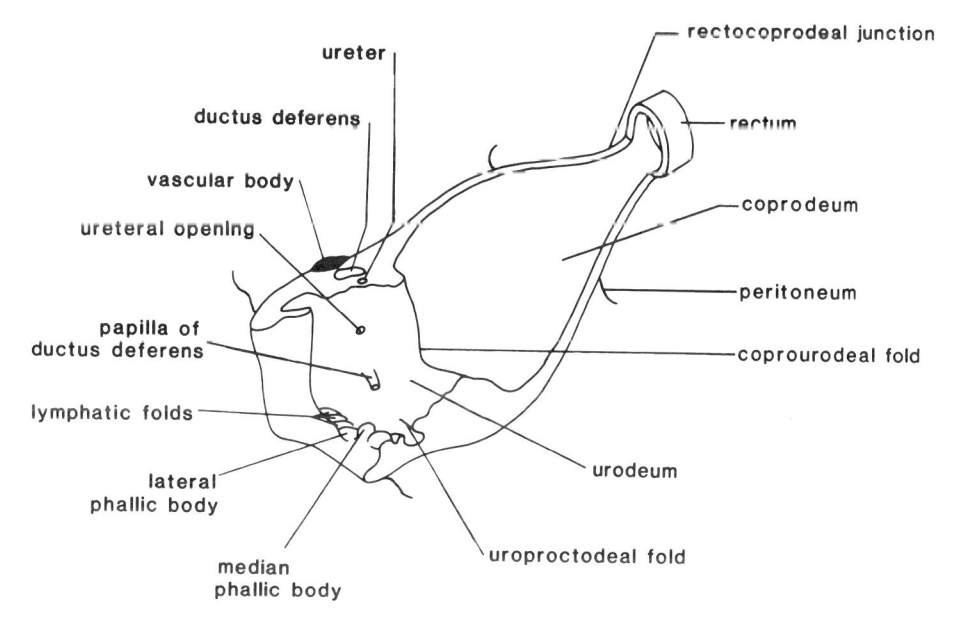

Fig 11–1 The cloaca of the male domestic fowl. The cloaca has been dissected to show only its left ventral quadrant. From King (1975), with kind permission of the publisher.

the cloaca on its dorsal aspect. The urogenital ducts traverse the dorsolateral surface of the cloaca and open into it dorsolaterally just caudal to its widest part. The cloaca is divided internally by two mucosal folds into three compartments, the coprodeum, urodeum and proctodeum, but these compartments are not distinguishable externally.

Coprodeum

The coprodeum is the most cranial compartment of the cloaca and is usually larger than the other two (Figs 11–1, 11–2, 11–3, 11–5 and 11–6). The internal

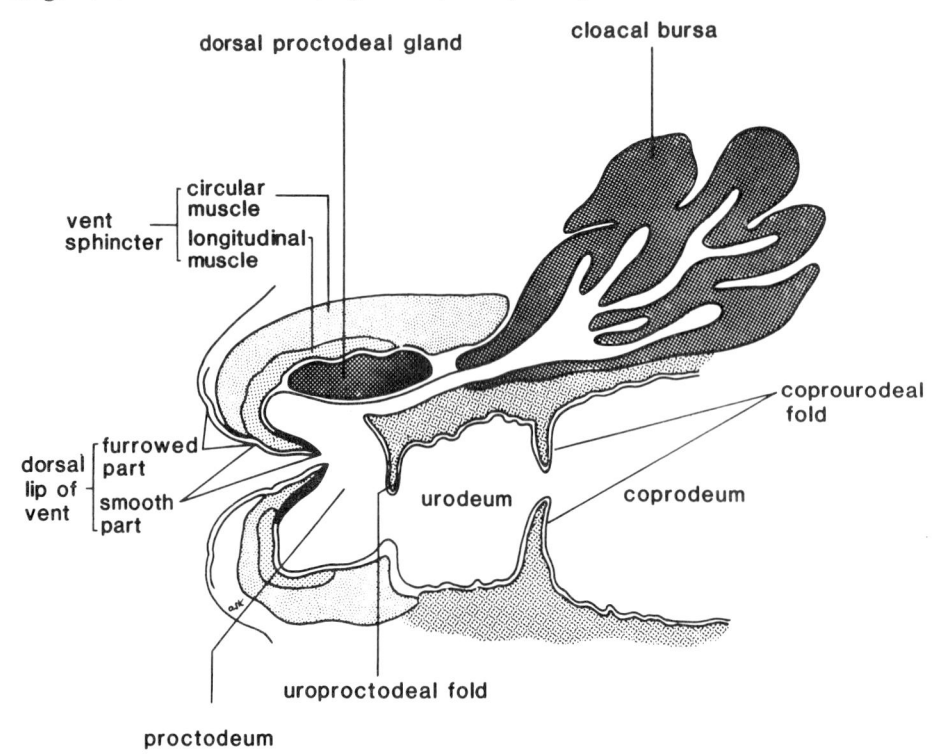

Fig 11–2 Median section of the cloaca of a four-month-old female domestic fowl. The black epithelial zone on the inner surface of the lips of the vent represents the extent of the stratified squamous epithelium. From King (1975), with kind permission of the publisher.

mucosal junction with the rectum is unmarked except (a) in the Ostrich (and possibly some other ratites) in which there consistently appears to be a true *rectocoprodeal fold* (Fig 11–3), and (b) in Anatidae in which there is an abrupt and conspicuous ridgelike change in the gross appearance of the mucous membrane caused by a sudden transition to a stratified squamous epithelium in the coprodeum (Figs 11–5 and 11–6). However, in most birds the boundary between the rectum and coprodeum is indicated only by an expansion in calibre at the beginning of the coprodeum. In the domestic fowl the mucosa is lined by villi which are similar to those of the rectum, apart from being somewhat lower and broader. Crypts and simple glands are present. The epithelium is tall columnar with goblet cells. Similar villi occur in some passerines (e.g. the Zebra Finch and Singing Honeyeater); in the Emu the surface area of the coprodeum is further increased by folds which carry villi. The villi and folds may be devices for cloacal reabsorption of water from urine in these desert species. On the other hand, in the xerophilic seed-eating Galah and in the Laughing Kookaburra there are no villi in the coprodeum. Only in the Anatidae among the species explored so far is the coprodeum (and all the rest of the cloaca) lined by stratified squamous epithelium.

Urodeum

The urodeum, which is the middle and smallest compartment of the cloaca, is usually partly separated from the other two compartments by two circular

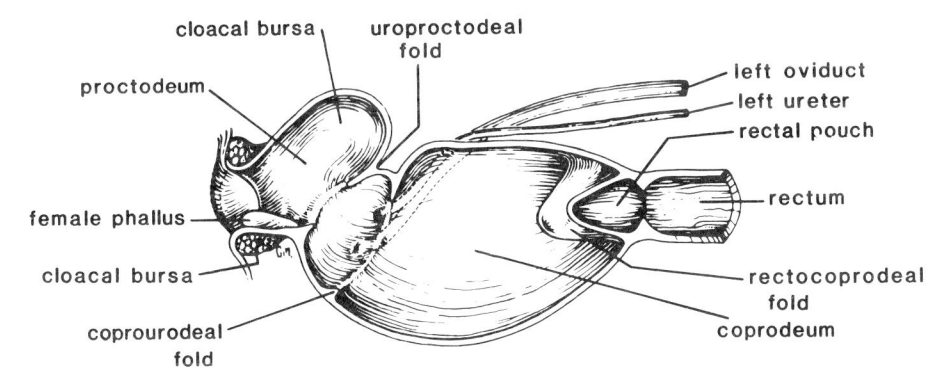

Fig 11–3 Longitudinal section of the cloaca of a female Ostrich seen from the right side. The right half of the cloaca has been removed, exposing the interior of the left half. From Saint-Hilaire (1822), *Mémoires du Muséum nationale d'histoire naturelle*, Paris, 9, 438.

mucosal folds (Figs 11–1, 11–2, 11–3, 11–5 and 11–6). The more cranial of these two folds, the *coprourodeal fold*, is an annular ridge between the coprodeum and urodeum (Figs 11–1, 11–2, 11–3 and 11–5). If the coprodeum is full of faeces this fold becomes a thin diaphragm with a central circular aperture. The presence of the faeces can eventually cause the aperture to be everted through the vent. Thus the faeces may not have to travel through the urodeum and proctodeum during defaecation but instead may be deposited via the everted coprodeum through the vent to the outside. Also, the fold may close during egg-laying, thus preventing simultaneous evacuation of the faeces from the coprodeum due to the increased intra-abdominal pressure. The diaphragm-like coprourodeal fold with its central aperture is also visible externally during ejaculation in the male bird, at least in the domestic fowl and turkey (Fig 11–4). The more caudal fold, the *uroproctodeal fold*, is a semicircular dorso-lateral fold between the urodeum and proctodeum which fades out ventrally

(Figs 11–1, 11–2, 11–3 and 11–5). It is everywhere lower than the coprourodeal fold.

The *urogenital ducts* open into the urodeum. Their openings are on the dorsolateral mucosal surfaces of the urodeum, the ureters usually being relatively dorsal and the genital ducts relatively lateral in position.

In the majority of birds each *ureter* appears to open by a simple orifice (Figs 11–1, 11–5 and 11–6) which is difficult to see in the fresh cloaca. In some species, however, including the Ostrich and penguins the opening of each ureter is situated on the summit of a small papilla (Fig 11–3). In the male each ductus deferens opens on the end of a slender conical projection, the *papilla of the ductus deferens* (Figs 11–1, 11–4, 11–5 and 11–6), the length of the papilla in the domestic fowl being about 2.5 mm and the diameter about 2–3 mm in the mature but detumescent bird. Each papilla points medially and caudally in detumescence. If the cloaca of the domestic fowl is laid open the tip of the papilla is about 1 cm from the opening of the related ureter. In the immature female domestic fowl, duck and goose, as well as in penguins, a small conical papilla which is probably the *female homologue of the papilla of the ductus deferens* is present on each side of the cloaca. In the domestic birds these papillae disappear as the bird matures, but in penguins they appear to persist in the adult. By means of a speculum the presence or absence of papillae has been used to sex mature pigeons and penguins. In penguins the papilla of the ductus deferens (or its homologue in the female) must be distinguished from the more dorsomedially situated ureteric papilla. In males the two papillae are approximately the same size. In females, however, the lateral papilla (the homologue of the ductus deferens) is only one-third the length of the papilla of the ureter.

Embedded in the ventrolateral wall of the urodeum is the paired red, egg-shaped *paracloacal vascular body* (Figs 11–1 and 11–7) about 5 mm in diameter and 7–10 mm long in the domestic fowl. This structure is the source of the lymph which enlarges the lateral phallic bodies and lymphatic folds during tumescence. It is known to occur in the domestic fowl, turkey, duck and grouse, and also a few other species; despite the fact that little is known of its occurrence in other species it is almost certainly a characteristic of birds in general. Within the paracloacal vascular body of the domestic fowl a cortex and medulla can be distinguished. The cortex is made up of many reteform tufts of blood capillaries, the *glomera*, each of which is invaginated into a *lymphatic vessel*, rather like a ball pressed into the surface of a balloon. The lymph vessels finally empty into subcapsular lymphatic sinuses which drain from the caudal end of the vascular body and join the erectile lymphatic cavities of the phallus. At the cranial end of the vascular body the subcapsular lymphatics open into the lymphatic vessels running with the pudendal artery, thus providing an escape route for lymph during detumescence. The medulla of the vascular body is formed by large blood vessels, and by nerves and connective tissue. In the paracloacal vascular body of the duck the subcapsular lymphatic sinuses are replaced by a large lymphatic (paired) cavity which connects with the lymphatic channels of the vascular body on the one hand and with the erectile tissue of the phallus on the other (Fig 11–7).

In the female, the *left oviduct* opens ventrally and laterally relative to the left ureter (Figs 8–6 and 11–3). The opening in the domestic fowl and turkey is generally situated on a slight domelike mound which becomes inconspicuous when the cloaca is cut open and flattened out. In young ducks, geese and swans the opening of the left oviduct is closed by a membrane which is not resorbed until the bird attains sexual maturity. In some species of duck this occluding membrane disappears before the bird is one year old; in other species of duck and in geese generally it probably persists until $1\frac{1}{2}$ years of age. The presence or absence of the membrane can be seen with the aid of a speculum and has been used to distinguish immature from mature females. A similar membrane occurs in the domestic fowl and is lost at the beginning of the first breeding season. The disappearance of the membrane is probably under the control of oestrogen. The vestigial *right oviduct* can quite often be found attached to the right side of the urodeum, but usually lacks a patent cloacal orifice (Fig 8–6). As already mentioned, in immature female Anatidae and domestic fowl a small papilla lies between the orifice of the ureter and the opening of the oviduct on each side of the cloaca, and is believed to be the female homologue of the papilla of the ductus deferens. It is absent in mature birds. The mucosa of the urodeum is smooth apart from a number of irregular folds and furrows. The epithelium in the domestic fowl is tall columnar with goblet cells. A few crypts and glands are present. In other species the epithelium is stratified squamous, and in yet others there is a mixture of epithelia.

Proctodeum

The proctodeum is a short compartment (about 1.0–1.5 cm in length in the domestic fowl) between the uroproctodeal fold and the lips of the vent (Figs 11–1, 11–2, 11–3, 11–5 and 11–6). In young birds an opening in the dorsal midline (Figs 11–2 and 11–3) leads into the globular *cloacal bursa* (see Chapter 14). In the midline, immediately caudal to the opening of the cloacal bursa the roof of the cloaca in some galliform birds carries an oval glandular mound about 1 cm long in the domestic fowl, the *dorsal proctodeal gland* (Figs 8–6 and 11–2). In the domestic fowl it consists of mucous glands invaded by lymphoid tissue. In sexually mature male domestic quail this gland is generally enlarged, reaching a size of about $10 \times 11 \times 12$ cm and causing the dorsal lip of the vent to protrude slightly. The gland in this species has numerous openings, each leading into one of the saclike glandular cavities which form a layer below the mucosa. The secretion is a white frothy fluid (hence the gland in the quail was formerly known as the 'foam gland') which is transferred to the oviduct during coitus, although no function has been found for it. In many species a number of glands, the *lateral proctodeal glands*, occur in the lateral wall of the cloaca (Figs 11–5 and 11–6). Again the function of these glands is unknown.

In the domestic cock the paired *lymphatic folds* lie on the floor of the proctodeum, just inside the ventral lip of the vent (Fig 11–1). In some Anatidae and ratites most of the floor of the proctodeum is taken up by the intromittent phallus. These structures are described below. In females of the species with an intromittent phallus the floor of the proctodeum has a phallic homologue analogous to the mammalian clitoris (Fig 11–3).

The mucosa of the proctodeum is smooth except for some irregular projections and furrows. The epithelium in most species examined is stratified squamous, but in the domestic fowl it is simple columnar except on the inner aspect of the lips of the vent where it becomes stratified squamous. The absence of stratified squamous epithelium in the proctodeum, as well as in the urodeum, suggests that in these compartments in the domestic fowl resorption of water may occur. In the domestic fowl large numbers of Herbst corpuscles, which are sensitive to vibration, are present under the epithelium of the lips of the vent.

The vent

At rest the vent is usually a transverse slit (Figs 6–3 and 6–10) guarded by *dorsal* and *ventral lips*, the lips being inverted into the cavity of the proctodeum (Fig 11–2). When a large mass of faeces is evacuated from the cloaca, the lips are partly everted exposing the orifice of the vent which then assumes a circular shape. In the domestic Anatidae the resting vent is U-shaped. In the domestic fowl the part of the caudal surface of each vent lip that is visible in the resting condition is keratinized and characterized by numerous radial furrows (the furrowed part, Fig 11–2). This is continued by a concealed part which is only slightly furrowed (the smooth part, Fig 11–2). The cranial (inner) surfaces of the lips are lined by stratified squamous epithelium. In many species (but not the Anatidae or the domestic fowl) the vent lips contain numerous mucous glands.

Spontaneous and vigorous sucking movements of the vent ('cloacal drinking') have been observed in domestic fowl chicks at hatching, and material applied to the lips of the vent is carried to both the cloacal bursa and the caeca. The experimental evidence strongly suggests that the domestic fowl acquires part of its basic immunity via this pathway.

Cloacal promontory

The terminal convolutions of the ductus deferens in breeding males of a number of passerine species form a conical projection of the cloacal region, the cloacal promontory. For details see Chapter 9.

Cloacal muscles

There are three striated cloacal muscles. The *sphincter muscle* of the cloaca surrounds the proctodeum and vent (Fig 11–2). The *transverse muscle* of the cloaca arises from the caudal part of the pelvic bone or the caudal vertebrae and passes ventral to the proctodeum and vent, its fibres interdigitating with those of the sphincter muscle. In the duck it pulls the proctodeum ventrocranially during coitus with the result that the tumescent phallus is directed cranially towards the cloaca of the female. The *levator muscle* of the cloaca extends from the ventral side of the tail lateroventrally around the proctodeum to insert on its ventral wall. In male birds it attaches to the phallus. In the domestic fowl and duck it acts after copulation, and possibly defaecation, to pull the ventral lip of the vent and the floor of the proctodeum back into the resting position.

The phallus

The ratites have a penislike protrusible intromittent phallus which resembles that of crocodiles, the contemporary reptiles most closely related to birds. The anseriform birds also have a protrusible intromittent phallus, but it is somewhat more advanced in having, for example, a spiral phallic sulcus. In birds generally, however, there may be a small non-protrusible structure on the ventral lip of the vent, as in the domestic fowl and turkey, or it may be absent altogether; there is almost a total lack of reliable anatomical knowledge about this region in the great majority of birds. The similarity between the phallus of ratites and crocodiles led to the suggestion that all ancestral birds had a protrusible phallus and that those which still retain it are relatively primitive forms. However, an alternative hypothesis proposes the opposite, that the phallus in ancestral birds was in fact simple and that only a few groups (ratites, tinamous and anseriforms) have developed a specialized type. It has been suggested that a fully intromittent phallus offers an adaptational advantage in the anseriforms by preventing water from entering the female cloaca during coitus and thus avoiding osmotic damage to the spermatozoa. However, a number of supposedly primitive anseriforms (screamers, Magpie Goose, whistling ducks, Coscoroba Swan and *Cereopsis*) copulate on land or while standing in shallow water.

The non-intromittent phallus. In the absence of information about wild birds the following account describes the domestic fowl. In this species the phallus is mounted on the crest of the ventral lip of the vent (Fig 11–1). It consists of the white *median phallic body* which in the detumescent male is about 1.5–3.5 mm in diameter, and a pair of flesh-coloured *lateral phallic bodies* which are 2 × 4 mm in the detumescent male, the larger measurement being mediolateral. In the classical method of sexing day-old chicks, which depends on observing the presence or absence of a phallus of male dimensions (the female has a smaller one), the ventral lip of the vent is everted by digital pressure to expose the phallic region on the crest of the lip. The paired spindle-shaped reddish *lymphatic folds* lie on the ventrolateral floor of the proctodeum, rather than on its crest. The median and lateral phallic bodies and the lymphatic folds consist of connective tissue and interconnecting lymphatic channels.

Tumescence is mainly due to the flow of lymph from the paired *paracloacal vascular bodies* to the phallic bodies and lymphatic folds. The lymphatic folds become greatly enlarged and merge with the less enlarged lateral phallic bodies from which they are now separated only by indistinct grooves (Fig 11–4A); the lymphatic folds are believed to contribute lymphlike fluid to the semen (p. 171). The median phallic body enlarges least (Fig 11–4A). These enlargements evert the ventral lip of the vent and adjacent proctodeal floor and thus extrude the phallus as a whole. The meeting in the midline of the right and left lymphatic folds and the right and left lateral phallic bodies forms a median groove with the median phallic body at its ventral tip (Fig 11–4A). Contraction of the levator and sphincter cloacal muscles just before ejaculation compresses the lymphatic channels carrying lymph to the phallus and causes the phallus to protrude still further. At this stage the urodeum and proctodeum are both more or less everted, so that the coprourodeal fold becomes

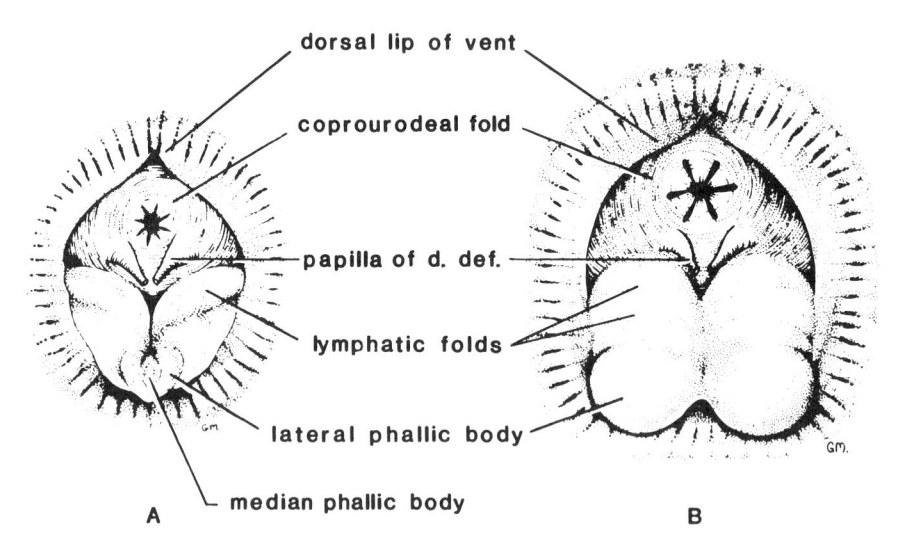

Fig 11–4 A, The erect phallus of the male domestic fowl as seen during collection of semen by manually-induced ejaculation. The degree of eversion of the floor of the cloaca is probably exaggerated by the squeezing of the cloaca between the finger and thumb. B, The erect phallus of the male domestic turkey as observed during collection of semen by manually-induced ejaculation. The degree of eversion of the cloaca is probably exaggerated by pressure from the fingers. d. def. = ductus deferens. From King (1981b), with kind permission of the publisher. These drawings are based on photographs by P.E. Lake and colleagues, Poultry Research Centre, Edinburgh.

visible like a diaphragm (Fig 11–4A). At the moment of ejaculation the protrusion of the phallus occurs with darting rapidity. Semen is discharged from the deferent papilla into the median groove, and at the same moment the ventral tip of the phallus is applied to the protruded oviduct of the female. *Detumescence* occurs in a few seconds and is due to the drainage of lymph into the general lymphatic system.

The phallus of the turkey is essentially similar to that of the domestic fowl, but the median phallic body although still small has a double apex, and the lateral phallic bodies are longer dorsoventrally and therefore protrude further from the vent in tumescence (Fig 11–4B).

The phallus in the female is a reduced structure consisting of a pair of folds separated by a shallow median groove.

The intromittent phallus. The anseriforms, tinamous and ratites possess a truly intromittent type of phallus of which two forms can be distinguished depending on whether or not it contains a blind tubular cavity.

An *intromittent phallus* with *no cavity* occurs in the Ostrich, kiwis and tinamous. In the Ostrich the base and body of the phallus consist essentially of right and left erectile *fibrolymphatic bodies*, the left body being larger than the

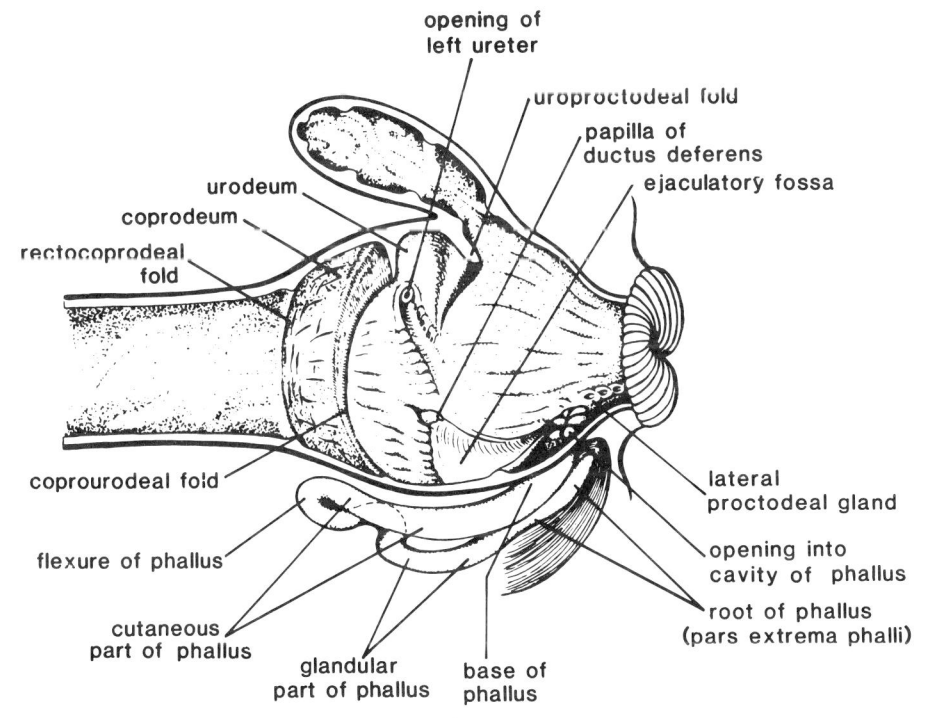

Fig 11-5 Longitudinal section of the cloaca of an immature male domestic goose viewed from the left side. The left wall of the cloaca has been removed showing the interior of the right side. The phallus is in the resting position. From Komárek (1969), with kind permission of the editor of *Acta Veterinaria* Brno.

right. On the dorsal surface of the phallus between the fibrolymphatic bodies, is the *phallic sulcus* into which the two deferent ducts eject semen. Along the ventral surface of the caudal free half of the phallus there is an *elastic vascular body* consisting of an outer layer of elastic tissue and an inner layer of erectile tissue. At rest the phallus lies bent in a pocket in the floor of the proctodeum. Since it takes up so much space in the cloaca it has to be partially protruded before micturition and defaecation can occur. Tumescence of the phallus occurs by lymphatic engorgement. The fully erect phallus is bright red and about 40 cm long in the mature Ostrich. It is directed ventrally and cranially, and because of the asymmetry of its fibrolymphatic bodies, is bent slightly to the left. This type of phallus resembles that of chelonian and crocodilian reptiles.

An *intromittent phallus* with a *cavity* occurs in the Emu, cassowaries, rheas and the anseriforms (Figs 11–5, 11–6 and 11–7). In the domestic duck and goose the resting phallus is a long blind-ending tube which lies coiled in a sac lined with peritoneum along the ventrolateral wall of the cloaca, like the invaginated finger of a glove (Fig 11–5). The deep blind-ending portion of the tube secretes mucus and is non-erectile. It is not everted in tumescence. The more superficial portion of the tube (i.e. in the resting phallus the part that opens on the cloacal floor) is lined internally by a keratinized stratified squamous epithelium. This portion of the tube is everted and protruded in tumescence (Fig 11–6). At the tip of the protruded phallus there is an opening,

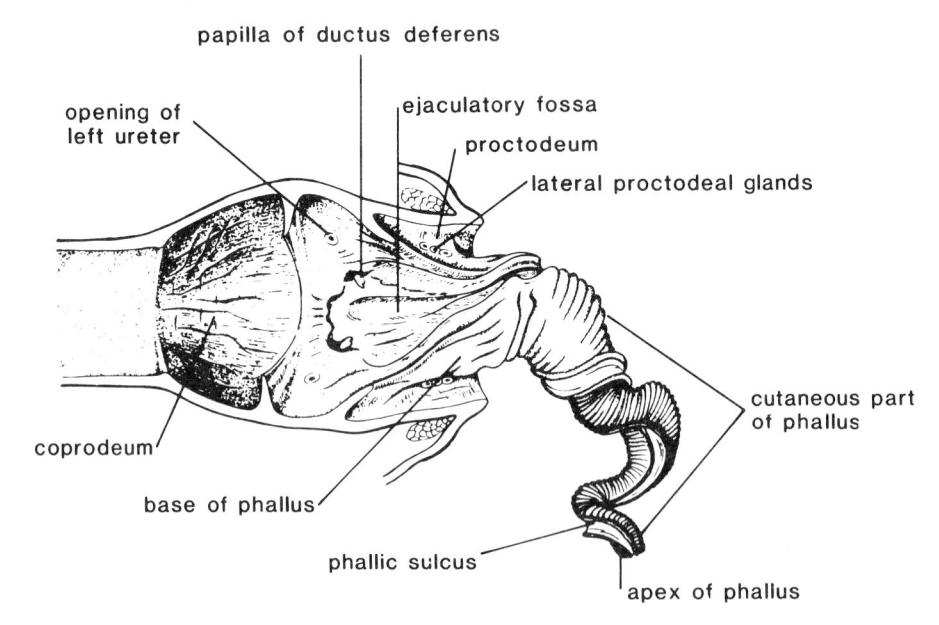

Fig 11–6 Horizontal section of the cloaca of the male domestic duck viewed from the dorsal aspect. The dorsal half of the rectum and cloaca have been removed to show the ventral floor of the cloaca. The phallus is erect. From Komárek (1969), with kind permission of the editor of *Acta Veterinaria* Brno.

which leads into the deep, non-everted blind end of the tube (Figs 11–6 and 11–7). The erectile mechanism is based on the right and left lymphatic cavities which surround the right and left paracloacal vascular bodies (Fig 11–7). Tumescence of the phallus is by engorgement with lymph which is pumped into the lymphatic cavities by the paracloacal vascular bodies. This forces the phallus out of its resting invaginated form. The right and left lymphatic cavities open directly into the cavities of the right and left fibrolymphatic bodies at the base of the erect phallus (Fig 11–7). There, the paired fibrolymphatic bodies become continuous with each other internally, forming a common erectile lymphatic cavity; on the surface, however, they are still divided by a groove housing the beginning of the *phallic sulcus*. The left and right fibrolymphatic bodies twist spirally round each other, with the phallic sulcus between them (Figs 11–6 and 11–7), the left body being much greater in diameter than the right; this asymmetry of the fibrolymphatic bodies occurs in all birds with a protrusible phallus. The bases of the fibrolymphatic bodies are supported by the trough-shaped *fibrocartilaginous body* which strengthens the ventral and lateral walls of the urodeum and proctodeum (Fig 11–7). An *elastic ligament* extends through the phallus from the fibrocartilaginous body to the apex. The erect phallus is about 8 cm long in the domestic drake (only about 4 cm long in the wild Mallard), is greyish yellow in colour and is directed cranially.

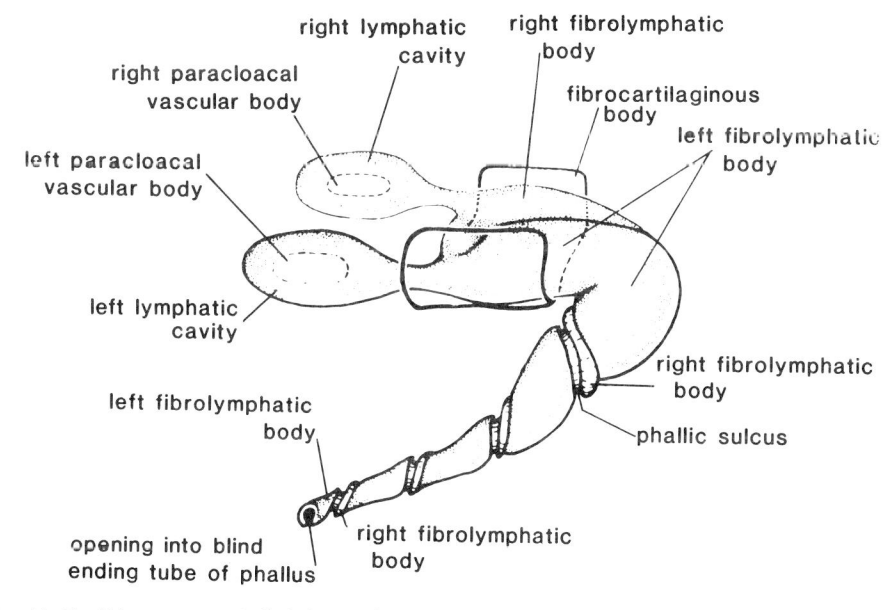

Fig 11–7 Diagrammatic left lateral view of the erect male phallus of the duck with the bird in the standing position, to show the lymphatic and vascular components of the phallus. The fibrocartilaginous body is drawn as though it is transparent. From King (1981b), with kind permission of the publisher. Based on Liebe (1914), *Jenaische Zeitschrift für Naturwissenschaft*, 51, 627, and Rautenfeld, Preuss and Fricke (1974).

The process of tumescence involves (a) eversion of the proctodeum so that the base of the phallus is brought to the vent opening; (b) contraction of the transverse cloacal muscle, turning the orifice of the vent cranially towards the female cloaca; (c) distension of the lymphatic cavities of the fibrolymphatic bodies with lymph (aided by peristaltic contraction of the cloacal sphincter muscle); and (d) sudden eversion of the engorged base and shaft of the phallus. In the fully erect phallus the phallic sulcus can become a closed tube. During manually induced ejaculation, however, semen drips from the *base* of the phallus not the tip, so that the sulcus is not then a functional seminal groove; whether or not it serves as a seminal groove in natural coition is not known. Detumescence results from draining of the lymphatic spaces of the phallus into the general lymphatic system. As the pressure falls, the tip of the phallus is pulled back by the elastic ligament and invaginates into its base, much as the finger of a glove can be pushed into the hand. In geese venereal infection transmitted at coitus can involve the phallus and cloaca, and may lead to infertility and reduced egg production.

Because of the elongated form of the erect phallus and its cranioventral direction, true intromission into the female cloaca occurs in anseriforms and ratites. The phallus in these birds is thus fully analogous to the mammalian penis. It is not, however, homologous to the mammalian penis: its erectile mechanism is lymphatic, whereas that of the mammal is vascular; the semen travels via the external surface, as opposed to the internal urethra of the mammalian penis; finally, the avian phallus is solely reproductive in function, while the mammalian penis is both reproductive and urinary.

Further reading

Burrows, W.H. & Quinn, J.P. (1937) The collection of spermatozoa from the domestic fowl and turkey. *Poult. Sci.*, 16, 19–24.

Coil, W.H. & Wetherbee, D.K. (1959) Observations on the cloacal gland of the Eurasian Quail, *Coturnix coturnix*. *Ohio J. Sci.*, 59, 268–270.

Fujihara, N., Nishiyama, H. & Nakashima, N. (1976) Studies on the accessory reproductive organs in the drake. 2. Macroscopic and microscopic observations on the cloaca of the drake with special reference to the ejaculatory groove region. *Poult. Sci.*, 55, 927–935.

Gilbert, A.B. (1980) Female genital organs. In *Form and Function in Birds* (Ed.) King, A.S. & McLelland, J. Vol. 1. London and New York: Academic Press.

Gunawardana, V.K. & Scott, M.G.A.D. (1978) On the structure of the vascular body in the domestic fowl. *J. Anat.*, 127, 447–457.

Guzsal, E. (1974) Erection apparatus of the copulatory organ of ganders and drakes. *Acta vet. hung.*, 24, 361–373.

Ikeda, K. & Taji, K. (1954) On the foamy ejaculate of Japanese Quail, *Coturnix coturnix japonica*. *Scient. Rep. Matsuyama agric. Coll.*, 13, 1–4.

Johnson, O.W. & Skadhauge, E. (1975) Structural–functional correlations in the kidneys and observations of colon and cloacal morphology in certain Australian birds. *J. Anat.*, 120, 495–505.

King, A.S. (1975) Aves urogenital system. In *Sisson and Grossman's The Anatomy of the Domestic Animals* (Ed.) Getty, R. Vol. 2, 5th Edn. Philadelphia: Saunders.

King, A.S. (1980) Systema urogenitale. In *Nomina Anatomica Avium* (Ed.) Baumel, J.J., King, A.S., Lucas, A.M., Breazile, J.E. & Evans, H.E. London and New York: Academic Press.

King, A.S. (1981a) Cloaca. In *Form and Function in Birds* (Ed.) King, A.S. & McLelland, J. Vol. 2. London and New York: Academic Press.

King, A.S. (1981b) Phallus. In *Form and Function in Birds* (Ed.) King, A.S. & McLelland, J. Vol. 2. London and New York: Academic Press.

Klemm, R. D., Knight, C.E. & Stein, S. (1973) Gross and microscopic morphology of the glandula proctodealis (foam gland) of *Coturnix c. japonica* (Aves). *J. Morph.*, 141, 171–184.

Knight, C.E. (1967) Gross and microscopic anatomy of the structures involved in the production of seminal fluid in the chicken. MSc Thesis, Michigan State University.

Knight, C.E. (1970) The anatomy of the structures involved in the erection–dilution mechanism in the male domestic fowl. PhD Thesis, Michigan State University.

Komárek, V. (1969) Die maännliche Kloake unserer Entenvögel. *Anat. Anz.*, 124, 434–442.

Komárek, V. (1970) The cloaca of the turkey-cock and of the cock. *Acta vet. Brno*, 39, 227–234.

Komárek, V. (1971) The female cloaca of anseriform and galliform birds. *Acta vet. Brno*, 40, 13–22.
Komárek, V. & Marvan, F. (1969) Beitrag zur mikroskopischen Anatomie des Kopulationsorganes der Entenvögel. *Anat. Anz.*, 124, 467–476.
Lake, P.E. (1981) Male genital organs. In *Form and Function in Birds* (Ed.) King, A.S. & McLelland, J. Vol. 2. London and New York: Academic Press.
McFarland, L.Z., Warner, R.L., Wilson, W.O. & Mather, F.B. (1968) The cloacal gland complex of the Japanese Quail. *Experientia*, 24, 941–953.
Miller, W.J. & Wagner, F.H. (1955) Sexing mature columbiformes by cloacal characters. *Auk*, 72, 279–285.
Nagra, C.L. (1959) Cloacal glands in Japanese Quail (*Coturnix coturnix japonica*): histogenesis and response to sex steroids. *Anat. Rec.*, 133, 415.
Nishiyama, H. (1950) Studies on the physiology of reproduction in the male fowl. I. On the accessory organs of the phallus. *Sci. Bull. Fac. Agric. Kyushu Univ.*, 12, 27–36.
Nishiyama, H. (1950) Studies on the physiology of reproduction in the male fowl. II. On the erection of the rudimentary copulatory organ (so called phallus) *Sci. Bull. Fac. Agric. Kyushu Univ.* 12, 37–46.
Nishiyama, H. (1955) Studies on the accessory reproductive organs in the cock. *J. Fac. Agric. Kyushu Univ.*, 10, 277–305.
Nishiyama, H. & Ogawa, K. (1961) On the function of the vascular body, and accessory reproductive organ, of the cock. *Jap. J. zootech. Sci.*, 32, 89–96.
Rautenfeld, D.B., Preuss, F. & Fricke, W. (1974) Neue Daten zur Erektion und Reposition des Erpelphallus. *Prakt. Tierarzt*, 10, 553–556.
Sladen, W.J.L. (1978) Sexing penguins by cloacascope. *Int. Zoo. Yb.* 18, 77–80.
Sorvari, R., Naukkarinen, A. & Sorvari, T.E. (1977) Anal sucking-like movements in the chicken and chick embryo followed by the transportation of environmental material to the bursa of Fabricius, caeca and caecal tonsils. *Poult. Sci.*, 56, 1426–1429.
Tamura, T. & Fujii, S. (1967) Studies on the cloacal gland of the quail. 1. Macroscopical and microscopical observations. *Jap. Poult. Sci.*, 4, 187–193.
Yamano, S., Sugimura, M. & Kudo, N. (1977) The lymphatic system of the corpus paracloacalis vascularis and the second fold in the male domestic fowl. *Jap. J. vet. Res.*, 25, 93–98.

Chapter 12

ENDOCRINE SYSTEM

HYPOPHYSIS

The hypophysis (pituitary gland) is a small organ attached to the ventral surface of the diencephalic part of the brainstem immediately caudal to the optic chiasma. It consists of two components, the adenohypophysis which arises from the embryonic stomodeum, and the smaller neurohypophysis which arises from the diencephalon (Fig 12–1). The anatomy of the avian hypophysis is complicated by minor species variations, and by an assortment of different names for the same structure. The following account is based on the domestic fowl.

Adenohypophysis

This comprises two parts: (a) the pars tuberalis and (b) the pars distalis (Fig 12–1). Unlike the mammalian organ there is no separate pars intermedia. The cells equivalent to those of the pars intermedia of mammals are incorporated within the pars distalis.

The pars tuberalis. This component of the adenohypophysis covers the median eminence of the neurohypophysis as far rostrally as the optic chiasma (Fig 12–1) and tends also to form a collar round the infundibular stalk of the neurohypophysis. Ventrally it connects with the pars distalis, and it is this connecting region which carries the portal vessels from the median eminence of the neurohypophysis to the pars distalis of the adenohypophysis.

The cells of the pars tuberalis are round or elongated, finely granular and faintly basophilic. They are arranged in cords containing a few colloidal acini.

The pars distalis. This forms the bulk of the adenohypophysis (Fig 12–1). It lies immediately ventral and rostral to the neurohypophysis. Its most caudo-dorsal part is in contact with the ventral surface of the pars nervosa of the neurohypophysis, except for a connective tissue sheath which separates them.

The secretory cells of the pars distalis of the adenohypophysis are arranged either in cords or follicles which have a central lumen and often contain a basophilic 'colloid'. On the basis of their staining affinities and ultrastructure, seven types of secretory cell have been identified (alpha, beta, gamma, delta, epsilon, eta and kappa cells). In addition, a chromophobe cell is present which appears to be non-secretory. The general distribution of these different cell types enables two regions to be distinguished in the pars distalis of the adenohypophysis, a rostral zone and a caudal zone (Fig 12–1).

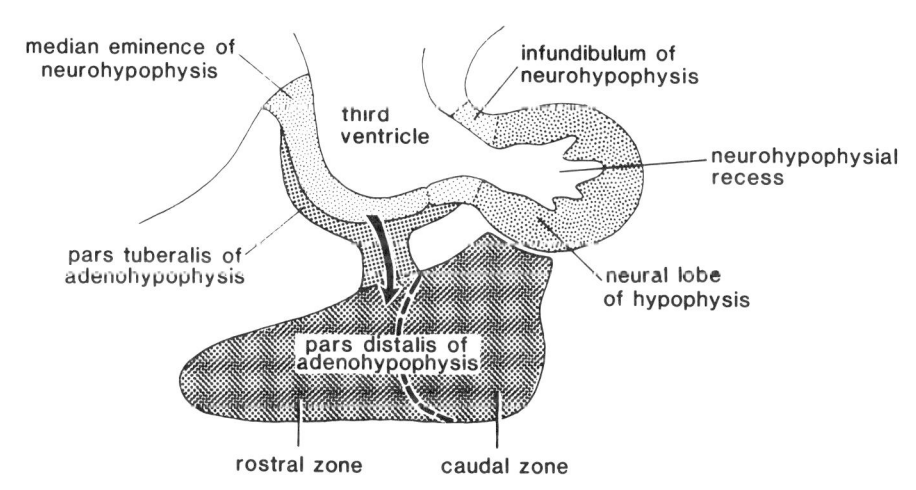

Fig 12–1 Diagrammatic sagittal section of the avian hypophysis. The two components of the adenohypophysis are its pars tuberalis and its pars distalis. They form the more ventral regions of the hypophysis, both being relatively darkly shaded in the diagram. From the distribution of its different cell types rostral and caudal zones can be distinguished in the pars distalis. The arrow represents the portal system, which is the pathway of blood vessels running from the primary capillary plexus in the median eminence of the neurohypophysis to the secondary capillary plexus in the pars distalis of the adenohypophysis; the portal system carries the neurosecretory releasing factors liberated from the hypothalamohypophysial tract to the pars distalis of the adenohypophysis. The neurohypophysis consists of the three more dorsal parts of the hypophysis, all relatively lightly stippled in this diagram. Of these, the median eminence continues directly into the infundibulum of the neurohypophysis, which presently becomes a hollow stalk carrying the enlarged neural lobe of the neurohypophysis.

Dorsally the median eminence is directly continuous with the hypothalamus.

Neurohypophysis

The neurohypophysis is a direct extension of the hypothalamus. It has three anatomical components, namely the median eminence, the infundibulum and the neural lobe (Fig 12–1). The first two of these are not clearly separable, the one running imperceptibly into the other. The neural lobe of the neurohypophysis is distinctly recognizable as a slightly enlarged lobe joined to the infundibulum by a somewhat constricted neck. Histologically the three components are essentially similar, consisting of three layers, i.e. an inner ependymal layer, a middle fibre layer and an outer palisade (the so-called 'glandular') layer.

The median eminence. The median eminence is the rostroventral region of the floor of the third ventricle being directly continuous, without definite demarcation, with the tuber cinereum of the hypothalamus. Of its three layers, the inner ependymal layer forms the epithelial lining of the cavity of the third ventricle of the brain; the middle fibre layer consists of axons which belong to the hypothalamohypophysial tract and project to the neural lobe of the neurohypophysis; the outer palisade layer consists of the processes of cells of the ependymal layer and the processes of some of the cells of the fibre layer passing outwards towards the external surface of the median eminence. Functionally, and in many species structurally, the median eminence can be divided into rostral and caudal zones.

The infundibulum. This is a ventrocaudal continuation of the median eminence. As it approaches the neural lobe it becomes a hollow tubular stalk (the infundibular stalk). The structure of the whole of the infundibulum of the neurohypophysis, including the dorsal and ventral walls of its stalk, is generally similar to that of the median eminence; however, the fibre layer has now reached its maximal development having received all its axons, and the palisade layer is correspondingly relatively reduced.

The neural lobe. This component of the neurohypophysis is carried on the caudal end of the stalklike infundibulum. The same three layers occur as in the median eminence and infundibulum. The fibre layer contains the nerve endings of the hypothalamohypophysial tract, which are surrounded at their ends by the cytoplasmic processes of pituicytes. Large masses of neurosecretory material accumulate around these cells. The lumen of the neural lobe, the neurohypophysial recess, is continuous with the cavity of the third ventricle.

The blood vessels of the hypophysis. The internal carotid arteries vascularize the hypophysis, including the primary capillary plexus of the median eminence. The supply to the pars distalis of the adenohypophysis begins at the dense primary capillary plexus of the median eminence. In some species the primary capillary plexus is clearly divided into rostral and caudal parts corresponding to the division of the median eminence. A group of veins, forming the *portal system* of the hypophysis, drains from the primary capillary plexus through the pars tuberalis of the adenohypophysis, and ends by discharging into a second capillary plexus in the pars distalis of the adenohypophysis. Like the primary capillary plexus, the portal system is divided into rostral and caudal groups of veins. Thus blood is transferred from the rostral and caudal zones of the median eminence to the rostral and caudal zones, respectively, of the pars distalis. The neurohypophysis receives an independent blood supply from its own branch of the internal carotid artery.

The nerve supply of the hypophysis. The hypophysis is innervated by the paired hypothalamohypophysial tract. Each tract consists of axons of the supraoptic and paraventricular nuclei, and of the infundibular nucleus, all three of which are nuclei of the hypothalamus. Most of these axons end in the neural lobe of the neurohypophysis, close to the basal lamina of the capillaries. Hypothalamic neurosecretory substances travel to the neurohypophysis via these axons. However, some of the axons of the hypothalamohypophysial tract end in the median eminence, in association with the capillaries of the primary plexus. There is evidence that these axons release neurosecretory substances which reach the pars distalis of the adenohypophysis via the portal blood vessels (see below).

The functions of the hypophysis. So intimate are the vascular and nervous interconnections between the hypothalamus and the hypophysis that it is difficult or impossible in some instances to separate the functions of these two structures.

Hormones of the pars distalis of the adenohypophysis. The pars distalis of the adenohypophysis secretes at least seven hormones including follicle stimulating hormone (FSH) attributed to beta cells, thyroid stimulating hormone (TSH) attributed to delta cells, luteinizing hormone (LH) attributed to gamma cells, prolactin attributed to eta cells, somatotropic hormone (STH) attributed to alpha cells, adrenocorticotropic hormone (ACTH) attributed to epsilon cells, and melanotropic hormone (MSH) attributed to kappa cells. Although each of these hormones has been assigned to a specific cell type it has not been conclusively established that such a one to one relationship exists.

FSH in females stimulates the development of ovarian follicles and the secretion of oestrogens by the ovaries; in males it stimulates tubular growth of the testes and spermatogenesis. TSH controls the function of the thyroid glands. LH in females appears to cause ovulation (although the details of the mechanism are still unknown), and in males stimulates the interstitial cells of the testes to produce androgens. Prolactin appears to be involved in broodiness, possibly by suppressing secretion of gonadotrophic hormone by the pars distalis of the adenohypophysis. In pigeons it stimulates the production of crop milk. STH in young birds appears to regulate body growth. ACTH controls the production of adrenal corticosteroids. The physiological role of MSH is unknown.

The hypothalamus appears to be important in the control of the pars distalis of the adenohypophysis. Evidently releasing factors (neurosecretory substances) are formed in the hypothalamic nuclei and travel to the median eminence in some of the axons of the hypothalamohypophysial tract. In the median eminence they are transferred to the primary capillary plexus, and are then carried by the portal vessels to the secondary capillary plexus in the pars distalis of the adenohypophysis. The action of these substances on the cells of the pars distalis of the adenohypophysis causes these cells to release their hormones.

The evidence for this hypothalamic control of the pars distalis of the adenohypophysis is supported by studies on the influence of photoperiodicity on the avian reproductive cycle. The poultry farmer exploits this phenomenon when using artificial light to induce the early onset of laying.

Hormones of the neurohypophysis. Vasotocin and oxytocin are produced in the hypothalamic nuclei and transported in the hypothalamohypophysial tract to the neurohypophysis where they are stored. Vasotocin has a hyperglycaemic action and is antidiuretic. It is probably involved in oviposition. The physiological role of oxytocin in the bird is unknown.

THYROID GLANDS

Anatomy

The paired thyroid glands are dark red, round or oval structures situated at the thoracic inlet (Figs 6–3, 6–9 and 14–2) medial to the jugular vein and just cranial to the origin of the subclavian and common carotid arteries (Fig 12–2). They arise from the floor of the embryonic pharynx at about the level of the first and second pharyngeal pouches. The size of the glands (about $10 \times 6 \times 3$ mm in the domestic fowl) appears to be affected by a number of factors including age, sex, diet, climate and secretory activity. As in all known vertebrates, the glands consist of follicles lined by a single layer of epithelial cells. Between the follicles are many blood vessels and only a few nerves. Each follicle contains thyroglobulin (colloid) in its lumen. During activity, the amount of colloid is reduced and the secretory cells become taller, as in mammals. During this process the thyroglobulin is hydrolysed by the follicular cells, releasing the thyroid hormones thyroxine and triiodothyronine which are then secreted into the blood stream.

Functions

In contrast to the thyroid hormones of mammals the physiological potencies of the two avian hormones, thyroxine and triiodothyronine, appear to be very similar in most aspects.

Among the various functions of the hormones in birds are the following.

1. Stimulation of general metabolism, to regulate heat production in response to changes in environmental temperature.
2. Regulation of the growth of the body as a whole and the reproductive organs in particular. Moderately increased availability of thyroid hormone accelerates growth and increases egg production.
3. Control of moulting. An increase in thyroid hormones precipitates moulting, possibly by stimulating the growth of new feathers.

PARATHYROID GLANDS

Anatomy

The parathyroid glands arise from the third and fourth pharyngeal pouches in the embryo. In the adult they comprise two pairs of small yellowish glands on

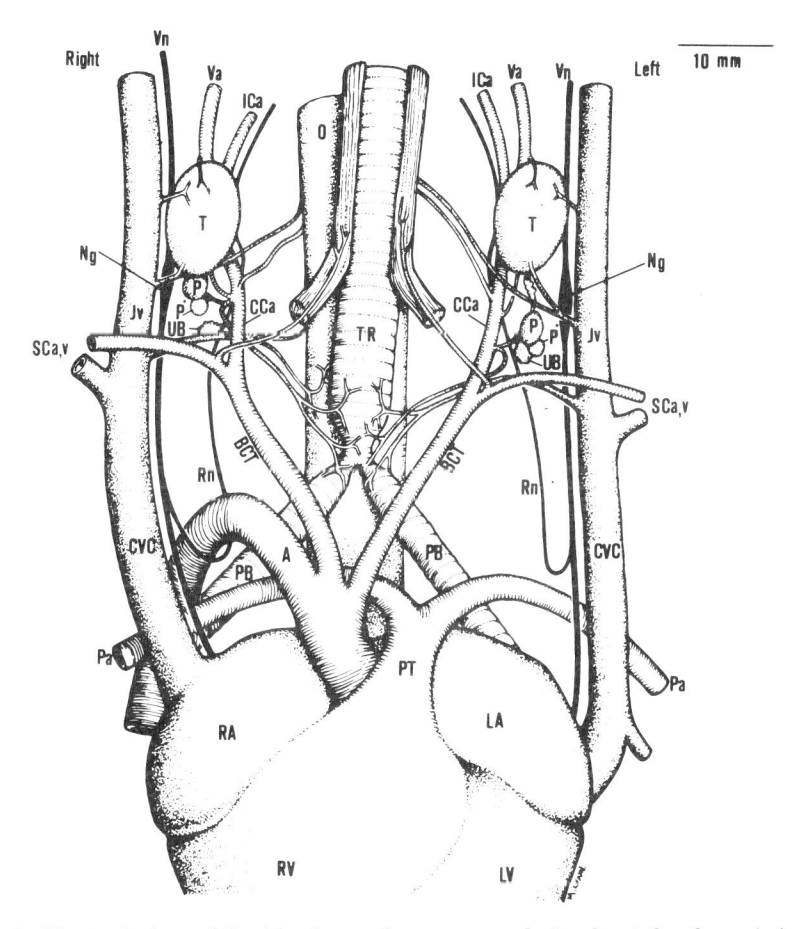

Fig 12–2 Ventral view of the blood vessels, nerves and glands at the thoracic inlet of the domestic fowl. The carotid bodies are drawn in black on the medial surfaces of the parathyroid glands. The drawing is approximately to scale. A = aorta; BCT = brachiocephalic trunk; CCa = common carotid artery; CVC = cranial vena cava; ICa = internal carotid artery; Jv = jugular vein; LA = left atrium; LV = left ventricle; Ng = nodose (distal vagal) ganglion; O = oesophagus; OAa = ascending oesophageal artery; OAv = ascending oesophageal vein; OTBa = oesophagotracheobronchial artery; OTBv = oesophagotracheobronchial vein; P = cranial parathyroid gland; P' = caudal parathyroid gland; Pa = pulmonary artery; PB = primary bronchus; PT = pulmonary trunk; RA = right atrium; Rn = recurrent nerve; RV = right ventricle; SCa = subclavian artery; SCv = subclavian vein; T = thyroid gland; TR = trachea; UB = ultimobranchial body or gland; Va = vertebral artery; Vn = vagus nerve. From Abdel-Magied and King (1978), with kind permission of the editor of the *Journal of Anatomy*.

each side of the body just caudal to the thyroid glands (Figs 12–2, 12–3 and 14–2). In the domestic fowl the glands measure about $2 \times 2 \times 0.9$ mm. The two members of the pair in the domestic fowl adhere to each other, the cranial parathyroid touching the thyroid on the right side but not on the left. The cranial parathyroid is usually the larger. Smaller accessory parathyroid glands have also been reported in the domestic fowl in several sites and

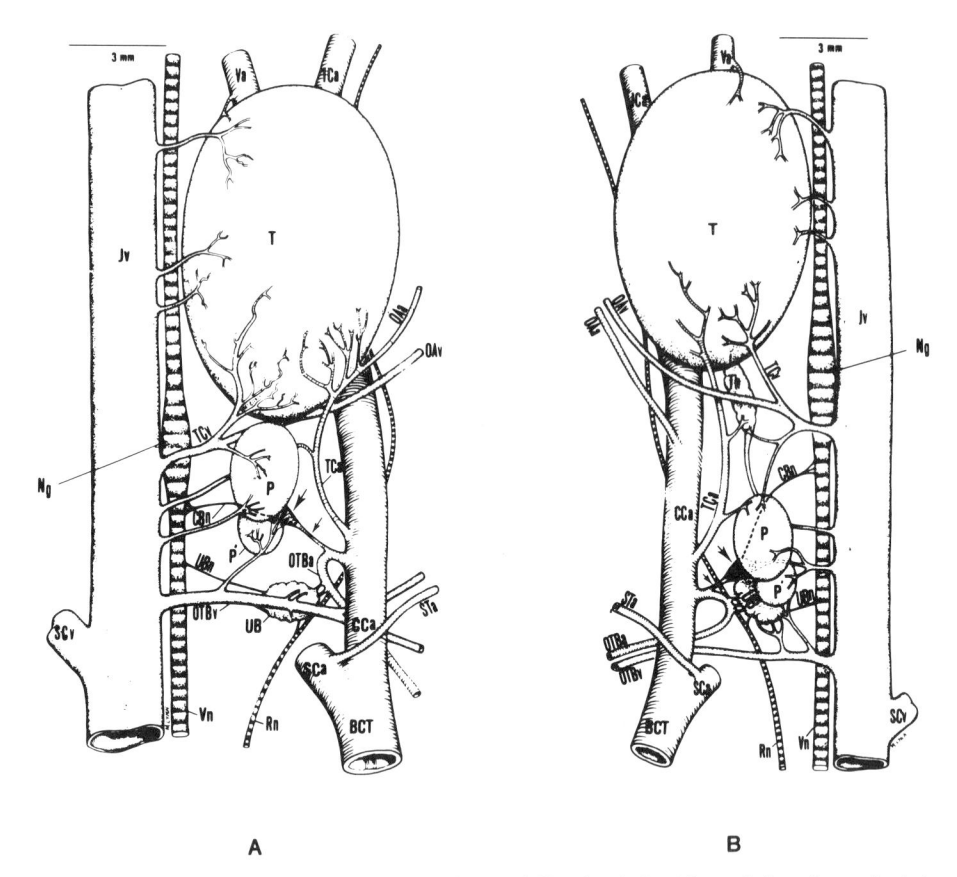

Fig 12–3 Ventral views of **A**, the right and **B**, the left sides of the thoracic inlet showing the endocrine glands and carotid body, together with their blood and nerve supply, at the thoracic inlet of the domestic fowl. The drawings are approximately to scale. The orientation of the two illustrations is the same as the left and right sides of Fig 12–2. The abbreviations are as in Fig 12–2, with the following additions: CBn = carotid body nerve; STa = sternotracheal artery; TCa = caudal thyroid artery; TCv = caudal thyroid vein; Th = thymus; UBn = nerve to ultimobranchial gland; large arrow = carotid body; small arrow = artery to carotid body. From Abdel-Magied and King (1978), with kind permission of the editor of the *Journal of Anatomy*.

especially in the ultimobranchial glands. Each gland consists of irregular groups and cords of columnar 'chief' cells, the oxyphil cells of the mammal being absent.

Function

The parathyroid glands secrete parathyroid hormone (PTH) which, as in mammals, resorbs bone and raises the level of blood calcium. During egg laying the hormone regulates the plasma level of ionic calcium, large amounts of calcium being shifted from the medullary bone to the egg shell. It seems likely that at the onset of rapid calcification of the shell there is a fall in the plasma ionic calcium, and this induces an increase in the secretion of PTH which stimulates the number and activity of the osteoclasts.

Parathyroid hormone also increases the renal excretion of phosphate by reducing tubular resorption.

ULTIMOBRANCHIAL GLANDS

Anatomy

The ultimobranchial glands are believed to arise from the sixth pharyngeal pouch (the fifth pouch disappearing). In the adult, the glands are small (2–3 mm in diameter in the domestic fowl), dorsoventrally flattened, irregularly-shaped pink structures lying on each side of the thoracic inlet, caudal to the parathyroid glands and usually just craniolateral to the origin of the common carotid arteries (Figs 12–2 and 12–3). In the domestic fowl the left ultimobranchial gland is usually adherent to the caudal parathyroid gland but on the right side the gland lacks this attachment, generally lying on the dorsal surface of the right oesophagotracheobronchial vein just lateral to the point where the vein is crossed by the recurrent nerve (Fig 12–3A). The ultimobranchial glands are unencapsulated and consist of four main components.

C-cells. These are eosinophilic cells arranged in scattered groups and cords. They are homologous to the C-cells of the mammalian thyroid gland.

Parathyroid nodules. The parathyroid nodules are encapsulated accumulations of parathyroid tissue. Cords of parathyroid cells grow from these nodules, penetrate between the C-cells, and link up with the vesicles.

Vesicles. These may form a large proportion of the gland in the adult domestic fowl. They are lined by a secretory epithelium and accumulate carbohydrate–protein secretion in their lumen. It is probable that the parathyroid nodules give rise to the ultimobranchial vesicles, since the cells of the strands that link the two structures have an appearance intermediate between that of the parathyroid and vesicular lining cells.

Lymphoid tissue. Foci of lymphoid cells and thymus tissue may be present.

Function

The C-cells secrete calcitonin. This hormone blocks the transfer of calcium from bone to blood, but its physiological role in birds is not clear. It may be relatively unimportant in regulating the level of plasma calcium but significant in preventing excessive bone resorption by parathyroid hormone. The physiological role of the vesicular secretion is unknown.

CAROTID BODIES

The principal cells of the carotid body belong to the APUD system of endocrine cells (see below under **Gastro–Entero–Pancreatic Endocrine System**) and may have endocrine functions. Their main role, however, is to participate in blood–vascular chemoreception. Nevertheless it is convenient to consider the carotid body alongside the endocrine organs which lie within the thoracic inlet, since these structures are all very closely related topographically.

The paired carotid bodies in the domestic fowl (Figs 12–2 and 12–3) lie on each side of the thoracic inlet in contact with the medial surface of one or both parathryoid glands. In this species the carotid body is a whitish, pear-shaped or ovoid structure about $0.8 \times 0.6 \times 0.5$ mm in size and therefore considerably smaller than the parathyroid glands. The carotid body in many passerines is partially or wholly embedded in one of the parathyroid glands. In some species, including the domestic fowl, there may also be contact between the carotid body and the ultimobranchial gland, especially on the left side. As well as the carotid body, accessory carotid body tissue has been observed in a number of sites including, in the domestic fowl, the wall of the common trunk of the oesophagotracheobronchial and carotid body arteries.

The carotid body consists of glomus (Type I) cells and sustentacular (Type II) cells embedded in a rich capillary network. The glomus cells form the bulk of the tissue, and are large cells with long processes which lie close to blood capillaries. Their most characteristic feature is the presence of numerous dense-cored granular vesicles about 120 nm in diameter. The glomus cells are innervated by afferent nerve fibres which are stimulated by a fall in the oxygen tension of the blood. These afferent fibres come from the vagus, in contrast to those of the mammalian carotid body which come from the glossopharyngeal nerve. The sustentacular cells are relatively small and invest the unmyelinated axons in a manner similar to that of Schwann cells. These cells are not innervated and are generally believed to have a supportive role.

ADRENAL GLANDS

Anatomy

The adrenal glands are generally paired yellow structures lying on each side of the midline at the cranial end of the kidney (Figs 6–10, 9–2 and 10–1), dorsal to the gonads. In the domestic fowl they are about $13 \times 8 \times 5$ mm in size and in the male bird are firmly attached to the appendix of the epididymis by connective tissue (Fig 9–1). In a relatively small number of species (e.g. the Greater Rhea and the Common Loon) the right and left adrenal glands are fused together. In the Jackdaw up to three small accessory adrenal glands are embedded in the epididymis. There is evidence in some species, including the domestic fowl, for an adrenal portal system between the glands and the muscles of the lateral abdominal wall.

As in mammals, the cells of the adrenal gland arise from two different embryonic sources, neural crest and mesoderm. Unlike those of mammals, however, these two groups of cells do not form a separate medulla and cortex but are extensively intermingled. The ectodermal cells of the neural crest form the medullary parts of the gland, and the mesodermal cells form the cortical parts.

Medullary tissue. The medullary parts of the gland consist of irregular clumps of basophilic polygonal cells that are larger than those of the cortical parts of the gland. They receive a sympathetic preganglionic innervation and fluoresce strongly in tests for biogenic monoamines.

Cortical (inter-renal) tissue. The cortical parts of the gland consist of anastomosing cords of granular vacuolated eosinophilic cells. They contain carotenoids, which account for the yellow colour of the glands as a whole. Ultrastructurally these cells resemble the cells of the mammalian adrenal cortex. In nearly all species zonation of the cortical tissue as in mammals is not apparent.

Functions

Adrenalin and noradrenalin are secreted by the medullary parts of the glands.
The hormones of the cortical parts, which include corticosterone and aldosterone, are important in electrolyte balance and carbohydrate and lipid metabolism. Bilateral removal of the adrenal glands is rapidly fatal if there is no steroid replacement.

GASTRO–ENTERO–PANCREATIC ENDOCRINE SYSTEM

This is a diffuse endocrine complex formed of several types of endocrine cell scattered throughout the gastrointestinal tract and pancreas and producing a variety of hormones which are involved in digestion or have no known function. The endocrine cells of this system, along with a large range of other widely dispersed endocrine cells such as the glomus cells of the carotid body, constitute the so-called APUD system of cells; such cells have the ability to take up the amino acid precursors of fluorogenic amines, to decarboxylate these, and to store the resulting amines in specific granules (the Amine Precursor Uptake, Decarboxylation and storage, or APUD system of cells). All these cells are characterized by the presence of dense-cored granular vesicles within their cytoplasm.

Gastrointestinal endocrine cells

Endocrine cells in the epithelium of the gastrointestinal tract can be demonstrated by silver impregnation techniques and are either argyrophil or argentaffin, or both argyrophil and argentaffin. The number of endocrine cells varies considerably between the different regions of the tract, in general by far the greatest concentration of the cells occurring in the pyloric part of the stomach. The hormones secreted by many of these cells have only recently been identified using immunocytochemical techniques. Amongst the types of endocrine cell which have so far been found are enterochromaffin cells, enteroglucagon cells, gastrin cells, D cells (somatostatin), D1 cells, secretin cells, avian pancreatic polypeptide cells, neurotensin cells and vasoactive intestinal polypeptide cells (see also p. 105).

Pancreatic islets

In contrast to the mammalian pancreas there are three types of islet; light, dark and mixed types. The dark islets (A or alpha islets) are composed mainly of A and D cells (in the domestic fowl 72 per cent being A cells and 28 per cent D cells), while the light islets (B or beta islets) are composed mainly of B and D cells (in the domestic fowl 86 per cent being B cells and 14 per cent D cells). Mixed islets contain A cells, B cells and some D cells. A islets are generally large, irregular in shape and indistinctly separated from the surrounding exocrine tissue. B islets, in contrast, are typically small, round or ellipsoid, and clearly set apart from the exocrine tissue by collagen.

The A cells of the pancreatic islets secrete glucagon. In birds the pancreatic levels of glucagon can be up to ten times greater than those in mammals while the plasma levels can be up to eight times greater. Glucagon has a lipolytic action since it increases the concentrations of free fatty acids, and it also raises the plasma glucose; it appears to play a major role in lipid and glucose homeostasis. The B cells secrete insulin. However, the importance of insulin in birds is not well understood, the level of the hormone in the pancreas being only about one-tenth that of mammals. In general, the mechanisms by which the islets in birds control lipid and carbohydrate metabolism are far from clear. The D cells secrete somatostatin.

Among other types of endocrine cell in the pancreas are the APP cells, which secrete avian pancreatic polypeptide (APP). These cells are scattered throughout the exocrine tissue of the pancreas. The physiological role of APP appears to be unknown. However, at high dose rates in the domestic fowl, APP induces hepatic glycogenolysis and hyperglycerolaemia, whilst at low dose rates it is a powerful gastric secretagogue.

PINEAL GLAND

Anatomy

The pineal gland is a median, conical, dorsally-directed projection of the diencephalon, lying in the space between the cerebral hemispheres and the cerebellum. The principal part of the gland, the body, is connected ventrally by the narrow pineal stalk to the roof of the third ventricle in the region of the choroid plexus. In the adult domestic fowl the gland is about 3.5 mm long by 2.0 mm thick. The structure of the gland in birds appears to be highly variable. In some species it is basically tubular or saccular, while in others it has a basically follicular structure; in yet other birds, including the domestic fowl, it has a solid lobular structure. In the domestic fowl the lobules are separated from each other by thin connective tissue strands and consist of parenchymal cells arranged as rosettes and follicles. The follicles are surrounded by tracts of fine nerve fibres.

With the electron microscope, the wall of the follicle is seen to be lined by receptor-type cells with cilia, and supportive cells with microvilli. Peripheral to these cell types are the so-called glial cells which have no connection with the lumen. The gland is well innervated especially by adrenergic fibres. The pineal stalk is solid in most species.

Function

Whilst no physiological role for the avian pineal gland has been conclusively established, it is generally believed that the gland is concerned in some way with photoreception, circadian rhythms and the control of reproduction (p. 253).

Further reading

Abdel-Magied, E.M. & King, A.S. (1978) The topographical anatomy and blood supply of the carotid body region of the domestic fowl. *J. Anat.*, 126, 535–546.

Adams, W.E. (1958) *The Comparative Morphology of the Carotid Body and Carotid Sinus.* Springfield, Illinois: C.C. Thomas.

Andrews, A. (1976) Intestinal endocrine cells of chicks around the time of hatching. *Cell Tissue Res.*, 172, 541–551.

Andrews, A. (1976) Endocrine cells of the stomach of chicks around the time of hatching. *Cell Tissue Res.*, 172, 553–561.

Assenmacher, I. (1973) The peripheral endocrine glands. In *Avian Biology* (Ed.) Farner, D.S. & King, J.R. Vol. III. New York and London: Academic Press.

Bhattacharyya, T.K. (1975) Fine structure of the interrenal cell in the quail and the pigeon. *Anat. Embryol.*, 146, 301–311.

Chan, A.S. (1977) Ultrastructure of the parathyroid glands in the chicks. *Acta anat.*, 97, 205–212.

Cronshaw, J., Holmes, W.N. & Loeb, S.L. (1974) Fine structure of the adrenal gland in the duck (*Anas platyrhynchos*). *Anat. Rec.*, 180, 385–406.

Falconer, I.R. (1971) The thyroid glands. In *Physiology and Biochemistry of the Domestic Fowl* (Ed.) Bell, D.J. & Freeman, B.M. Vol. 1. London and New York: Academic Press.

Freeman, B.M. (1983) The adrenal gland. In *Physiology and Biochemistry of the Domestic Fowl* (Ed.) Freeman, B.M. Vol. 4. London and New York: Academic Press.

French, E.I. & Hodges, R.D. (1977) Fine structural studies on the thyroid gland of the normal domestic fowl. *Cell Tissue Res.*, 178, 397–410.

Ghosh, A. (1977) Cytophysiology of the avian adrenal medulla. *Int. Rev. Cytol.*, 49, 253–284.

Gilbert, A.B. (1979) Glandulae endocrinae. In *Nomina Anatomica Avium* (Ed.) Baumel, J.J., King, A.S., Lucas, A.M., Breazile, J.E. & Evans, H.E. London and New York: Academic Press.

Gould, R.P. & Hodges, R.D. (1971) Studies on the fine structure of the avian parathyroid glands and ultimobranchial bodies. *Mem. Soc. Endocr.*, 19, 567–604.

Harrison, F. (1978) Ultrastructural study of the adenohypophysis of the male Chinese Quail. *Anat. Embryol.*, 154, 185–211.
Hodges, R.D. (1970) The structure of the fowl's ultimobranchial gland. *Annls Biol. anim. Biochim. Biophys.*, 10, 255–279.
Hodges, R.D. (1974) *The Histology of the Fowl.* London and New York: Academic Press.
Hodges, R.D. (1980) Endocrine organs. In *Form and Function in Birds* (Ed.) King, A.S. & McLelland, J. Vol. 2. London and New York: Academic Press.
Holmes, W.N. & Phillips, J.G. (1976) The adrenal cortex of birds. In *General, Comparative and Clinical Endocrinology of the Adrenal Cortex* (Ed.) Chester Jones, I. & Henderson, I.W. London and New York: Academic Press.
Isler, H. (1973) Fine structure of the ultimobranchial body of the chick. *Anat. Rec.*, 177, 441–459.
Knouff, R.A. & Hartman, F.A. (1951) A microscopic study of the adrenal gland of the Brown Pelican. *Anat. Rec.*, 109, 161–187.
Kobayashi, H. & Wada, M. (1973) Neuroendocrinology in birds. In *Avian Biology* (Ed.) Farner, D.S. & King, J.R. Vol. III. New York and London: Academic Press.
Larsson, L.-I., Sundler, F., Hakanson, R., Rehfeld, J.F. & Stadil, F. (1974) Distribution and properties of gastrin cells in the gastrointestinal tract of chicken. *Cell Tissue Res.*, 154, 409–421.
Menaker, M. & Oksche, A. (1974) The avian pineal organ. In *Avian Biology* (Ed.) Farner, D.S. & King, J.R. Vol. IV. New York and London: Academic Press.
Mikami, S. & Mutch, K. (1971) Light- and electron-microscopic studies of the pancreatic islet cells in the chicken under normal and experimental conditions. *Z. Zellforsch. mikrosk. Anat.*, 116, 205–227.
Okamoto, T., Sugimura, M. & Kudo, N. (1976) Distribution of endocrine cells in duck digestive tracts. *J. Fac. Fish. Anim. Husb. Hiroshima Univ.*, 15, 127–134.
Oksche, A., Kirchstein, H., Kobayashi, H. & Farner, D.S. (1972) Electron microscopic and experimental studies of the pineal organ in the White-crowned Sparrow, *Zonotrichia leucophrys gambelii*. *Z. Zellforsch. mikrosk. Anat.*, 124, 247–274.
Pearce, R.B., Cronshaw, J. & Holmes, W.N. (1978) Evidence for the zonation of interrenal tissue in the adrenal gland of the duck (*Anas platyrhynchos*). *Cell Tissue Res.*, 192, 363–379.
Piezzi, R.S. & Entierrez, L.S. (1975) Electron microscopic studies on the pineal organ of the Antarctic Penguin (*Pygoscelis papua*). *Cell Tissue Res.*, 164, 559–570.
Ralph, C.L. (1970) Structure and alleged function of avian pineals. *Am. Zool.*, 10, 217–235.
Ringer, R.K. (1976) Thyroids. In *Avian Physiology* (Ed.) Sturkie, P.D. 3rd Edn. New York: Springer-Verlag.
Ringer, R.K. (1976) Adrenals. In *Avian Physiology* (Ed.) Sturkie, P.D. 3rd Edn. New York: Springer-Verlag.
Ringer, R.K. & Meyer, D.C. (1976) Parathyroids, ultimobranchial glands, and the pineal. In *Avian Physiology* (Ed.) Sturkie, P.D. 3rd Edn. New York: Springer-Verlag.
Simkiss, K. & Dacke, C.G. (1971) Ultimobranchial glands and calcitonin. In *Physiology and Biochemistry of the Domestic Fowl* (Ed.) Bell, D.J. & Freeman, B.M. Vol. 1. London and New York: Academic Press.
Singh, K.B. & Dominic, C.J. (1975) Anterior and posterior groups of portal vessels in the avian pituitary: incidence in forty nine species. *Archs. Anat. microsc. Morph. exp.*, 64, 359–374.
Sivaram, S. (1965) Structure and development of the adrenal gland of *Gallus domesticus*. *Can. J. Zool.*, 43, 1021–1031.
Smith, P.H. (1974) Pancreatic islets of the Coturnix Quail. A light and electron microscopic study with special reference to the islet organ of the splenic lobe. *Anat. Rec.*, 178, 567–586.
Stockell-Hartree, A. & Cunningham, F.J. (1971) The pituitary gland. In *Physiology and*

Biochemistry of the Domestic Fowl (Ed.) Bell, D.J. & Freeman, B.M. Vol. 1. London and New York: Academic Press.

Stoeckel, M.E. & Porte, A. (1970) A comparative electron microscope study on the fowl, the pigeon, and the turtle dove of the C cells localized in the ultimobranchial body and the thyroid. In *Calcitonin 1969. Proceedings of the Second International Symposium* (Ed.) Taylor, S. & Foster, G. London: Heinemann Medical Books.

Sturkie, P.D. (1976) Hypophysis. In *Avian Physiology* (Ed.) Sturkie, P.D. 3rd Edn. New York: Springer-Verlag.

Taylor, T.G. (1971) The parathyroid glands. In *Physiology and Biochemistry of the Domestic Fowl* (Ed.) Bell, D.J. & Freeman, B.M. Vol. I. London and New York: Academic Press.

Tixier-Vidal, A. & Follett, B.K. (1973) The adenohypophysis. In *Avian Biology* (Ed.) Farner, D.S. & King, J.R. Vol. III. New York and London: Academic Press.

Unsicker, K. (1973) Fine structure and innervation of the avian adrenal gland. I. Fine structure of adrenal chromaffin cells and ganglion cells. *Z. Zellforsch. mikrosk. Anat.*, 145, 389–416.

Unsicker, K. (1973) Fine structure and innervation of the avian adrenal gland. IV. Fine structure of interrenal cells. *Z. Zellforsch. mikrosk. Anat.*, 145, 385–402.

Wells, J.W. & Wight, P.A.L. (1971) The adrenal glands. In *Physiology and Biochemistry of the Domestic Fowl* (Ed.) Bell, D.J. & Freeman, B.M. Vol. 1. London and New York: Academic Press.

Wight, P.A.L. (1971) The pineal gland. In *Physiology and Biochemistry of the Domestic Fowl* (Ed.) Bell, D.J. & Freeman, B.M. Vol. I. London and New York: Academic Press.

Wingstrand, K.G. (1966) Comparative anatomy and evolution of the hypophysis. In *The Pituitary Gland* (Ed.) Harris, G.W. & Donovan, B.T. Vol. 1. London: Butterworths.

Yamada, J., Kayamori, T., Okamoto, T., Yamashita, T. & Misu, M. (1978) Endocrine cells in the pyloric region of the Japanese Quail (*Coturnix coturnix japonica*). *Archvm histol. jap.*, 41, 41–52.

Chapter 13

CARDIOVASCULAR SYSTEM

Heart

The avian heart is relatively much larger than that of mammals, forming about 1.02–1.38 per cent of the body weight in the sparrow compared to about 0.5 per cent of the body weight in the mouse; in hummingbirds the heart may account for about 2.4 per cent of the total body weight, and the rate can reach over 1000 beats per minute. The combination of a relatively large size and fast

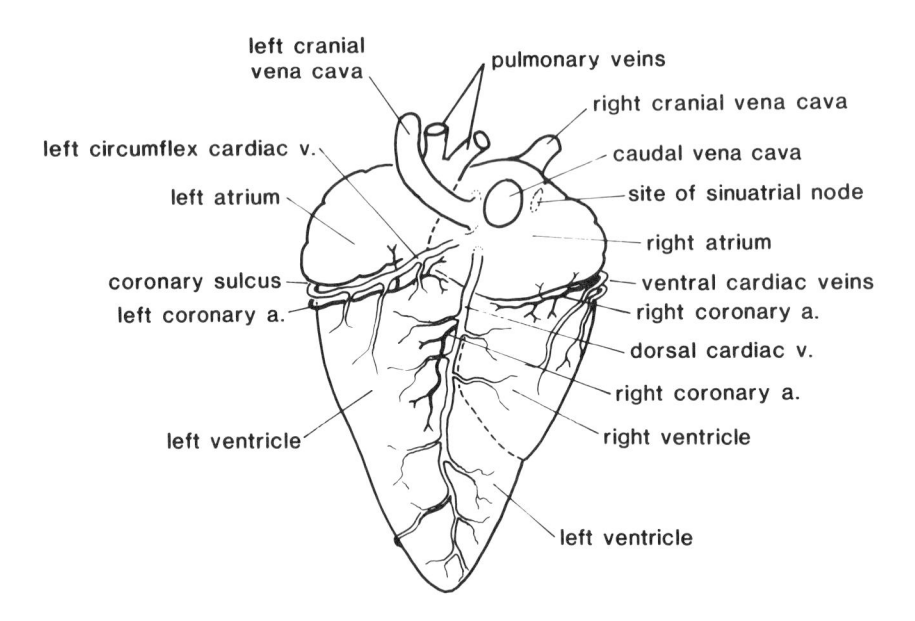

Fig 13–1 The dorsal surface of the heart of the domestic fowl to show the four chambers, the great vessels entering and leaving the heart, and the coronary arteries and cardiac veins. The left and right coronary arteries possess large deep branches as well as circumflex branches, but only the latter are shown, except for a small branch of the deep ramus of the right coronary artery which accompanies the dorsal cardiac vein in the centre of the dorsal surface. The broken lines represent the interventricular sulcus and the border between the atria. Based on Lindsay and Smith (1965) and Lindsay (1967).

rate leads to a relatively great cardiac output. The cardiac output (expressed as litres per kg body weight per minute) has been estimated to be about seven times greater in a flying Budgerigar than in a man or dog at maximum exercise. The total peripheral resistance in birds is slightly lower than in mammals. Nevertheless the high cardiac output requires a correspondingly high arterial pressure (140 to 250 mmHg in various species) to drive the high rates of flow. These substantial differences in circulatory function in birds and mammals comprise an efficient system of oxygen transport which contributes to the remarkable exercise capacity of birds (p. 5).

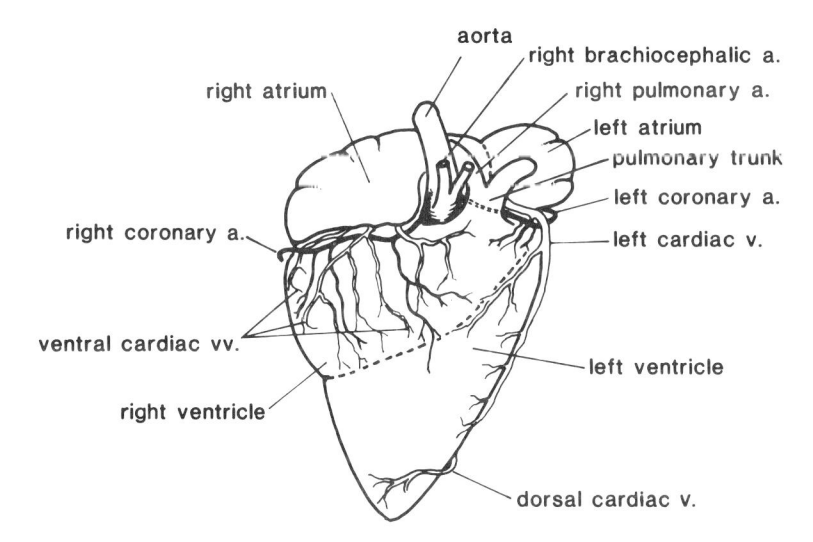

Fig 13–2 The ventral surface of the heart of the domestic fowl to show the four chambers, the great vessels entering and leaving the heart, and the coronary arteries and cardiac veins. Only the circumflex branches of the left and right coronary arteries are shown; the much larger deep rami of these arteries supply the interventricular septum and other deep regions of the heart wall. The single broken lines represent the interventricular sulcus and the boundary between the atria. The double broken line represents the root of the left coronary artery. Based on Lindsay and Smith (1965) and Lindsay (1967).

The heart lies in the midline within the thoracic cage (Figs 6–3, 6–10 and 14–2). Since the lungs are dorsal in position, the liver not the lungs encloses the heart on each side. The anatomy of the avian heart (Figs 13–1 and 13–2) resembles that of typical mammals except for a few special features. In the domestic fowl and some other species of bird the sinus venosus has not been fully incorporated into the wall of the right atrium; consequently, the right cranial vena cava and the single caudal vena cava open into a recognizable but not very clearly defined *sinus venosus* guarded by delicate but muscular left and right *sinuatrial valvules*. The sinus septum separates the mouth of the left cranial vena cava from the orifices of the other two veins. The *right atrioventricular valve* takes the form of a muscular flap devoid of chordae tendineae. The *left atrioventricular valve* and the *valves* of the *aorta* and *pulmonary artery* resemble their mammalian counterparts. The left and right *pulmonary veins* open into the left atrium either together or separately (Fig 13–1), the latter being more usual in the domestic fowl. Within the atrium the veins combine to form a single vessel the opening of which protrudes into the left atrioventricular orifice and is guarded by a flap, the *valve* of the *pulmonary vein*. This valve appears to direct the pulmonary flow towards the left ventricle and prevent regurgitation. It is much reduced in passerines. The wall of the left ventricle is two to three times thicker than that of the right.

The cardiac muscle cells of birds have a diameter (1–10 μm) which is only one-fifth to one-tenth that of mammals. The internal structure of the avian cell is also much simpler. Thus the system of T-tubules, which in mammals forms tubular invaginations of the sarcolemma carrying the extracellular space deep inside each muscle fibre, are absent in birds. As in mammals, however, the sarcoplasmic reticulum is well developed. Evidently the small diameter of the avian cardiac muscle cell, with a correspondingly large surface area relative to volume, obviates the need for internal tubules in excitation–contraction coupling; instead, the contacts between the sarcolemma and the sarcoplasmic reticulum at the surface of the cell provide adequate peripheral coupling. A further difference between the avian and the mammalian cardiac muscle cell is the absence in birds of the M line, which in mammals is caused by the delicate cross-connections holding the myosin filaments in register at the middle of the A band.

In cold-blooded vetebrates the musculature of the atria is freely continuous with that of the ventricles, but in birds, as in mammals, rings of fibrous tissue surround the atrioventricular and aortic and pulmonary openings, thus separating the atria from the ventricles and preventing the generalized spread of excitation from the atria to the ventricles. Consequently, in birds and mammals a specialized conducting system is required to carry the impulse from the atria to the ventricles, piercing the fibrous ring at a single point, i.e. at the atrioventricular node. The conducting system begins at the *sinuatrial node*, which lies at the base of the right sinuatrial valve, between the openings of the right cranial vena cava and the caudal vena cava (Fig 13–3). It then extends into subendocardial ramifications which make contact with ordinary myocardial cells in the atrial walls. As in mammals, the cells of the SA node have a higher frequency of depolarization than any other cardiac muscle cells, and therefore they control the rate at which the heart beats. The *atrioventricular node*, which is situated in the right caudodorsal part of the interventricular septum, gives rise to the *atrioventricular bundle* (Fig 13–3). In the domestic fowl the latter pierces the fibrous tissue separating the atria from the ventricles, and divides into *right* and *left crura* which continue down the interventricular septum. The a–v node also gives off a fascicle, the *right atrioventricular ring* (AAD in Fig 13–3), which is peculiar to birds. This ring encircles the right atrioventricular opening (presumably controlling the activity of the muscular valve) and joins the variably developed *truncobulbar node* (FTB in Fig 13–3) at the base of the aorta. The truncobulbar node in turn gives off a fascicle, the *truncobulbar bundle* (FTB in Fig 13–3), which extends through the aortic fibrous ring and connects with the atrioventricular bundle

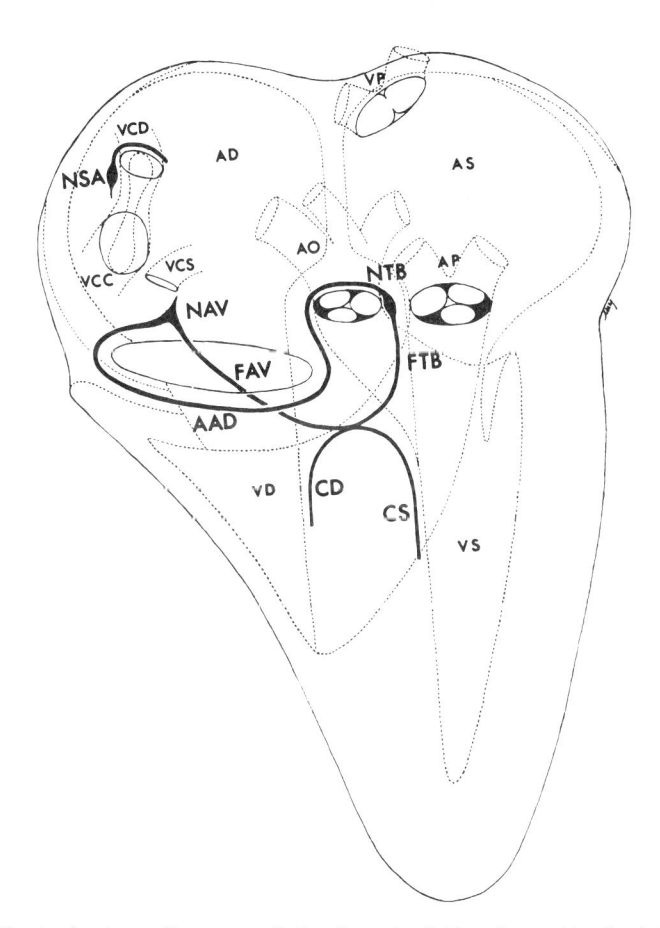

Fig 13-3 Ventral view diagram of the heart of the domestic fowl to show the conducting system. AAD = right atrioventricular ring; AD = right atrium; AO = aorta; AP = pulmonary artery; AS = left atrium; CD = right crus of atrioventricular bundle; CS = left crus of atrioventricular bundle; FAV = atrioventricular bundle; FTB = truncobulbar bundle; NAV = atrioventricular node; NSA = sinuatrial node; NTB = truncobulbar node; VCC = caudal vena cava; VCD = right cranial vena cava; VCS = left cranial vena cava; VD = right ventricle; VP = pulmonary vein; VS = left ventricle. From Kim and Yasuda (1979), with kind permission of the editor of *Zentralblatt für Veterinärmedizin, C.*

at its bifurcation into the left and right crura. The heart is innervated by tonic accelerator sympathetic fibres and by vagal fibres which have a tonic inhibitory influence on heart rate (p. 275).

The atrioventricular bundle and its branches consist of Purkinje cells. In the domestic fowl the diameter of these cells is about five times greater than that of the ordinary myocardial cells. As would be expected from their relatively large diameter, Purkinje cells have relatively fast conduction velocities compared with ordinary myocardial cells. Consequently, they play an important role in integrating myocardial contraction to achieve the optimum efficiency of the pump. As in mammals electrical coupling between myocardial and Purkinje cells is believed to occur across gap junctions in the intercalated discs, although the latter seem to be smaller and scarcer than in mammals.

Arteries

In contrast to mammals, the aorta is derived embryologically from the *right* fourth arterial arch and *right* dorsal aorta. The *ascending aorta* therefore curves to the *right* (Fig 12-2) to supply the heart, neck, head and wing. As in mammals the first branches of the ascending aorta are the left and right *coronary arteries* (Figs 13-1 and 13-2). In the domestic fowl each of these has a superficial branch which follows the coronary sulcus (Figs 13-1 and 13-2), and a larger deep branch which is the main supply to the ventricular myocardium including the interventricular septum; the left artery also forms an interatrial ramus which is important because it supplies the nodal tissue of the conducting system. Of the two coronary arteries, the right is usually dominant. Arising after the coronary arteries are the left and right *brachiocephalic trunks* (Figs 12-2, 13-4 and 14-2). Each brachiocephalic trunk divides into a common carotid (Figs 6-3 and 13-4) and a subclavian artery (Figs 12-2, 13-4 and 14-2). The *subclavian artery* distributes many branches to the wing, including a large *pectoral trunk* (Fig 13-4) to the pectoral musculature and incubation patch, a *brachial artery* (Fig 13-4) supplying mainly the region of the humerus, and *radial* and *ulnar arteries* (Fig 13-4) forming *metacarpal* and *digital* arteries (Fig 13-4) to the distal regions of the wing. The *common carotid* is a relatively short artery compared to that of mammals; it divides at the base of the neck to form mainly the *vertebral trunk* (Fig 13-4) and the internal carotid artery (Figs 12-2, 12-3, 13-4 and 14-2). The *internal carotid* is as long as the common carotid is short. It ascends the neck in a bony groove along the ventral midline of the cervical vertebrae. In the domestic fowl both the left and right internal carotid arteries persist, lying in contact with each other side by side. In other species the two vessels fuse, or one drops out. At the base of the skull the internal carotid gives off the *external ophthalmic artery* (Fig 13-4) which supplies the eyeball including the pecten, and the muscles and glands of the orbit. The internal carotid arteries are continued rostrally by the *cerebral carotid arteries* (Fig 13-4). These supply blood to the brain by branches which run in the subarachnoid space and send small penetrating rami into the substance of the brain. Immediately caudal to the hypophysis the left and right cerebral carotid arteries are connected by the large transverse *intercarotid anastomosis* (Fig 13-4), which is the functional analogue of the mammalian cerebral arterial circle (of Willis). In the domestic fowl the external ophthalmic branch of the internal carotid artery (of the external carotid in some species) forms a network of anastomosing arterioles, the *rete mirabile ophthalmicum*, situated in the temporal fossa of the skull. The arterial rete is enmeshed within a venous rete which receives cooled blood from the rostral regions of the face and pharyngeal wall where there is heat loss by evaporation. Some of the arteries which emerge from the arterial rete anastomose with intracranial arteries and may supply the brain with blood. The entangled arterial and venous retia may therefore be a heat exchanger for

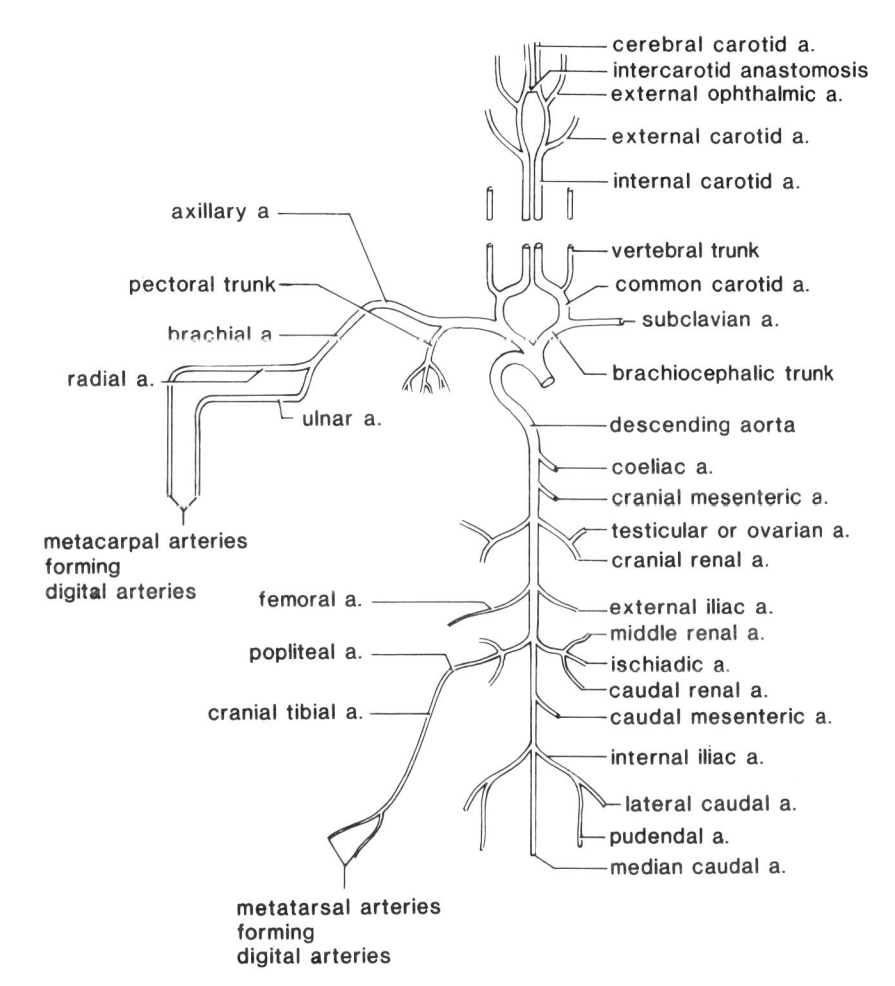

Fig 13–4 Ventral view diagram of the main arteries of a typical bird. The arteries of the right wing and right leg are shown on the left side of the diagram.

cooling arterial blood en route to the brain. The *external carotid artery* (Fig 13–4) arises from the internal carotid artery near the base of the skull. It vascularizes all the extensive extracranial regions of the head which are not supplied by the external ophthalmic artery, i.e. the lower jaw and its muscles, the tongue, palate, oropharynx, upper jaw and larynx. It also contributes vessels to the nasal cavity and orbit.

The *descending aorta* (Fig 13–4) supplies the gastrointestinal tract, urogenital system and other nearby viscera, and the trunk, tail and leg. Close to the origin of the coeliac artery the tunica intima of the aorta contains aggregations of smooth muscle. They may be of pathological interest, since this part of the aorta is a predilection site for atheromatous plaques in growing turkeys. Combined with the exceptionally high blood pressure characteristic of this species, such plaques can lead to a dissecting aneurysm and death by rupture of the aorta. The *coeliac artery* (Fig 13–4) supplies the proventriculus, gizzard, duodenum, ileum, caeca and pancreas (Fig 6–12). The *cranial mesenteric artery* (Fig 13–4) distributes branches to the duodenum, jejunum, ileum and caeca (Fig 6–12). This artery is unusual in possessing outside the tunica muscularis a longitudinal layer of smooth muscle which can shorten the isolated vessel by up to 60 per cent of its resting length and may adjust the vessel to changes in the position of the intestines. The *caudal mesenteric artery* (Fig 13–4) supplies the ileum and rectum, anastomosing with the cranial mesenteric artery at the caeca. The kidneys are supplied by the *cranial, middle* and *caudal renal arteries* (Figs 10–5 and 13–4). In the male the left and right *testicular arteries* (Fig 13–4) arise from the left and right cranial renal arteries. In the female the left gonad is supplied by an *ovarian artery* or *arteries*, which generally arise from the left cranial renal artery (Fig 10–5) but quite often also arise directly from the aorta. The oviduct is supplied by five arteries (Fig 13–5): (1) the *cranial oviductal artery* arises from the left cranial renal artery and supplies the infundibulum and magnum; (2) the *accessory cranial oviductal artery* arises from the left external iliac artery and supplies the magnum; (3) the *middle oviductal artery* arises from the left ischiadic artery and supplies the magnum and uterus; (4) the *caudal oviductal artery* arises from the pudendal branch of the internal iliac artery and supplies the uterus; and (5) the *vaginal artery* also arises from the pudendal artery and supplies the vagina. The cranial oviductal and accessory cranial oviductal arteries are variable. The domestic fowl typically has the cranial oviductal artery and only occasionally possesses the accessory artery as well. In the turkey and duck only the accessory cranial oviductal artery is present. The goose generally has both vessels. The pigeon sometimes has neither of these two arteries and then the middle oviductal artery takes over their territories. The ovarian and oviductal arteries greatly hypertrophy during egg laying.

The trunk is supplied by paired intercostal and segmental synsacral arteries. There is no common iliac artery in birds, since the external and internal iliac arteries arise separately from the aorta. The pelvic limb is supplied by the *external iliac artery* (Figs 10–1 and 13–4) which continues into the thigh as the *femoral artery* reaching the region of the knee joint (Fig 13–4). The leg is also supplied by the *ischiadic artery* (Figs 10–1 and 13–4), which is far larger than the femoral artery. The ischiadic artery is directly continued down the distal part of the limb by the *popliteal* and *cranial tibial arteries* (Fig 13–4). The cranial tibial artery forms *metatarsal arteries* at the ankle (hock) joint, and in the foot these give rise to *digital arteries* (Fig 13–4). In some species including aquatic and wading birds the arteries and veins in the proximal feathered part of the leg form a counter-current *tibiotarsal rete*. The function of the rete is probably to transfer heat from the warm downward flowing arterial blood to the cool upward flowing venous blood returning from the feet, and thereby prevent excessive heat loss at the extremities. However, this heat exchange device has only been found in a limited number of species, and in most birds heat loss from the distal part of the limb and feet is probably controlled simply by regulating the rate of blood flow to the feet. In very cold conditions, damage by frost-bite is countered by intermittent brief periods of very great flow through the feet. During heat stress the blood flow to the extremity is greatly increased, so that the unfeathered parts of the pelvic limbs and feet can be used as radiators to get rid of heat. During steady flight, when heat production goes up by as much as 10 times, transfer from the feet accounts for about 80 per cent of the heat which is being produced; this amounts to an increase in heat transfer from the feet by 10 to 15 times above the transfer

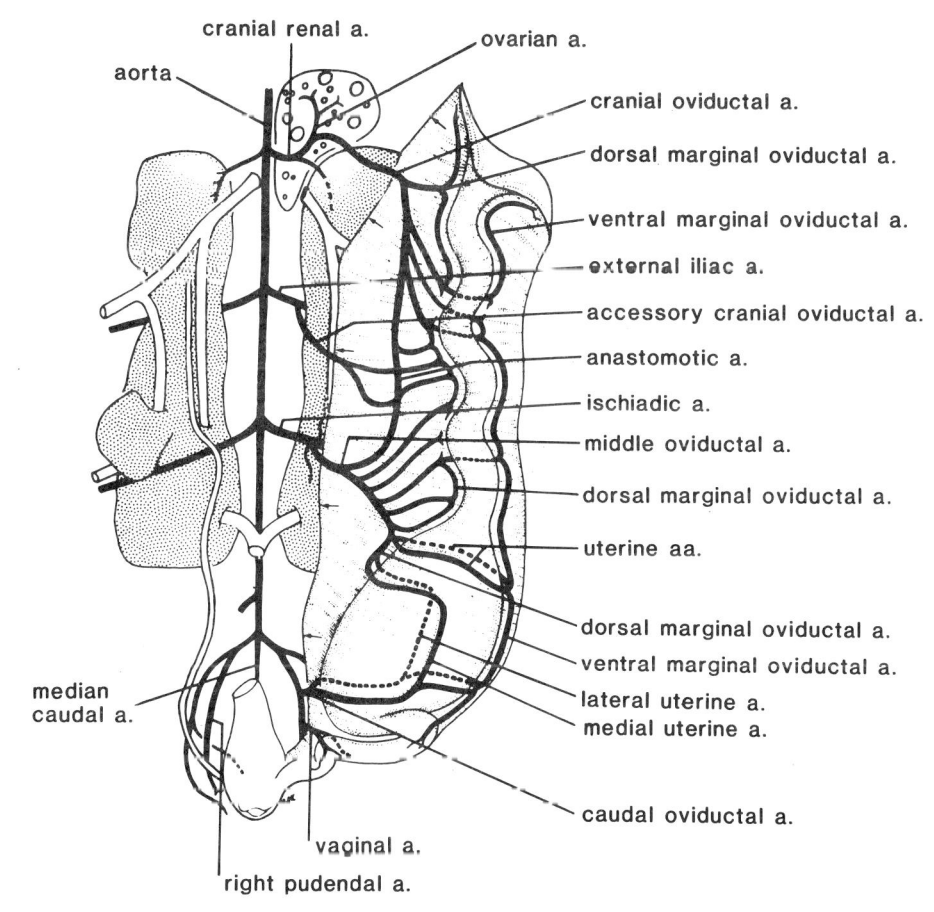

Fig 13–5 Ventral view diagram of the main arteries supplying the ovary, oviduct and cloaca of the domestic fowl. The accessory cranial oviductal artery is only occasionally present in this species. From King (1975), with kind permission of the publisher.

while at rest. It is likely that the regulation of perfusion of the pelvic limbs and feet is achieved by the vasomotor control of arterioles and arteriovenous anastomoses.

The *internal iliac artery* (Fig 13–4) arises at the end of the aorta, and supplies the vagina and cloacal region via the *pudendal artery* (Fig 13–4) and its caudal oviductal and vaginal branches (Fig 13–5), and the tail region via the *lateral caudal artery*.

Veins

As already stated, in the domestic fowl the left and right *pulmonary veins* open separately into the left atrium (Figs 13–1 and 13–3). There are five groups of *cardiac veins*: the dorsal cardiac vein, the left cardiac vein, the left circumflex cardiac vein, the ventral cardiac veins and the small cardiac veins (Figs 13–1 and 13–2). Of these, the last two groups consist of a number of veins with separate openings, but the other three vessels are single. The dorsal cardiac vein is the largest. The cardiac veins open into the right atrium; the small cardiac veins (Thebesian veins) drain into all four chambers of the heart as in mammals, their ostia being abundant in all chambers except the left ventricle.

The left and right *cranial venae cavae* (Figs 12–2 and 13–1) are formed on each side essentially by the union of the subclavian and jugular veins, there being no brachiocephalic vein on either side. The left and right *jugular veins* (Figs 6–3 and 14–2) are the main drainage of the head and neck. In most birds including the domestic fowl the right jugular vein is much larger than the left and it appears to receive blood from the left jugular vein via an oblique anastomosis at the cranial end of the neck. Cutting the neck in the region of the anstomosis is the method used in slaughter houses to bleed poultry. The right jugular vein can be used for taking blood samples. The *subclavian vein* (Fig 12–2) drains the wing by means of tributaries which are, in principle, satellites of the arteries. One of its branches on the lateral body wall, the *common pectoral vein*, is reported to be the best vessel for repeated intravenous injections. The *deep ulnar vein* is the largest vein of the forearm. At the elbow where it becomes subcutaneous it is accessible (like its central continuation, the *basilic vein*) for intravenous injections and blood sampling.

The *caudal vena cava* receives the *left* and *right hepatic veins* both of which are large, and several small *middle hepatic veins*. More caudally it receives the left and right *adrenal veins*, the *ovarian* or *testicular veins* and a large left and right *common iliac vein* (Fig 10–5). Opening into the adrenal veins are veins draining the body wall thus forming an adrenal portal system. The common iliac vein drains the kidney by means of the *caudal renal vein* and several *cranial renal veins* (Fig 10–5). It is also involved in the *renal portal system* since it contains the renal portal valve and receives the cranial renal portal and caudal renal portal veins peripheral to the valve (Fig 10–5). The common iliac vein also drains the cranial part of the oviduct.

The large *external iliac vein* (Fig 10–5) empties directly into the common iliac vein and is ultimately the main drainage of the pelvic limb; the external iliac vein is itself the direct intrapelvic continuation of the *femoral vein* (Fig 13–6). The femoral vein is initially a relatively small vein draining the cranial region of the thigh, but as it approaches the hip joint it is reinforced by a very large anastomotic vein from the *ischiadic vein* (Fig 13–6). The anastomotic vein transfers most of the flow from the ischiadic vein into the femoral vein. Running caudal to the femur and cranial to the ischiadic nerve, the ischiadic vein is the main venous drainage for most of the limb. Proximal to the anastomotic vein it continues as a small vein which enters the pelvic cavity alongside the ischiadic nerve and artery by passing through the ischiadic foramen, and empties into the caudal renal portal vein (Fig 10–5). In the region of the knee the ischiadic vein is the direct continuation of the *popliteal vein*, which in turn is formed mainly by the *cranial tibial vein* (Fig 13–6). The cranial tibial vein continues the *dorsal metatarsal vein*, which receives *digital veins* from the lateral aspect of the foot (Fig 13–6). The largest vein in the tibial region is the *caudal tibial vein* (Fig 13–6). This passes over the mediodorsal aspect of the hock, where it is visible through the skin in the domestic fowl, and can be used for venipuncture; it is the continuation of the *superficial plantar metatarsal vein* (Fig 13–6), a large vein which drains digital veins from the medial aspect of the foot.

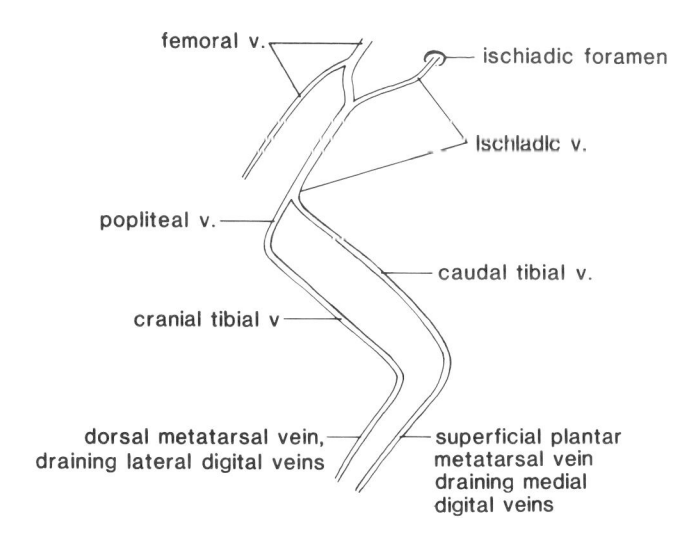

Fig 13–6 Lateral view diagram of the principal veins of the left leg of the domestic fowl.

The *internal iliac vein*, which drains the caudal part of the oviduct and the cloacal region, continues directly into the caudal renal portal vein (Fig 10–5); the latter is in fact a part of the embryonic internal iliac vein.

There are two *hepatic portal veins* (Fig 6–14). The smaller left vein drains the stomach region, and the much larger right vein drains the stomach, spleen, and the small and large intestines. The main venous drainage of the large intestine is by the *caudal mesenteric vein* (Fig 10–5), which is commonly known as the coccygeomesenteric vein. This vein empties into both the hepatic portal system and the renal portal system; cineradiography has shown that the direction of flow often changes, sometimes being directed to the liver and then being reversed and flowing to the kidney (p. 182). In addition to this major anastomosis between the hepatic portal and renal portal systems, there is an extensive anastomosis between the hepatic portal and systemic circulations; this occurs through the venous drainage of the proventriculus, which empties extensively into the left cranial vena cava as well as into the left hepatic portal vein.

As in mammals an extensive system of venous sinuses is associated with the dura mater inside the cranial cavity and neural canal. The *intracranial venous sinuses* include olfactory, sagittal, petrosal, transverse, sphenotemporal, cavernous and occipital sinuses. The flow in these sinuses converges mainly on the occipital region and then has an exit from the cranial cavity into the occipital veins, and thence into the jugular vein. The sinuses in the region of the medulla oblongata have a much more extensive contact with the inner surface of the skull in birds than in mammals, so it is impossible to expose the medulla oblongata in the live bird (at least in the domestic fowl) without substantial haemorrhage. The intracranial dural sinuses continue through the foramen magnum into the *internal vertebral venous sinuses* within the neural canal. This system is not confined to the ventral region of the neural canal as in mammals, but lies on both the dorsal and the ventral aspect of the spinal cord. Consequently, laminectomy (certainly in the domestic fowl) causes profuse haemorrhage. The system empties into the systemic veins mainly via the vertebral veins and thence into the jugular vein in the neck.

Blood cells

The blood cells of birds are made up of erythrocytes, thrombocytes, lymphocytes, monocytes, heterophils, eosinophils and basophils.

Erythrocytes are flattened ellipsoidal cells, which in contrast to those of mammals are nucleated and very large in size. In the domestic fowl they have a long axis of about 10.7 to 13.0 μm and a short axis of about 6.5 to 7.2 μm. In birds generally they range from about 9.75×5.0 μm in a warbler to 15.9×9.7 μm in the Ostrich. As a rule their size is smaller in species possessing relatively high metabolic rates (e.g. passerines and hummingbirds) thus facilitating efficient gaseous exchange. The number of erythrocytes in the domestic fowl is approximately the same in both sexes immediately after hatching (2×10^6 per mm^3), but in adult birds males generally have more red cells (about 3.23×10^6 per mm^3) than females (about 2.72×10^6 per mm^3). Typically, mature red cells under the light microscope appear to have a clear homogeneous cytoplasm. In electron micrographs the cytoplasm is packed with electron-dense haemoglobin, which almost completely conceals a few mitochondria, ribosomes and even a Golgi complex; the presence of these organelles is presumably the reason why the avian erythrocyte has a metabolic rate that is about 10 times that of the mammalian red cell. In most birds there are two types of haemoglobin, type II being more acidic than type I and having a very different amino acid composition. The affinity of the avian haemoglobins for oxygen is often said to be less than that of mammals, but there is evidence for species variations and for individual variations reflecting for example the hormonal status or physical fitness; the organelles of the avian cell suggest that its oxygen affinity is under a more elaborate set of controls than that of mammals. The life span of the avian erythrocyte is only 20 to 35 days (120 days in man). Erythropoiesis takes place in both the yolk sac and bone marrow in the embryo, and in the bone marrow after hatching.

Thrombocytes are fragile nucleated cells which are homologous with the platelets of mammals. While superficially similar to red cells they are usually smaller (long axis in the domestic fowl about 6.1 to 11.5 μm, and short axis 3.0 to 6.1 μm), have a larger and more rounded nucleus, and a cell outline which is less regularly oval. In addition, the cytoplasm under the light microscope has a reticulated appearance and possesses one or more acidophilic granules containing 5-hydroxytryptamine. These granules correspond to the large osmiophilic granules (0.2 to 0.3 μm in diameter) seen with the electron microscope. Mitochondria, smooth and rough endoplasmic reticulum, and a Golgi complex are also present. In the domestic fowl the number of thrombocytes ranges from 35 000 to 40 000 per mm^3. Thrombocytes are involved in blood coagulation, although their precise role in this has not been established. Since they contain little thromboplastin they do not appear to be involved in the initiation of the clotting process. Thrombocytes are also phagocytic.

Lymphocytes are rounded cells which in the domestic fowl range in diameter from 6.0 to 12 μm. Large, medium and small cells are often arbitrarily distinguished, the large lymphocytes occurring relatively infrequently and probably being immature. Medium-sized cells have more cytoplasm than small cells. Under the light microscope the nuclear chromatin is especially strongly condensed in small lymphocytes, while in all cell sizes the cytoplasm appears homogeneous. Ultrastructurally the cytoplasm is seen to contain relatively few organelles and inclusions including some mitochondria, rough endoplasmic reticulum, ribosomes and dense membrane-bound granules (maximum diameter 1.0 μm). In differential leucocyte counts lymphocytes form between 59 and 80 per cent in the normal domestic fowl but the proportion is lower in several other species (e.g. 46 per cent in the Canada Goose and 27 per cent in the Ostrich). Lymphocytes derived from the thymus (T lymphocytes) have a long life span and colonize thymus-dependent zones. They are concerned with the development of cellular immunity (e.g. delayed immune reactions and graft rejections). Lymphocytes derived from the cloacal bursa (B lymphocytes) have a short life span and are concerned with humoral immune responses. Under antigenic stimulation they differentiate into plasma cells which secrete specific antibodies.

Monocytes are rounded cells, generally, with an average diameter in the domestic fowl of 12 μm. They possess an elongated indented nucleus and cytoplasm which has a characteristic reticular appearance, especially near the nuclear indentation where it forms the 'Hof'. While monocytes are sometimes difficult to distinguish from lymphocytes with the light microscope, the electron microscope shows that the nuclear chromatin is more diffusely distributed in monocytes and that the cytoplasm contains more endoplasmic reticulum. Throughout the cytoplasm are distributed a number of membrane-bound granules (maximum diameter 0.5 μm). The well developed Golgi complexes are concentrated close to the nuclear indentation and correspond to the 'Hof' seen with the light microscope. Monocytes form about 1 to 10 per cent of the total number of leucocytes in the domestic fowl and other species. As in mammals, avian monocytes exist in the blood stream for only a short time before passing into the tissues and becoming macrophages.

Heterophils are rounded cells with a diameter in the domestic fowl of about

8.0 to 10 μm. They are functionally equivalent to the neutrophils of mammals and like the neutrophils have a polymorphic nucleus (usually bi- or tri-lobed). Their granules under the light microscope, however, differ from those of neutrophils in being spindle-shaped and in having an acid staining reaction. Three types of membrane-bound granules have been identified in the domestic fowl with the electron microscope: small dense granules (maximum diameter 0.2 μm), large spindle-shaped dense granules (maximum diameter 3.5 μm), and smaller, round to oval, pale granules. In the domestic fowl there are between 3000 and 12 000 heterophils per mm^3. They form from about 10 to 35 per cent of the total number of leucocytes in the domestic fowl, and from 20 to 60 per cent in other species. Heterophils are highly motile and phagocytic and have an important role in the defence mechanism.

Eosinophils are rounded cells with a diameter in the domestic fowl of about 4.0 to 11.0 μm (average 7.3 μm). Characteristically, they have a bilobed nucleus and their cytoplasm is packed with relatively large spherical granules which are so numerous sometimes that they mask the nucleus. With the electron microscope the granules appear as round, oval or elongated, densely granulated structures (maximum diameter 1.6 μm); in the domestic duck and goose the granules have a crystalline core. Eosinophils form about 1 to 3 per cent of the total number of leucocytes in the domestic fowl, and 10 to 12 per cent in other species. Their function in birds has not been established.

Basophils are rounded cells with a diameter in the domestic fowl of about 8.0 μm. They possess an indented nucleus with a single lobe and many spherical basophilic and metachromatic granules which frequently mask both the cytoplasm and nucleus. The membrane-bound granules seen with the electron microscope in the domestic fowl have a maximum diameter of 0.8 μm and consist of four types depending on their internal structure which may be dense, granular, webbed or with myelin whorls. In differential counts of leucocytes, basophils average about 2 to 4 per cent in the domestic fowl, and 1 to 10 per cent in other species. Avian basophils are motile cells which in contrast to those of mammals are involved in inflammatory reactions associated with both immunogenic and non-immunogenic agents.

Further reading

Abdalla, M.A. & King, A.S. (1975) The functional anatomy of the pulmonary circulation of the domestic fowl. *Resp. Physiol.*, 23, 267–290.

Abdalla, M.A. & King, A.S. (1976) The functional anatomy of the bronchial circulation of the domestic fowl. *J. Anat.*, 121, 537–550.

Akester, A.R. (1967) Renal portal shunts in the kidney of the domestic fowl. *J. Anat.*, 101, 569–594.

Akester, A.R. (1971) The Heart. In *Physiology and Biochemistry of the Domestic Fowl* (Ed.) Bell, D.J. & Freeman, B.M. Vol. 2. London and New York: Academic Press.

Baudinette, R.V., Loveridge, J.P., Wilson, K.J., Mills, C.O. & Schmidt-Nielsen, K. (1976) Heat loss from feet of Herring Gulls at rest and during flight. *Am. J. Physiol.*, 230, 920–924.

Baumel, J.J. (1967) The characteristic asymmetrical distribution of the posterior cerebral artery of birds. *Acta anat.*, 67, 523–549.

Baumel, J.J. (1975) Heart and blood vessels. Aves. In *Sisson and Grossman's The*

Anatomy of Domestic Animals (Ed.) Getty, R. Vol. 2, 5th Edn. Philadelphia: Saunders.

Baumel, J.J. (1979) Systema cardiovasculare. In *Nomina Anatomica Avium* (Ed.) Baumel, J.J., King, A.S., Lucas, A.M., Breazile, J.E. & Evans, H.E. London and New York: Academic Press.

Baumel, J.J. & Gerchman, L. (1968) The avian intercarotid anastomosis and its homologue in other vertebrates. *Am. J. Anat.*, 122, 1–18.

Bhaduri, J.L., Biswas, B. & Das, S.K. (1957) The arterial system of the domestic pigeon (*Columba livia* Gmelin). *Anat. Anz.*, 104, 1–14.

Bolton, T.B. (1968) Electrical and mechanical activity of the longitudinal muscle of the anterior mesenteric artery of the domestic fowl. *J. Physiol., Lond.*, 196, 272–281.

Davies, F. (1930) The conducting system of the bird's heart. *J. Anat.*, 64, 129–147.

Freedman, S.L. & Sturkie, P.D. (1963) Blood vessels of the chicken's uterus (shell gland). *Am. J. Anat.*, 113, 1–7.

Freeman, B.M. (1971) The corpuscles and the physical characteristics of blood. In *Physiology and Biochemistry of the Domestic Fowl* (Ed.) Bell, D.J. & Freeman, B.M. Vol. 2. London and New York: Academic Press.

Fukuta, K., Nishida, T. & Yasuda, M. (1969) Comparative and topographical anatomy of the fowl. LVI. Blood vascular supply of the spleen in the fowl. *Jap. J. Vet. Sci.*, 31, 179–185.

Glenny, F.H. (1951) A systematic study of the main arteries in the region of the heart. Aves XII. Galliformes, Part 1. *Ohio J. Sci.*, 51, 47–54.

Gobeil, R.E. (1970) Arterial system of the Herring Gull (*Larus argentatus*). *J. Zool., Lond.*, 160, 337–354.

Harris, J.R. (1971) The ultrastructure of the erythrocyte. In *Physiology and Biochemistry of the Domestic Fowl* (Ed.) Bell, D.J. & Freeman, B.M. Vol. 2. London and New York: Academic Press

Hartman, F.A. & Lessler, M.A. (1963) Erythrocyte measurements in birds. *Auk*, 80, 467–473.

Hodges, R.D. (1965) The blood supply to the avian oviduct with special reference to the shell gland. *J. Anat.*, 99, 485–506.

Hodges, R.D. (1974) *The Histology of the Fowl*. London and New York: Academic Press.

Hodges, R.D. (1977) Normal avian (poultry) haematology. In *Comparative Clinical Haematology* (Ed.) Archer, R.K. & Jeffcott, L.B. Oxford: Blackwell Scientific.

Hodges, R.D. (1979) The blood cells. In *Form and Function in Birds* (Ed.) King, A.S. & McLelland, J. Vol. 1. London and New York: Academic Press.

Johansen, K. & Millard, R.W. (1974) Cold-induced neurogenic vasodilatation in the skin of the Giant Fulmar, *Macronectes giganteus*. *Am. J. Physiol.*, 227, 1232–1235.

Jones, D.R. & Johansen, K. (1972) The blood vascular system of birds. In *Avian Biology* (Ed.) Farner, D.S. & King, J.R. Vol. II. New York and London: Academic Press.

Kaku, K. (1959) On the vascular supply in the brain of the domestic fowl. *Fukuoka Acta Med.*, 50, 4293–4306.

Kim, Y. & Yasuda, M. (1979) The cardiac conducting system of the fowl. *Zbl. vet. Med.*, C, 8, 138–150.

King, A.S. (1975) Aves urogenital system. In *Sisson and Grossman's The Anatomy of the Domestic Animals* (Ed.) Getty, R. Vol. 2, 5th Edn. Philadelphia: Saunders.

Kitoh, J. (1962) Comparative and topographical anatomy of the fowl. XII. Observations on the arteries with their anastomoses in and around the brain in the fowl. *Jap. J. vet. Sci.*, 24, 141–150.

Kitoh, J. (1964) Comparative and topographical anatomy of the fowl. XVI. Arterial supply of the spinal cord. *Jap. J. vet. Sci.*, 26, 169–175.

Lindsay, F.E.F. (1967) The cardiac veins of *Gallus domesticus*. *J. Anat.*, 101, 555–568.

Lindsay, F.E.F. & Smith, H.J. (1965) Coronary arteries of *Gallus domesticus*. *Am. J. Anat.*, 116, 301–314.

Lucas, A.M. & Jamroz, C. (1961) *Atlas of Avian Haematology*. Agriculture Monograph No. 25. Washington: US Department of Agriculture.

Malinovsky, L. (1965) Contribution to the comparative anatomy of the vessels in the abdominal part of the body cavity in birds. III. Nomenclature of branches of the a. coeliaca and of tributaries of the v. portae. *Folia Morph.*, 13, 252–264.

Malinovsky, L. & Visnanska, M. (1975) Branching of the coeliac artery in some domestic birds. II. The domestic goose. *Folia Morph.*, 23, 128–135.

Manning, P.J. & Middleton, C.C. (1972) Atherosclerosis in wild turkeys: morphological features of lesions and lipids in serum and aorta. *Am. J. vet. Res.*, 33, 1237–1246.

Maxwell, M.H. (1973) Comparison of heterophil and basophil ultrastructure in six species of domestic birds. *J. Anat.*, 115, 187–202.

Maxwell, M.H. (1974) An ultrastructural comparison of the mononuclear leucocytes and thrombocytes in six species of domestic bird. *J. Anat.*, 117, 69–80.

Maxwell, M.H. & Trejo, F. (1970) The ultrastructure of white blood cells and thrombocytes of the domestic fowl. *Br. vet. J.*, 126, 583–592.

Miyaki, T. (1978) The afferent venous vessels to the liver and the intrahepatic portal distribution in the fowl. *Zbl. vet. Med.*, C, 7, 129–139.

Nishida, T. (1960) Comparative and topographical anatomy of the fowl. II. On the blood vascular system of the thoracic limb in the fowl. Part 1. The artery. *Jap. J. vet. Sci.*, 22, 223–331.

Nishida, T. (1963) Comparative and topographical anatomy of the fowl. X. The blood vascular system of the hind limb in the fowl. Part 1. The artery. *Jap. J. vet. Sci.*, 24, 93–106.

Nishida, T. (1964) Comparative and topographical anatomy of the fowl. XLII. Blood vascular system of the male reproductive organs. *Jap. J. vet. Sci.*, 26, 211–221.

Nishida, T. & Mochizuki, K. (1976) The venous system of the proventriculus of duck (*Anas domesticus*). *Jap. J. vet. Sci.*, 38, 255–262.

Olson, C. (1959) Avian hematology. In *Diseases of Poultry* (Ed.) Biester, H.E. & Schwarte, L.H. 4th Edn. Ames, Iowa: The Iowa State University Press.

Paik, Y.K., Nishida, T. & Yasuda, M. (1969) Comparative and topographical anatomy of the fowl. LVII. The blood vascular system of the pancreas in the fowl. *Jap. J. vet. Sci.*, 31, 241–251.

Powell, F.L. (1983) Circulation. In *Physiology and Behaviour of the Pigeon* (Ed.) Abs, M. New York and London: Academic Press.

Richards, S.A. (1967) Anatomy of the arteries of the head in the domestic fowl. *J. Zool., Lond.*, 152, 221–234.

Richards, S.A. (1968) Anatomy of the veins of the head in the domestic fowl. *J. Zool., Lond.*, 154, 223–234.

Rigdon, R.H. & Frolich, J. (1970) The heart of the duck. *Zbl. Vet. Med.*, A, 17, 85–94.

Scott, T.M. (1971) The ultrastructure of ordinary and Purkinje cells of the fowl heart. *J. Anat.*, 110, 259–273.

Siller, W.G. & Hindle, R.M. (1969) The arterial blood supply to the kidney of the fowl. *J. Anat.*, 104, 117–135.

Simon, J.R. (1960) The blood vascular system. In *Biology and Comparative Physiology of Birds* (Ed.) Marshall, A.J. Vol. 1. New York and London: Academic Press.

Steen, I. & Steen, J.B. (1965) The importance of the legs in the thermoregulation of birds. *Acta physiol. scand.*, 63, 285–291.

Vitums, A., Mikami, S.-I. & Farner, D.D. (1965) Arterial blood supply to the brain of the White-crowned Sparrow, *Zonotrichia leuchophrys gambelii*. *Anat. Anz.*, 116, 309–326.

West, N.H., Langille, B.L. & Jones, D.R. (1980) Cardiovascular system. In *Form and Function in Birds* (Ed.) King, A.S. & McLelland, J. Vol. 2. London and New York: Academic Press.

Westpfahl, U. (1961) Das Arteriensystem das Haushuhnes (*Gallus domesticus*). *Wiss. Z. Humboldt-Univ. Berlin, Math.-Nat. R.*, 10, 93–124.

Yousuf, N. (1965) The conducting system of the heart of the House Sparrow, *Passer domesticus indicus*. *Anat. Rec.*, 152, 235–250.

Chapter 14

LYMPHATIC SYSTEM

The lymphatic vessels are mainly concerned with returning extravascular fluids to the blood. The lymphoid tissue is responsible for adaptive immunity.

LYMPHATIC VESSELS

Lymphatic vessels are present, as in mammals, but are relatively less numerous. With very few exceptions they closely follow the blood vessels, usually accompanying the arteries within the trunk (Fig 14–1) and the veins outside the trunk. Typically, there are two lymphatic vessels to each blood vessel although only one is shown in Fig 14–1. Retrograde flow in the lymphatic vessels is prevented by valves; since the valves are relatively less frequent than in mammals the lymphatic vessels of birds do not have a beaded appearance. A pair of *lymphatic hearts* with walls lined by striated muscle fibres occurs in the caudal abdominal region of a few species including some ratites and anseriforms, although only in the ratites do the hearts appear to be contractile. A pair of such lymphatic hearts is located on the lymphatic vessels accompanying the internal iliac vein in the chick embryo, but they disappear with maturity in the domestic fowl, as in most other species.

There is usually a pair of thoracoabdominal trunks (thoracic ducts), with frequent anastomoses between them (Fig 14–1). These receive the lymphatic drainage from the hindlimb via the external iliac and the ischiadic lymphatic vessels; they also drain the abdominal viscera via the coeliac, cranial mesenteric and caudal mesenteric lymphatic vessels; finally, they drain the pelvic region and cloacal bursa via the sacral, pudendal and cloacal lymphatic vessels (Fig 14–1). The jugular lymphatic vessel, which accompanies the jugular vein, drains the head and neck. The wing is drained by the subclavian lymphatic vessel, emptying into the subclavian vein (Fig 14–1). After draining the deep pulmonary lymphatics, the lymphatic vessels of each lung follow the left and right pulmonary veins and unite to form the single common pulmonary lymphatic vessel which discharges into the left cranial vena cava (Fig 14–1). The left and right cardiac lymphatic vessels are not associated with the coronary arteries or veins but combine into a common cardiac vessel opening into the right cranial vena cava (Fig 14–1). Other lymphatics named accord-

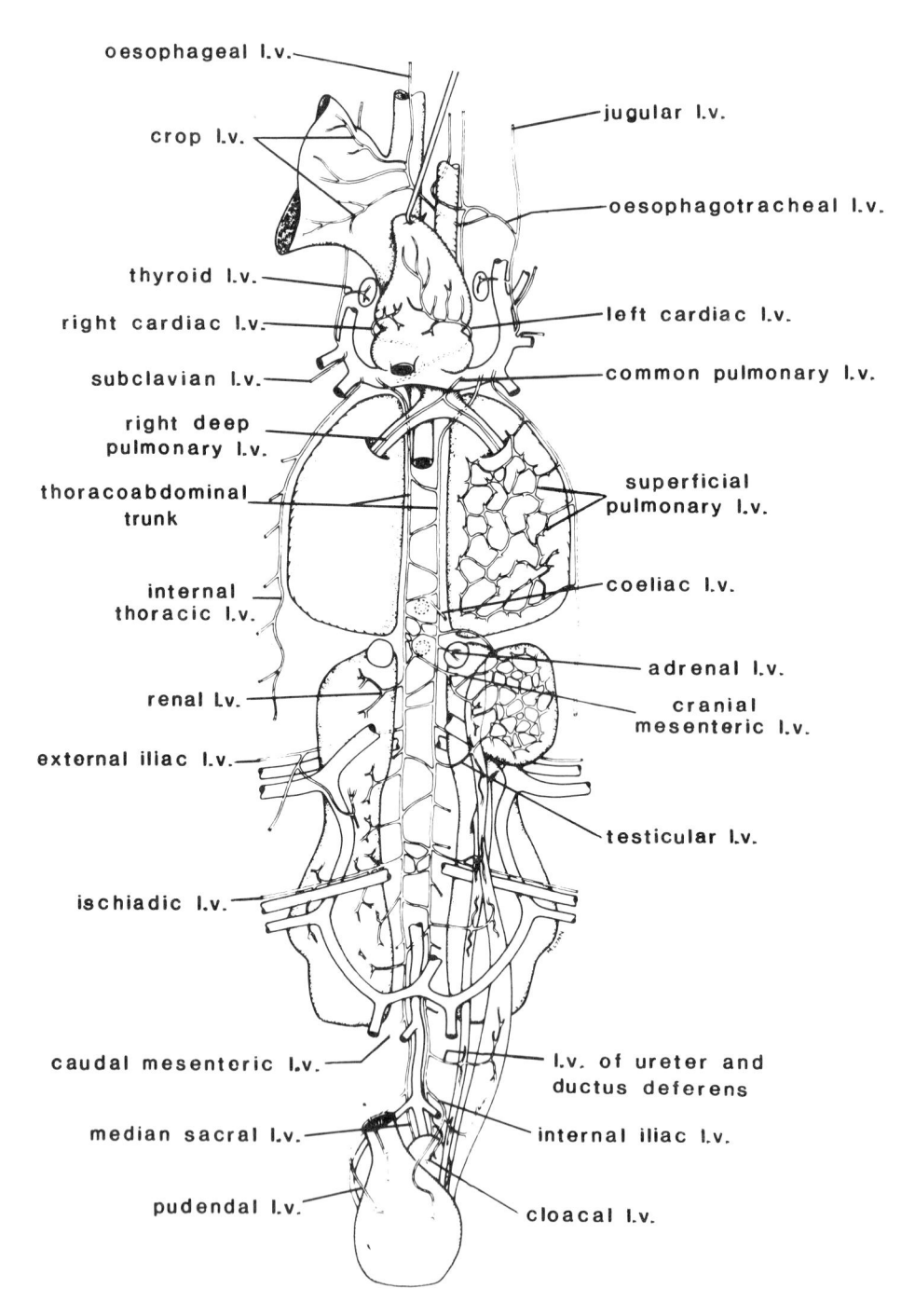

Fig 14–1 Ventral view diagram of the principal lymphatic vessels (l.v.) of the domestic fowl. From Payne (1979), with kind permission of the publisher.

ing to the organs they drain are shown in Fig 14–1. The ultimate drainage of all lymphatic vessels is into the left and right cranial venae cavae.

LYMPHOID TISSUE

Thymus

The thymus consists of three to eight pale-pink, flattened, irregularly shaped lobes (each about 1 cm long in the domestic fowl) strung along each side of the neck close to the jugular vein (Fig 14–2). In the domestic fowl these reach their

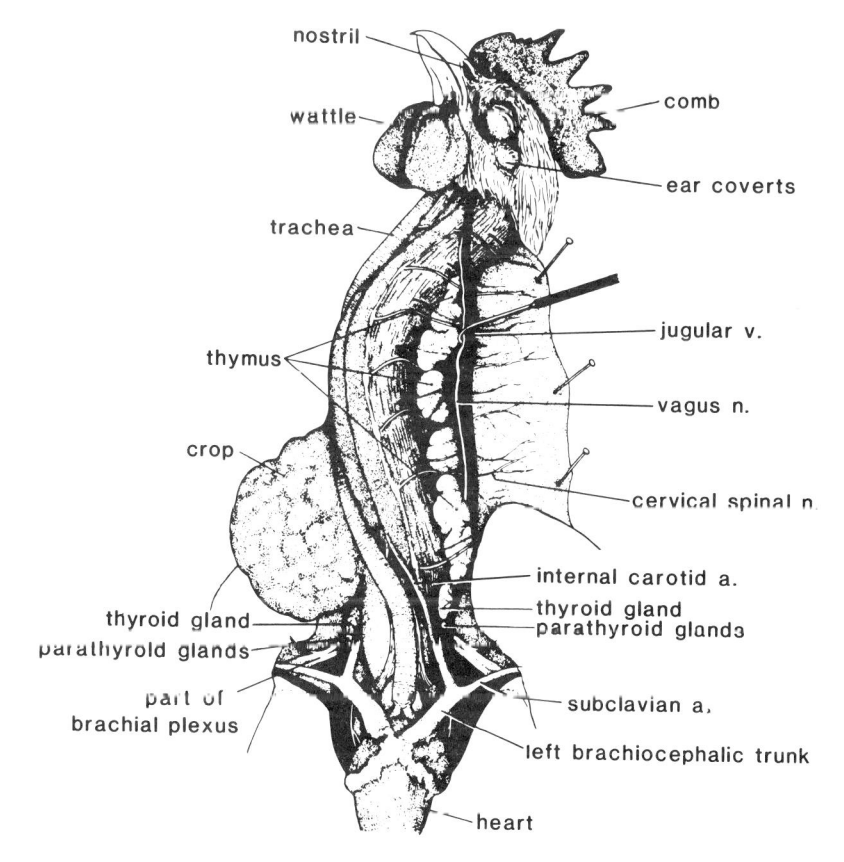

Fig 14–2 The neck of a young male domestic fowl. From Nonidez and Goodale (1927).

maximum size between 4 and 17 weeks of age and then begin to involute. In wild birds the thymus may re-enlarge after the annual breeding cycle. Its histological structure resembles that of the mammal. Each lobe consists of lobules which are partly separated by connective tissue. The lobule consists of an outer dark cortex and an inner pale medulla; both of these components are formed from a framework of scattered epithelial reticular cells and their reticular fibres, containing masses of small lymphocytes which are less densely packed in the medulla than in the cortex. Islands of reticular cells (thymic or Hassal's corpuscles) occur in the medulla; unlike those of mammalian corpuscles, the cells are usually vacuolated rather than laminated. Thymic (T) lymphocytes are derived from stem cells which in the early embryo are produced in the yolk sac, circulate in the blood and are attracted to the thymus by its developing epithelium; progenitor cells have also been found in the bone marrow. Having reached the thymus the stem cells then differentiate into immunologically competent T lymphocytes. However, the thymus does not consist exclusively of T lymphocytes, but contains a small number of B lymphocytes also.

Cloacal bursa (bursa of Fabricius)

The cloacal bursa is unique to birds. It consists of a dorsal median diverticulum of the proctodeum (Figs 6–10 and 11–2), and often contains traces of faecal material. In most birds the bursa is pear-shaped, as in the domestic fowl and the passeriform species that have been examined. In other birds, for example the anseriform species, it is an elongated spindle-shaped structure. In the turkey it is also more or less spindle-shaped. The bursa reaches its absolute maximum size before the bird is fully mature; in the domestic fowl this size is attained at between 4 and 12 weeks of age depending on the strain, with measurements up to $3 \times 2 \times 1$ cm and a weight of nearly 4 g. The wall of the fully developed bursa is usually thick as in the domestic fowl, but in passeriforms and psittaciforms it is thin and saclike. In most species the internal cavity extends the whole length of the bursa.

The internal surface either consists of thick longitudinal folds as in the domestic fowl, or is riddled with fossae separated by broad septa as in pigeons, some passeriforms and the ratites. In the domestic fowl the internal wall shows about 12 folds. In each fold the lymphoid tissue is separated by connective tissue into follicles each of which consists, as in most birds, of an outer cortex and inner medulla. The cortex stains more deeply than the medulla because it contains a relatively larger number of closely packed small lymphocytes. Separating the cortex from the medulla is a narrow and inconspicuous layer of epithelial cells. The internal lumen is lined by columnar epithelium. The cloacal bursa is the site of differentiation of immunologically competent bursal (B) lymphocytes. The initial source of B lymphocytes seems to be mainly from blood-borne stem cells derived from the yolk sac. Progenitor cells may also arise from the embryonic spleen, and from the bone marrow at a later stage. A few T lymphocytes are also found in the bursa.

Involution of the bursa begins at about two to three months of age in the domestic fowl and is advanced by the time of sexual maturity, the weight of the bursa being reduced to about 0.5 g in the domestic fowl at five months. During involution lymphocytes are lost from the cortex and medulla, and the epithelial layer becomes more conspicuous. Cavities form within the follicles and become continuous with the internal lumen of the bursa. In the goose involution is slower, since sexual maturity may not be reached until two years. In all species a remnant of the bursa persists for a long while after involution, as a tiny globule or a thin band depending on whether the initial form was pear-shaped or spindle-shaped.

In the Ostrich and other ratites the neck of the bursa has a very wide lumen, in contrast to the narrow lumen in all the other species. Consequently, in the ratites the proctodeum and cloacal bursa form a single large cavity. It has long been believed that the ratites differ from all other birds in retaining the bursa at its full size for life. This supposition, combined with the unusually wide entrance of the bursa, probably explains the commonly held view that the ratites have a urinary bladder. However, the bursa does in fact involute to the point of disappearance in these birds just as in the other species, and the wide entrance to the bursa narrows as the bursa atrophies.

Lymph nodes

True lymph nodes occur only in certain aquatic birds including the duck and goose. In these species there are two pairs, one pair (cervicothoracic nodes) lying near the thyroid gland, and the other pair (lumbar) near the kidneys. Each node is spindle-shaped and about 1 to 3 cm long. The microscopic structure of these nodes differs from that of mammals. The main afferent vessel continues directly into a central sinus which runs through the length of the node. At the efferent end of the node the central sinus subdivides into many small sinuses which finally empty into the efferent vessel. The central sinus opens into a network of lymphatic spaces lined by a discontinuous endothelium. Loose lymphoid tissue containing small and medium-sized lymphocytes surrounds the lymphatic spaces. Filtration occurs in these spaces but is probably less efficient than in the mammalian lymph node, although the flow of lymph is faster in the avian node.

Mural lymphoid nodules

These are microscopic nodules of lymphoid tissue situated at irregular intervals (a few millimetres to several centimetres apart in the domestic fowl) in the walls of all the lymph vessels. Each nodule is embedded in one side of the wall of the vessel and sometimes protrudes into the lumen. Therefore there is essentially a central sinus and peripheral lymphoid tissue, as in the true nodes of aquatic birds. Each larger nodule has three or four germinal centres. Mural lymphoid nodules can have little or no capacity for filtration. They have been found in a variety of domestic and wild species.

Solitary and aggregated lymphoid nodules

Variable amounts of lymphoid tissue including both solitary and aggregated nodules occur in virtually all avian tissues and organs.

Solitary lymphoid nodules are not encapsulated around their periphery, but some contain germinal centres which are lightly encapsulated. The normality of these solitary nodules has been much argued, and it has been suggested that they represent an abnormal response to lymphomatosis virus. However, the consensus of opinion is that they should be regarded as normal, in the sense that they represent the natural response of the bird to its day-to-day contact with the environment.

Aggregated lymphoid nodules are very numerous in the digestive tract. The most prominent are the paired 'caecal tonsils' in the wall of each caecum near the junction with the rectum. Each consists of numerous large clearly circumscribed germinal centres and a diffuse mass of dense lymphoid tissue in which small lymphocytes predominate and many plasma cells are present. Relatively prominent aggregated nodules also occur in the oropharynx around the choanal opening and the pharyngeal opening of the auditory tubes, in the small intestine, and possibly also in the caudal end of the oesophagus in some species, including the domestic duck and goose.

Spleen

The dark-red spleen lies in all birds on the right side of the junction between the proventriculus and gizzard (Fig 6–12). In the domestic fowl it is a nearly spherical organ about 2 cm in diameter, but in the domestic duck and goose it is more triangular with a flat dorsal surface and a rounded ventral surface. Small accessory spleens have been found in the domestic fowl adherent to the coeliac artery and mesentery. Histologically the avian spleen resembles that of the mammal, although its capsule is thin and it is not subdivided by well defined trabeculae. Its basic component is a scaffolding of reticular fibres and reticular cells. The white pulp consists of typical lymphoid tissue surrounding the arteries. The red pulp consists of venous sinuses separated by cords of cells which include lymphocytes, macrophages and the elements of the circulating blood. The distinction between red and white pulp is less obvious than in mammals. The splenic circulation in birds is believed to be 'open', no direct vascular connection between the arteries and veins being apparent. The functions of the adult spleen include phagocytosis of worn out erythrocytes in the red pulp, lymphopoiesis in the white pulp and antibody production in both the red and white pulps. In contrast to mammals, the spleen seems not to be a significant reservoir of blood.

Adaptive immunity

There are two components of adaptive immunity in birds, one bursa-dependent and the other thymus-dependent. The cloacal bursa and thymus constitute the primary sites for the development of lymphocytes, B lymphocytes in the bursa and T lymphocytes in the thymus. The immunologically competent B and T lymphocytes of the primary lymphoid tissue migrate to the secondary lymphoid tissue in the spleen, bone marrow, lymph nodes, mural lymphoid nodules, and the solitary and aggregated lymphoid nodules. Secondary lymphoid tissue, unlike primary tissue, is antigen-driven and is therefore the effector tissue.

The thymus and the thymus-dependent areas of secondary lymphoid tissue are responsible for the cell-mediated immune response, i.e. delayed and contact hypersensitivity, graft rejection and graft-versus-host reactions. The bursa and the bursa-dependent areas of secondary lymphoid tissue are responsible for the synthesis of circulating antibodies and are therefore the main defence against invading microorganisms. Besides sharing the general immune responses, the caecal aggregations and other intestinal aggregated nodules seem to have a local immunological function against bacteria and other antigenic agents in the gut.

Further reading

Aitken, I.D. (1974) Avian Immunology. In *Progress in Immunology, II* (Ed.) Brent, L. & Holborrow, J. Vol. 2. Amsterdam: North-Holland Publishing Company.

Bacchus, S. & Kendall, M.D. (1975) Histological changes associated with enlargement and regression of the thymus glands of the Red-billed Quelea *Quelea quelea* L. (Ploceidae: weaver birds). *Phil. Trans. R. Soc., B*, 273, 65–73.

Bell, R.G. & Lafferty, K.J. (1972) The flow and cellular composition of cervical lymph from unanaesthetized ducks. *Aust. J. exp. Biol. med. Sci.*, 50, 611–623.

Biggs, P.M. (1957) The association of lymphoid tissue with the lymph vessels in the domestic chicken (*Gallus domesticus*). *Acta anat.*, 29, 36–47.

Dransfield, J.W. (1945) The lymphatic system of the domestic fowl. *Br. Vet. J.*, 101, 171–179.

Durkin, H. G., Theis, G.A. & Thorbecke, G. J. (1972) Bursa of Fabricius as site of origin of germinal centre cells. *Nature, Lond.*, 235, 118–119.

Firth, G.A. (1977) The normal lymphatic system of the domestic fowl. *Vet. Bull.*, 47, 167–179.

Frazier, J.A. (1973) Ultrastructure of the chick thymus. *Z. Zellforsch. mikrosk, Anat.*, 136, 191–205.

Frazier, J.A. (1974) The ultrastructure of the lymphoid follicles of the chick bursa of Fabricius. *Acta anat.*, 88, 385–397.

Freeman, B.M. (1971) The endocrine status of the bursa of Fabricius and the thymus gland. In *Physiology and Biochemistry of the Domestic Fowl* (Ed.) Bell, D.J. & Freeman, B.M. Vol. 1. London and New York: Academic Press.

Fukuta, K., Nishida, T. & Yasuda, M. (1969) Comparative and topographical anatomy of the fowl. LVI. Blood vascular system of the spleen in the fowl. *Jap. J. Vet. Sci.*, 31, 179–185.

Fukuta, K., Nishida, T. & Yasuda, M. (1969) Comparative and topographical anatomy of the fowl. LXIII. Structure and distribution of the fine blood vascular system in the spleen. *Jap. J. Vet. Sci.*, 31, 303–311.

Glick, B. (1977) The bursa of Fabricius and immunoglobulin synthesis. In *International Review of Cytology* (Ed.) Bourne, G.H. & Danielli, J.F. Vol. 48. New York and London: Academic Press.

Glick, B., Holbrook, K.A., Olah, I., Perkins, W.D. & Stinson, R. (1978) A scanning electron microscope study of the caecal tonsil: the identification of a bacterial attachment to the villi of the caecal tonsil and the possible presence of lymphatics in the caecal tonsil. *Poult. Sci.*, 57, 1408–1416.

Glick, B., Holbrook, K.A. & Perkins, W.D. (1977) Scanning electron microscopy of the bursa of Fabricius from normal and testosterone-treated embryos. *J. Develop. comp. Immunol.*, 1, 41–46.

Glick, B. & Sato, K. (1964) Accessory spleens in the chicken. *Poult. Sci.*, 43, 1610–1612.

Hashimoto, Y. & Sugimura, M. (1976) Histological and quantitative studies on the postnatal growth of the thymus and the bursa of Fabricius of White Pekin ducks. *Jap. J. Vet. Sci.*, 24, 65–76.

Hodges, R.D. (1974) *The Histology of the Fowl*. London and New York: Academic Press.

Höhn, E.O. (1961) Endocrine glands, thymus and pineal body. In *Biology and Comparative Physiology of Birds* (Ed.) Marshall, A.J. Vol. II. New York and London: Academic Press.

King, A.S. (1975) Aves lymphatic system. In *Sisson and Grossman's The Anatomy of the Domestic Animals* (Ed.) Getty, R. Vol. 2, 5th Edn. Philadelphia: Saunders.

King, A.S. (1980) Cloaca. In *Form and Function in Birds* (Ed.) King, A.S. & McLelland, J. Vol. 2. London and New York: Academic Press.

Miyaki, T. & Yasuda, M. (1977) On the thoracic duct and the lumbar lymphatic vessel in the fowl. *Jap. J. Vet. Sci.*, 39, 559–570.

Mueller, A.P., Sato, K. & Glick, B. (1971) The chicken lacrimal gland, gland of Harder, caecal tonsil, and accessory spleens as sources of antibody-producing cells. *Cell. Immunol.*, 2, 140–152.

Nonidez, J.F. & Goodale, H.D. (1927) Histological studies on the endocrines of chickens deprived of ultraviolet light. *Am. J. Anat.*, 38, 319–341.

Panigraphi, D., Waxler, G.L. & Mallman, V.H. (1971) The thymus in the chicken and its anatomical relationship to the thyroid. *J. Immun.*, 107, 289–292.

Payne, L.N. (1971) The lymphoid system. In *Physiology and Biochemistry of the Domestic Fowl* (Ed.) Bell, D.J. & Freeman, B.M. Vol. 2. London and New York: Academic Press.

Payne, L.N. (1979) Systema lymphaticum et splen. In *Nomina Anatomica Avium* (Ed.) Baumel, J.J., King, A.S., Lucas, A.M., Breazile, J.E. & Evans, H.E. London and New York: Academic Press.

Potworowski, E.F. (1972) T and B lymphocytes: organ and age distribution in the chicken. *Immunology*, 23, 199–204.

Rose, M.E. (1980) Lymphatic system. In *Form and Function in Birds* (Ed.) King, A.S. & McLelland, J. Vol. 2. London and New York: Academic Press.

Sorvari, R., Naukkarinen & Sorvari, T.E. (1977) Anal sucking-like movements in the chicken and chick embryo followed by the transportation of environmental material to the bursa of Fabricius, caeca and caecal tonsils. *Poult. Sci.*, 56, 1426–1429.

Waltenbaugh, C.R. & Van Alten, P.J. (1974) The production of antibody by bursal lymphocytes. *J. Immun.*, 113, 1079–1084.

Ward, P. & Kendall, M.D. (1975) Morphological changes in the thymus of young and adult red-billed queleas (*Quelea quelea*) (*Aves*). *Phil. Trans. R. Soc.*, 273, 55–64.

Warner, N.L. (1967) The immunological role of the avian thymus and bursa of Fabricius. *Folia biol., Praha*, 13, 1–17.

Weber, W.T. & Mausner, R. (1977) Migration patterns of avian embryonic bone marrow cells and their differentiation to functional T and B cells. In *Avian Immunology* (Ed.) Benedict, A.A. *Advances in Experimental Medicine and Biology*, 88. New York and London: Plenum Press.

Chapter 15

NERVOUS SYSTEM

SPINAL CORD

Meninges

The avian spinal cord is enclosed, protected and supported by the three meninges, i.e. the dura, arachnoid and pia mater, essentially as in mammals. The *dura mater* is the tough, relatively thick, outermost layer. In the cervical and thoracic regions it is separated from the periosteal lining of the vertebral canal by an epidural space. This space is said to be filled by a gelatinous substance which enables the spinal cord and dura mater to adapt themselves to the great mobility of the neck. Towards the caudal end of the thoracic region the dura fuses with the periosteum, and they remain fused as far as the caudal end of the vertebral canal. The dura is also fused to the periosteum at the foramen magnum. The internal vertebral venous sinus runs the length of the vertebral canal, except for the region of the lumbar enlargement of the spinal cord where it is absent. It lies in the dorsal epidural space in the cervical and thoracic regions, and elsewhere it intervenes between the periosteum and dura mater. The *arachnoid mater* is a delicate membrane in more or less close contact with the dura mater. The *pia mater* firmly coats the spinal cord and contains many blood vessels. Between the pia and arachnoid is the *subarachnoid space* which is traversed by a network of fine filaments of the pia-arachnoid. The pia mater is thickened laterally and in the ventral midline into lateral and ventral longitudinal ligaments. The paired lateral ligaments project like a shelf along each side of the spinal cord, separating the dorsal and ventral roots of the spinal nerves. In the intervals between these roots they attach to the dura by toothlike *denticulate ligaments* as in mammals. The left and right denticulate ligaments thus suspend the spinal cord from the dura mater.

Macroscopic anatomy of the spinal cord

Unlike the mammalian spinal cord, the avian spinal cord is almost exactly the same length as the neural canal. Consequently, the spinal nerves pass laterally rather than caudally to their intervertebral foramina and there is no cauda equina. The fine dorsal and ventral rootlets of each spinal nerve pierce the dura separately and come together in the intervertebral foramen.

There are two distinct enlargements of the spinal cord, cervical and lumbosacral, that are associated with the brachial plexus and lumbosacral plexus, respectively. In flying birds the *cervical enlargement* is greater than the lumbar, but in the large flightless birds, some of which can run very energetically, the *lumbosacral enlargement* tends to exceed the cervical. In the dorsal midline of the lumbosacral enlargement is the *rhomboidal sinus*, a structure that is unique to birds. In the region of this structure the left and right dorsal columns of the spinal cord separate in the midline leaving a cleft which is occupied by the *gelatinous body* (or glycogen body). The gelatinous body consists of glial cells which are rich in glycogen and innervated by unmyelinated nerve fibres, but are of unknown function. The ventral part of the gelatinous body encloses the central canal and is continuous with neural tissues of the spinal cord. It has been claimed that there is also a brachial 'glycogen' body and that a central glycogen-rich zone may extend throughout the length of the spinal cord. In the region of the lumbosacral enlargement the ventral horn of grey matter is increased in size, thus causing a series of bulges on the ventrolateral surface of the cord from which the ventral rootlets arise. In the lumbosacral enlargement there are also small lateral projections from the lateral column, just dorsal to the denticulate ligament, which are caused by the marginal nuclei.

A ventral longitudinal *median fissure* is present throughout the length of the spinal cord, but only a barely perceptible *dorsal sulcus* is visible in the dorsal midline except at the rhomboidal sinus.

Internal anatomy of the spinal cord

The general anatomy of the grey and white matter is the same as in mammals, with a central 'butterfly' of grey matter surrounded by white matter, except for the presence of the outlying masses of grey matter known as the marginal nuclei.

White matter. The white matter is divided into three columns. The *dorsal column* lies between the dorsal median septum on the one hand, and the dorsal horn and the line of attachment of the dorsal rootlets of the spinal nerves on the other hand (Fig 15–1). The *lateral column* is the area of white matter between the dorsal and ventral rootlets, and the dorsal and ventral horns. The *ventral column* lies between the ventral horn and ventral rootlets, and the ventral median fissure. In general the ventral and lateral columns of white matter are relatively very large (accounting for about 60–70 per cent of the total cross-sectional area of the cord) and the dorsal column is relatively very small, compared to mammals. The dorsal column has much the same cross-sectional area in the regions between the lumbosacral and cervical enlargements, and between the cervical enlargement and the brain. This suggests that in birds many of the axons in the dorsal column must be quite short and travel in it for only a few segments; in contrast, the mammalian dorsal column gets progressively more bulky cranially as it receives more and more tactile and kinaesthetic axons on their way to the cuneate and gracile nuclei in the medulla oblongata.

The dorsal median septum and the ventral median fissure do not reach the grey matter. Consequently, there is a *dorsal white commissure* dorsal to the grey matter, and a *ventral white commissure* ventral to the grey matter. Each of these commissures allows nerve fibres to cross from one side of the spinal cord to the other. The dorsal commissure contains fibres from primary afferent neurons in the dorsal roots, as well as fibres arising from nerve cell bodies in the grey matter. The ventral commissure contains nerve fibres which are crossing over to form ascending pathways such as spinothalamic and spinoreticular tracts.

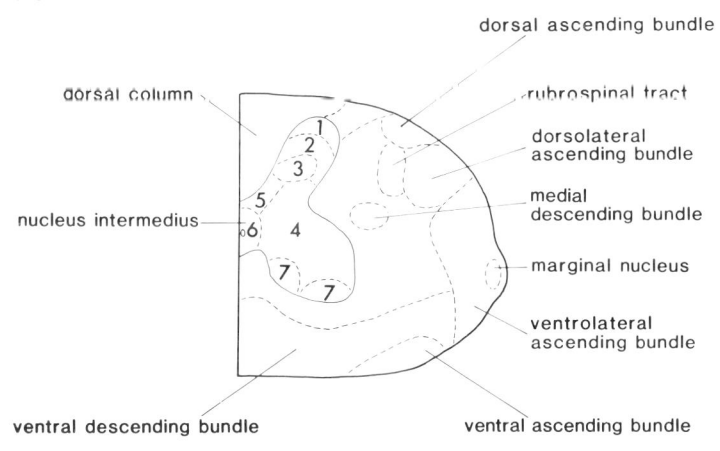

Fig 15–1 Diagrammatic transverse section of the right half of the spinal cord of the pigeon. The ascending bundles consist of afferent nerve fibres, and the descending bundles are efferent nerve fibres. 1–7 = areas of the grey matter. Based on Akker (1970).

Grey matter. In transverse sections of the two enlargements of the spinal cord the ventral horn has a much greater area than the dorsal horn (Fig 15–1), especially in the lumbosacral enlargement of the large flightless species, and it also projects further laterally into the white matter.

In the pigeon, where the grey matter has been relatively well explored, seven 'areas' of grey matter can be recognized (Fig 15–1). As knowledge of the avian spinal cord grows these areas begin to show some resemblance to Rexed's laminae in the cat. For example, area 5 in the pigeon includes large cells and is particularly prominent in the cranial regions of the two enlargements; it appears to correspond to Rexed's lamina VI which contains Clarke's column in segments T1 to L2. Both area 5 in the bird and Clarke's column in mammals contribute to the spinocerebellar pathways. Area 6 of the pigeon includes the nucleus intermedius (previously the column of Terni) which, despite its medial position (Fig 15–1), appears to be the cell station of the preganglionic sympathetic outflow extending from segments 14 to 22, i.e. essentially thoracolumbar segments (as in mammals). The marginal nuclei of the spinal cord form an almost continuous column of grey matter but are much more bulky between the rootlets of the spinal nerves. They consist of multipolar neurons like ventral horn cells but smaller. They may be displaced ventral horn motor neurons, but are more likely to be dislocated ventral commissural neurons projecting from one side of the cord to the other.

Ascending pathways. The anatomy of the ascending pathways in the avian spinal cord has not yet been extensively worked out, and the little that is known is largely based on the pigeon. However, there are at least five regions of the white matter (Fig 15–1) which have been shown by degeneration experiments to contain long ascending fibres. The connections of three of these, the dorsal column, the dorsolateral ascending bundle and the ventrolateral ascending bundle, are relatively well known, so that their probable homologies to mammalian ascending tracts can be suggested.

The *dorsal column* contains ascending fibres from primary afferent neurons with their cell bodies in the dorsal root ganglia. These fibres are arranged somatopically essentially as in mammals, those from caudal levels tending to lie medial to those from more cranial levels. However, as already stated, the uniform size of the dorsal column throughout the length of the spinal cord indicates that many of its axons must be quite short. Of those that do arrive at the brainstem many only reach the medulla oblongata where they end in the cuneate nucleus and gracile nucleus, with a synaptic relay to the thalamus. The ascending fibres of the dorsal column represent a part of the somatosensory system of birds. The exact range of modalities which it projects has not been fully established. Nevertheless, the general anatomical resemblance of the dorsal column and of its two brainstem nuclei in birds to the cuneate and gracile tracts and the cuneate and gracile nuclei of mammals strongly suggests that these avian pathways have the same function as in mammals, i.e. to transmit the modalities of touch, pressure and kinaesthesia (joint proprioception).

The *dorsolateral ascending bundle* in Fig 15–1, with a contribution from the ventrolateral ascending bundle, has been shown to contain a *dorsal spinocerebellar tract*, which is activated by muscle receptors. Its fibres arise from nerve cell bodies in area 5 of the grey matter, ascend ipsilaterally and enter the cerebellum through the (caudal) cerebellar peduncle; if area 5 of the bird corresponds to lamina VI in the mammal and is therefore related to Clarke's column, then these are all features suggesting homology with the mammalian dorsal spinocerebellar tract. However, in the bird this pathway is confined to the wing region of the spinal cord, whereas in mammals the dorsal spinocerebellar pathway is confined to the hindlimb and trunk, so the homology is not complete. In the mammalian spinal cord the spinocuneocerebellar pathway is the forelimb equivalent of the hindlimb dorsal spinocerebellar tract, but there seems to be no evidence that such a pathway serves the avian wing muscles.

The avian *ventrolateral ascending bundle* (Fig 15–1) evidently contains a *ventral spinocerebellar tract*. In both birds and mammals this tract is activated by muscle afferents and arises only in the hindlimb region of the spinal cord; it decussates in the cord, projects monosynaptically (or perhaps polysynaptically in birds) to the cerebellum, and finally enters a cerebellar peduncle (the rostral peduncle in mammals). At least some of these fibres in the bird, like many in the mammal, appear to decussate a second time in the commissure of the cerebellum, thus amounting functionally to an ipsilateral system. If the avian cerebellum is to exert an essentially ipsilateral regulation of motor activity in the wings and legs as it does in mammals, either there must be a double decussation or no decussation at all. In mammals the cranial spinocerebellar tract is the forelimb equivalent of the ventral spinocerebellar tract, but its presence in birds has not been directly demonstrated.

A distinct group of fine fibres caps the tip of the avian dorsal horn and is known as the *dorsolateral fasciculus*. The tip of the dorsal horn is the *nucleus of the substantia gelatinosa*. In mammals this fasciculus (also known as Lissauer's tract) consists of short intersegmental fibres and is integrated with the substantia gelatinosa in the transmission of pain.

Degeneration experiments on pigeons have shown that ascending fibres arise from the dorsal horn, decussate and then ascend in the lateral column to project directly to the thalamus. This suggests the existence of an avian homologue of the mammalian *spinothalamic tract*. As such it should mediate pinprick pain, temperature and touch, but evidence for only tactile transmission seems to be available for birds. There is also evidence from degeneration experiments for well-developed *spinoreticular pathways*, ascending bilaterally from spinal levels to the reticular formation of the medulla oblongata, pons and mesencephalon. From a knowledge of mammalian spinoreticular pathways it must be supposed that this apparently homologous system in birds would project somatosensory modalities and especially the sensation of true (as opposed to pinprick) pain. Ascending pathways in the ventral column appear to be restricted to short intersegmental fibres of the *propriospinal system* (fasciculus proprius).

Descending pathways. Even less is known about the descending pathways. Several regions shown in Fig 15–1 are known from degeneration experiments to contain fairly long descending fibres in the pigeon. The majority of them appear to be spinospinal. The lateral column contains rubrospinal and lateral reticulospinal pathways. The *rubrospinal tract* is probably homologous to its mammalian counterpart. Its fibres appear to originate in the red nucleus in the midbrain, to descend the whole length of the spinal cord, and to end in the middle and ventral region of area 4 in Fig 15–1 (equivalent to Rexed's laminae V, VI, VII in mammals). These endings are near motoneurons innervating mainly flexor muscles. The fibres of the *lateral reticulospinal pathway* end in the nucleus intermedius (Fig 15–1), and should therefore have visceral motor functions (as would the similarly situated lateral reticulospinal tract in mammals).

Other experimental studies have suggested the presence of a *cerebrospinal system* of long fibres descending from the archistriatum of the forebrain and projecting at least to the cervical spinal cord; in its course and distribution this avian motor pathway resembles the pyramidal tract of ungulates, which decussates in the pyramid, descends in the ventral column and dorsal column and ends in the neck.

Descending pathways form most of the ventral column and include vestibulospinal, reticulospinal and tectospinal tracts. There are indications that *vestibulospinal fibres* project from the medial longitudinal bundle of the brainstem into the avian spinal cord, as would be expected from the great phylogenetic age of this system. These fibres lie in two groups in the ventral column, one medial and the other lateral. Both run the whole length of the spinal cord and end mainly in relationship to extensor motoneurons. The medial group is the larger of the two. It appears to be homologous to the mammalian *ventral vestibulospinal tract*. The more lateral group evidently represents the mammalian *lateral vestibulospinal tract*. It regulates alpha and gamma motoneurons, and therefore must be an important factor in somatic motor control of locomotion. The reticulospinal pathway in the ventral column appears to be the *medial reticulospinal tract*. As in mammals it arises mainly from the ipsilateral pontine reticular formation. Its function in birds is not known, but from mammalian parallels it is assumed to play a major role in somatic motor, and perhaps visceral motor, activity. An ill-defined *interstitiospinal* pathway is believed to arise in the mesencephalon and project fibres into the ventral column. It is believed to contain mesencephalic and vestibular components which influence somatic and visceral motor control systems. The *tectospinal tract* appears to project from the optic tectum to the upper segments of the cord as well as to the oculomotor nucleus, and is presumably involved in the movements of the eye and neck in tracking moving objects; its fibres lie in the ventral column.

The main reference books frequently mention a descending cerebellospinal tract in birds. Similar suggestions have often been made for mammals also. *Indirect* pathways from the cerebellum to the spinal cord are commonplace enough. Thus appropriate electrical stimulation of the cerebellum will induce movements of the limbs, but this can be explained by cerebellar feedback to the higher motor centres (e.g. to the red nucleus or the descending reticular formation). *Direct* cerebellospinal pathways far down the spinal cord are inherently improbable, however, since it is axiomatic that the role of the cerebellum is not to initiate movement but only to regulate movement which is already in progress.

Despite stories of decapitated chickens racing round the farmyard, no true locomotor movements of the hindlimbs occur in the spinal bird. Any walking movements occur only immediately after decapitation and last only a short time. Nevertheless the spinal bird does have a relatively great reflex capacity compared to that of mammals, and all the indications are that the cervical and lumbosacral enlargements of the avian spinal cord do have a high degree of autonomy in the coordination of locomotion.

BRAIN

The hindbrain (rhombencephalon) and midbrain (mesencephalon) have evolved from the reptilian brain in a similar manner in birds and mammals. Consequently, in these more caudal parts of the brain many homologous structures can be recognized in birds and mammals among the motor and sensory nuclei and in the main subdivisions of the reticular formation. In contrast, the forebrain (telencephalon and diencephalon) has followed entirely divergent lines of evolution in birds and mammals, so that it is extremely difficult to identify homologous structures in this region. In mammals the sudden emergence of the neocortex on the surface of the cerebral hemisphere, with inevitable repercussions in the thalamus, dominates the architecture of the forebrain: in birds the neuronal population which is homologous to the mammalian neocortex appears to have taken up a totally different position deep within the cerebral hemisphere instead of on the surface (p. 256).

The meninges consist of the dura, arachnoid and pia mater, and these membranes have essentially the same form as in mammals. In the cranial cavity the outer surface of the dura mater fuses with the periosteum, except where the two layers are separated by the dural venous sinuses. Between the forebrain and optic lobes the dura extends as a transverse fold, the tentorial fold. A similar dural fold separates the optic lobes from the cerebellum.

The subarachnoid space lies between the arachnoid and the pia, and expands into a cistern between the cerebellum and the dorsal surface of the medulla oblongata. It is filled with cerebrospinal fluid, which is difficult to collect from the live bird; about 0.5 ml can be obtained from the cistern at the foramen magnum in adult domestic fowl, but only at the risk of copious haemorrhage from the venous sinuses which lie there. In mammals the cavity of the fourth ventricle connects with the subarachnoid space by a pair of lateral foramina, and in some species a median foramen also. In the pigeon and chick there are no such foramina, transfer of water and solutes between the two compartments being by diffusion through the delicate and extensive roof of the fourth ventricle. Arachnoid granulations, enabling drainage of cerebrospinal fluid into the dural venous sinuses, have been observed in pigeons and chicks.

Medulla oblongata and pons: external structure

At the foramen magnum the spinal cord bends sharply in a ventral direction as it expands into the wide medulla oblongata (Fig 15–2a), which is rather globular when seen from the ventral view (Fig 15–2b). The ventral fissure is conspicuous, but the mammalian pyramid and its decussation are lacking. Although there is not an obvious pons, at the rostral end of the hindbrain it is just possible to make out a broad band of transversely directed pontine fibres, which belong to a pontocerebellar pathway. The trapezoid body is not prominent on the ventral surface, but comparable fibres from the cochlear nuclei decussate below the surface.

The cranial nerves from XII to V inclusive arise from the medulla oblongata along two lines, one ventromedial and the other ventrolateral (Fig 15–2b). The ventromedial nerves are XII and VI; these are somatic motor nerves equivalent to the ventral roots of spinal nerves. Their ventromedial line of origin from the brainstem (Fig 15–2b) is continued rostrally by nerve III, another

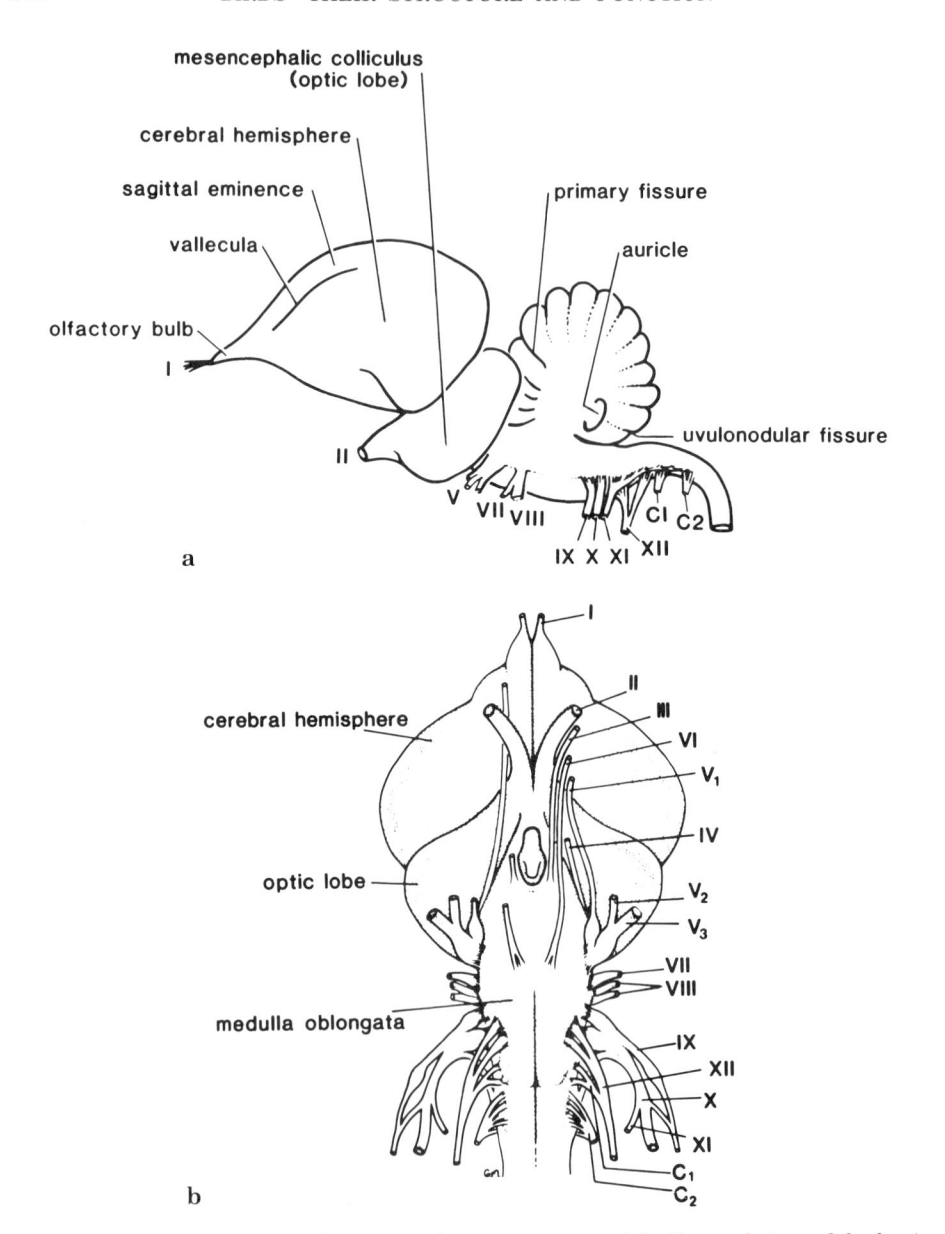

Fig 15–2 a, Lateral view of the brain of the domestic fowl. b, Ventral view of the brain of the domestic fowl. The roman numerals indicate the cranial nerves. C1 and C2 = first and second cervical spinal nerves.

ventral root nerve. The ventrolateral cranial nerves are XI, X, IX, VIII, VII and V. Except for VIII, which is a special somatic afferent nerve, these are the equivalent of the dorsal roots of the spinal nerves of primitive vertebrates and potentially contain somatic afferent, visceral afferent, visceral efferent and branchial (special visceral) efferent fibres.

The nuclei of the cranial nerves

Since the hindbrain and midbrain are a continuation of the spinal cord, their cranial nerve nuclei have features in common with the grey matter of the spinal cord. Thus the nuclei of the cranial nerves occur in the following five rostrocaudal columns, each column having its own particular function. (1) Somatic efferent: the nucleus of the hypoglossal nerve and the nucleus intermedius of the medulla oblongata; the nucleus of the abducent nerve and its accessory nucleus; the nucleus of the trochlear nerve; and the nucleus of the oculomotor nerve. (2) Branchial, or special visceral efferent: the nucleus ambiguus; the nucleus of the facial nerve; and the motor nucleus of the trigeminal nerve. (3) Visceral efferent: the dorsal motor nucleus of the vagus, and possibly the nucleus intermedius of the medulla oblongata; the nucleus of the glossopharyngeal nerve; and the accessory part of the nucleus of the oculomotor nerve. (4) Visceral afferent: the nucleus of the solitary tract. (5) Somatic afferent: the nucleus of the spinal tract of the trigeminal nerve, the principal sensory nucleus of the trigeminal nerve, and the mesencephalic nucleus of the trigeminal nerve. The discerning reader will wonder what happened to the visceral efferent nucleus of the facial nerve, i.e. the rostral salivatory nucleus of mammals? The site of this nucleus has not been established in birds.

Medulla oblongata and pons: internal structure

Nuclei of the cranial nerves. The XIIth nerve arises from two nuclei, both of which can be regarded as the direct continuation of the ventral horn. One of these is the nucleus of the hypoglossal nerve and the other is the nucleus intermedius of the medulla oblongata (Fig 15–3). These nuclei are somatic motor, and supply postotic somites forming, for example the muscles of the tongue and trachea. The nucleus of the abducent nerve projects somatic motor fibres via the VIth nerve to the third preotic somite forming the lateral rectus muscle; the accessory nucleus of the abducent nerve innervates the muscles of the nictitating membrane.

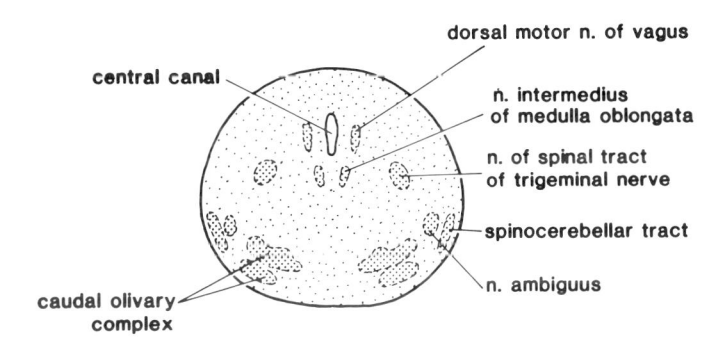

Fig 15–3 Diagrammatic transverse section of the medulla oblongata of the domestic fowl. n. = nucleus.

The motor fibres of the vagus arise from three nuclei. It appears likely from degeneration and stimulation studies that the *dorsal motor vagal nucleus* has the expected autonomic motor function, giving rise, for example to efferent cardiac inhibitory fibres. The function of the *intermediate nucleus of the medulla oblongata* is ill-understood but it seems to have an intimate association with both the vagus and the hypoglossal nerves. The *nucleus ambiguus* (Fig 15-3) distributes special visceral efferent fibres to the striated muscles derived from the embryonic pharyngeal arches, including the muscles of the pharynx and larynx. The *nucleus of the glossopharyngeal nerve* merges with the dorsal motor vagal nucleus and is presumably visceral motor in function

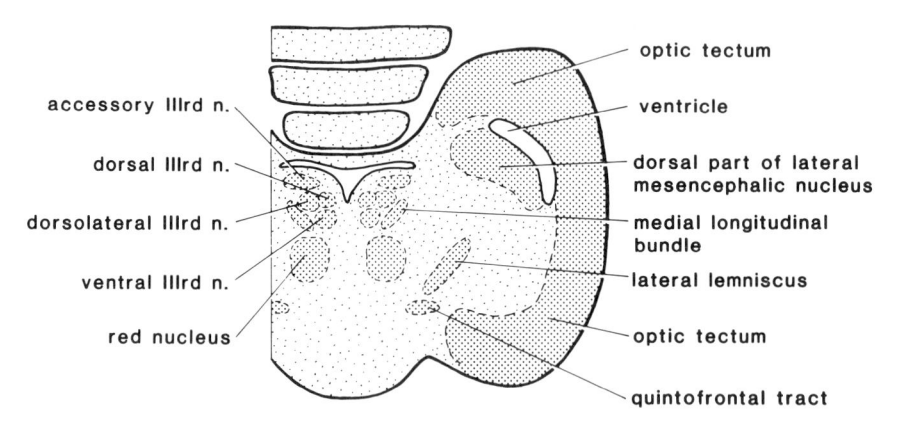

Fig 15-4 Diagrammatic transverse section of the midbrain of the domestic fowl. IIIrd n. = nucleus of the oculomotor nerve.

(corresponding perhaps to the caudal salivatory nucleus of mammals). Most of the afferent fibres entering the brainstem through the IXth and Xth nerves, including those of taste, project into the *nucleus of the solitary tract*, which therefore serves a visceral afferent function as in mammals. The *motor nucleus of the accessory nerve* is somewhat obscure, but it appears likely that motor accessory fibres arise somewhere in the region of the junction of the medulla oblongata with the first four segments of the spinal cord.

The vestibular nerve projects to six main *vestibular nuclei* (nucleus descendens, dorsolateralis, dorsomedialis, rostralis, tangentialis and ventrolateralis). The projections of these nuclei are similar to those of the vestibular nuclei of mammals, notably those to the motor nuclei of the IIIrd, IVth and VIth nerves, via the *medial longitudinal bundle* (Fig 15-4), thus coordinating the movements of the eye in relation to those of the head; they also project to the spinal cord via the *vestibulospinal tracts* hence activating the postural musculature, and to the cerebellum thus providing for coordination of the motor control of posture.

There are two primary *auditory nuclei* (the angular nucleus and the magnocellular cochlear nucleus) which receive their input from specific topographical regions of the cochlear ganglion much as in mammals. The efferent fibres from these nuclei decussate in the *trapezoid body* or *dorsal cochlear decussation* to form the *lateral lemniscus* (Fig 15–4); some of them end or form collaterals in the *rostral olivary nucleus*.

The motor fibres of the Vth nerve arise from the *motor trigeminal nucleus* which is divided into lateral, medial and ventral components. The sensory fibres of the trigeminal nerve project into three nuclei according partly to their topographical source and partly to their modality, much as in mammals. Thus the *principal trigeminal nucleus* receives its input from the somatic afferent cutaneous receptors of the face, beak and palate, in response to stimuli of touch and pressure. Hence it has much in common with the principal trigeminal nucleus of mammals. The *nucleus of the spinal tract of the trigeminal nerve* extends caudally to the cervical spinal cord, where it becomes continuous with the nucleus of the substantia gelatinosa which forms the dorsal tip of the dorsal horn. This nucleus therefore resembles the spinal trigeminal nucleus of mammals, but whether or not it is restricted to pain and temperature pathways from the face as in mammals is not known. The third afferent trigeminal nucleus, the *mesencephalic trigeminal nucleus*, belongs to the midbrain as its name implies, but is considered here for convenience. It receives a proprioceptive input from the muscles of the eye and jaw as in mammals, the receptors being muscle spindles in at least the jaw muscles. Of the various efferent tracts which project from the trigeminal sensory nuclei, a large one of particular interest is the *quintofrontal tract* (Fig 15–4) projecting from the principal trigeminal nucleus to the telencephalon (p. 253).

The *nucleus of the facial nerve*, which can be divided into dorsal, intermediate and ventral parts, is close to the motor nucleus of the trigeminal nerve as would be expected since both are special visceral (branchial) motor nerves innervating musculature derived from the embryonic pharyngeal arches. The afferent component of the VIIth nerve projects into poorly developed sensory areas which, however, appear to correspond to the nucleus of the solitary tract as in mammals.

Other nuclei and associated tracts. One of the main reasons for the swollen appearance of the avian medulla oblongata is the relatively large size of the *caudal olivary complex* (Fig 15–3). It is a massive structure with three major divisions each of which has several subdivisions. It projects to the cerebellum, and in view of this similarity to mammals is presumably involved in feedback pathways regulating the activity of higher somatic motor centres such as the red nucleus and probably the corpus striatum itself. Although somewhat rudimentary, lateral and medial *pontine nuclei* are also present and project by *pontocerebellar fibres* to the cerebellum via the intermediate cerebellar peduncle (brachium pontis). The resemblance of this pathway to the pontocerebellar fibres of mammals suggests that this is another motor feedback system.

The reticular formation of the hindbrain is rather extensive, especially the pontine reticular nuclei; knowledge of the comparable areas of the mammalian hindbrain suggests that the medullary reticular formation is likely to be extensively involved in the regulation of basic functions like respiration and circulation, but experimental data are scarce.

As already stated (p. 238) the dorsal column of the spinal cord is less well developed than in mammals, but cuneate and gracile nuclei are present in the medulla oblongata. There is also evidence for a slender spinothalamic tract in the spinal cord (p. 241). In mammals the cuneate and gracile nuclei project rostrally to the thalamus via the medial lemniscus, and so do ascending projections from the sensory trigeminal nuclei. Thus in mammals the medial lemniscal system is a massive ascending pathway for the modalities of touch, pressure, kinaesthesia, pain and temperature from the whole of the body and the head. In birds the *medial lemniscus* is present but only weakly developed. Moreover, only a few of its fibres reach the thalamus, and these appear to belong to the spinothalamic tract.

Cerebellum: external structure

The cerebellum attaches to the dorsal aspect of the medulla oblongata by *rostral* and *caudal cerebellar peduncles* (brachium conjunctivum and restiform bodies). Because of the relatively poor development of the pontocerebellar pathway, and hence the small development of the pons, it is difficult to identify a middle or *intermediate peduncle* (brachium pontis), but this peduncle also is recognized in the literature.

Externally the cerebellum consists of a single large median structure named the *body of the cerebellum* or *vermis*, and a small *cerebellar hemisphere* on each side of the vermis. The main components of each hemisphere are the paraflocculus and flocculus which project laterally from the caudal end of the vermis and together constitute the cerebellar auricle (Fig 15–2a). The *vermis* is divided by the primary and uvulonodular fissures into three main lobes (Fig 15–2a). The *rostral lobe* begins at the rostral end of the vermis and ends at the primary fissure. The *caudal lobe* lies between the primary and uvulonodular fissures. The *flocculonodular lobe* begins at the uvulonodular fissure and is largely concealed by the caudal part of the vermis; it gives off the cerebellar auricle. The three main lobes are themselves divided by transverse fissures into ten primary lobules (folia); these are numbered rostrocaudally I to X, but several of them have secondary subdivisions (e.g. lobule VIa, VIb and VIc). These fissures and lobules vary among the avian species and moreover are often difficult to distinguish. The flocculus is formed by a lateral extension of lobule X, and the paraflocculus by a lateral extension of lobule IX.

The ten primary lobules of the avian cerebellum have been partly related to the homologous parts of the mammalian cerebellum, but these homologies have not yet been fully worked out. Attempts have also been made to deduce the functions of the various lobules from their degree of development in various species of bird. This line of reasoning suggests, for example, that lobules II and III control the undercarriage, i.e. the legs. However, peripheral electrical stimulation and central recording of activation potentials indicate somewhat different conclusions, for instance that the legs are controlled by IV and V, or III, IV and V, or V and VI depending on the species. On the other hand, both lines of approach suggest that lobules IV, V and VI are involved with the wing, and that VII is concerned in hearing.

Cerebellum: internal structure

Histological structure. The histological structure of the cerebellar cortex is believed to be similar in principle to that of mammals. There are three layers. The incoming afferent fibres consist of climbing and mossy fibres. The climbing fibres presumably arise from the caudal olivary complex and project on the dendritic fields of Purkinje cells in the outermost layer, the molecular layer. The mossy fibres should originate from the spinocerebellar tracts and from the pontine and vestibular nuclei. They end in the dendritic fields of the neurons of the granular layer. The neurons of the granular layer, which is the deepest layer, project peripherally into the outermost layer, the molecular layer. There, their axons form T-shaped branches running at right angles to the long axis of the brainstem (parallel with the long axis of the folium). These T-shaped branches in turn project on the dendritic fields of the Purkinje cells which, as a single row of cells, form the middle layer. The dendrites of the Purkinje cells form branches parallel with the long axis of the brain, i.e. at right angles to the axonal fields of the granular cells. This arrangement allows extensive interaction throughout the cerebellar cortex. The Purkinje cells are the only efferent neurons of the cerebellar cortex. Their axons project to the cerebellar nuclei below the cerebellar cortex.

Cerebellar nuclei. There are three main cerebellar nuclei, the internal, intermediate and lateral nuclei.

Afferent projections to the cerebellum. The afferent projections consist of descending and ascending pathways. Of the descending pathways, the pontocerebellar projections and the projections from the caudal olivary complex and vestibular nuclei have already been mentioned (pp. 246–247); the olivocerebellar fibres are crossed and go to all parts of the cerebellar cortex. All these descending pathways, on the basis of comparisons with mammals, ought to be the afferent side of feedback pathways from the higher motor centres to the cerebellum. Besides these, there are also exteroceptive afferent projections which descend to the cerebellum. These include tectocerebellar pathways projecting visual stimuli to lobules VII and VIII of the cerebellar cortex, and cochleocerebellar fibres projecting auditory stimuli to the same lobules. There is also evidence for trigeminocerebellar tracts from the principal trigeminal nucleus to the cerebellar cortex, which presumably account for the presence of a facial tactile area in the cerebellar cortex (in lobule VI).

The ascending afferent pathways to the cerebellum consist mostly of proprioceptive fibres from the muscles of the limbs, i.e. the spinocerebellar tracts (p. 240) which end mainly in the rostral lobules (II to VI inclusive but also in VIII and IX). Again, however, there are exteroceptive fibres which project tactile stimuli from the tail, leg and wing (to lobules III, IV, V and VI), although their route is unknown.

Efferent projections from the cerebellum. The cerebellar nuclei give rise to efferent pathways which have not been extensively studied. However, there are some efferent projections which resemble those of mammals. These include projections to the vestibular nuclei and to the pontomedullary reticular formation. These areas are involved in regulating somatic motor activities. Therefore, as in mammals, such efferent cerebellar projections presumably represent the efferent side of feedback pathways by which the cerebellum regulates the motor activities of posture and locomotion. Cerebellospinal tracts have been suggested but remain dubious (p. 242).

Lesions of the cerebellum. Lesions of the cerebellum produce variable effects but, as in mammals, disorders of posture and movement are the main feature. There tends to be a more or less marked increase in muscle tone especially during movement, leading to strong extension of the wings, legs, tail and neck, with severe swaying and marked nystagmus. There may be a slight but persistent tremor.

Midbrain: external structure

The main feature of the midbrain is the mesencephalic (optic) tectum (Figs 15–2a and 15–4), which takes the form of the massive *mesencephalic colliculus*. This structure is commonly called the *optic lobe* and is also known as the *mesencephalic tectum* or *optic tectum*. Its huge size indicates the extent to which birds are visual animals. The mesencephalic colliculus is the homologue of the rostral colliculus of mammals, yet it lies not dorsally but ventrolaterally on the avian brainstem. This position is forced on it during embryonic development by competition for space with the forebrain and cerebellum. There is no externally obvious homologue of the caudal colliculus, but beneath the dorsolateral surface of the mesencephalon there is a homologous structure, i.e. the dorsal part of the *lateral mesencephalic nucleus* (Fig 15–4). Two cranial nerves emerge from the midbrain, the IIIrd and IVth. Both belong to the ventral root series, but as in mammals the IVth nerve emerges dorsolaterally between the cerebellum and optic lobe; however, this is because its fibres decussate dorsally over the cerebral aqueduct before leaving the brainstem. Caudally the midbrain is joined to the cerebellum by the rostral cerebellar peduncles.

Midbrain: internal structure

Cranial nerve nuclei. The single trochlear nucleus is continuous with the oculomotor nuclei, and they all lie ventrolateral and close to the central canal (Fig 15–4), a position comparable to the ventral horn of the spinal cord. As has just been stated, the efferent axons of the trochlear nucleus decussate over the dorsal aspect of the central canal before emerging.

There are four parts to the oculomotor nucleus. Three of these (the dorsal, ventral and dorsolateral parts) are presumably somatic motor and innervate extrinsic eye muscles. The fourth component (the accessory part) is believed to be the homologue of the mammalian Edinger–Westphal nucleus, innervating the intrinsic eye muscles (including presumably the striated constrictor of the iris and the striated sclerocorneal muscles).

The mesencephalic trigeminal nucleus has already been discussed with the other sensory trigeminal nuclei (p. 247). As in mammals the neurons of this nucleus appear to be primary afferent neurons which have failed to migrate out of the neuraxis during embryonic growth, thus breaking the otherwise almost inviolable rule that primary afferent neurons have their cell stations in a dorsal root ganglion. Evidence for this failure to migrate is the fact that the peripheral axonal process of each neuron in the nucleus ends directly in a muscle spindle.

Optic components. The *optic tectum* (optic lobe) can be divided into six main strata, of which the deepest adjoins the dorsolateral aspect of the lateral ventricle (Fig 15-4). However, these strata can be further subdivided, and under optimum conditions 15 layers can be made out, in the form of more or less alternating cellular and fibrous zones. Incoming retinal fibres reach the outer strata. Since only a few retinal fibres are uncrossed, the representation of the retina on the optic tectum is very largely contralateral. The middle grey layers of the optic tectum give rise to *tectospinal fibres* which project to the motor nuclei controlling the extrinsic eye muscles and at least to the upper segments of the spinal cord; thus they control the eye and neck movements which track moving objects.

The *isthmo-optic nucleus* lies near the oculomotor and trochlear nuclei and projects efferent fibres to the retina which presumably regulate receptor sensitivity.

Auditory and vestibular components. The dorsal part of the *lateral mesencephalic nucleus* (Fig 15-4) is the apparent homologue of the mammalian caudal colliculus. Consistent with this principle, this nucleus receives projections from the cochlear nuclei via the *lateral lemniscus* (Fig 15-4); it then projects rostrally to the *ovoidal nucleus* of the diencephalon.

The projections of the vestibular nuclei to the oculomotor and abducens nuclei via the *medial longitudinal bundle* (Fig 15-4), coordinating eye and head movements, have been mentioned on pages 242 and 246.

The red nucleus. The red nucleus lies on the ventromedial aspect of the oculomotor nuclei (Fig 15-4). It gives rise to the rubrospinal tract (p. 241).

The intercollicular nucleus. This nucleus appears to coordinate the motor control of the respiratory and syringeal muscles during vocalization.

Diencephalon: external structure

Dorsally the diencephalon carries the *pineal gland*. In the domestic fowl this is a conical pink structure lying in the triangular space between the cerebral hemispheres and cerebellum, about 3.5 mm long and 2.0 mm wide. Ventrally the main feature of the diencephalon is the *optic chiasma*, leading to the left and right optic tracts. Immediately caudal to the chiasma is the ventral surface of the hypothalamus which carries the hypophysis (see Chapter 12).

Diencephalon: internal structure

The diencephalon includes the thalamus, hypothalamus and epithalamus.

Thalamus. Judged by the mammalian thalamus, the avian thalamus should be the final relay station of afferent pathways ascending to the cerebral hemisphere.

Several optic components have been identified. The *rotund nucleus* (Fig 15–5), which is one of the largest thalamic nuclei, receives tectothalamic fibres

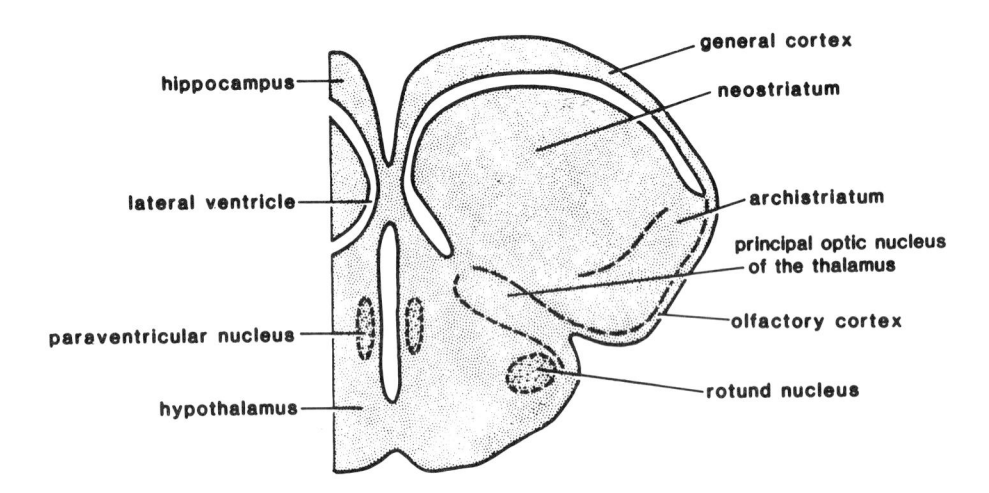

Fig 15–5 Diagrammatic transverse section of the cerebral hemisphere and diencephalon of the domestic fowl. The level of the section is slightly caudal to Fig 15–6. The region labelled 'general cortex' is the parahippocampal area.

from the optic tectum, and in turn projects to the ectostriatum of the cerebral hemisphere. The avian thalamic region which receives direct retinal projections has now been identified as the *principal optic nucleus* of the thalamus (Fig 15–5); this nucleus is therefore the homologue of the mammalian lateral geniculate body, or more precisely of the dorsal or main part of the mammalian lateral geniculate nucleus. The principal optic nucleus in turn projects to the rostral part of the sagittal eminence (the Wülst) of the cerebral hemisphere. The main auditory component is the *ovoidal nucleus*, a possible homologue of the mammalian medial geniculate body, which projects to the neostriatum. In contrast to mammals, there appears to be no well-defined area in the thalamus acting as the final relay station of ascending spinal somatic and visceral pathways, such as the ventroposterior thalamic nucleus of mammals. Indeed, of the whole *medial lemniscal system* in birds, only a few spinothalamic fibres seem to reach the thalamus, and even these are only distributed diffusely in the caudal regions of the thalamus. There is a possibility, however, that the *quintofrontal tract* (Fig 15–4), which arises from the principal trigeminal nucleus, might represent the trigeminal contribution to the medial lemniscus in mammals; however, this tract projects to the *basal nucleus* (nucleus basalis), which lies in the forebrain near the ventral surface and has no known direct anatomical connections to the thalamus.

Hypothalamus. Several nuclear masses have been identified in the hypothalamus (Fig 15–5), including preoptic, paraventricular, supraoptic and infundibular nuclei. The last three of these nuclei contribute to the hypothalamohypophyseal tract which innervates the hypophysis. Ventrally the hypothalamus is continuous, through its tuber cinereum, with the median eminence of the neurohypophysis (Fig 12–1). Caudally the hypothalamus continues directly into the reticular formation of the midbrain. As in mammals it dominates virtually all autonomic functions, including thermoregulation, respiration, circulation, eating and drinking, reproduction, and defensive and aggressive reactions, probably via the descending reticular formation.

Epithalamus. The epithalamus contains a number of nuclei and tracts and includes also the *pineal gland* (p. 210). Some of the cells of the avian pineal gland have a structure that is suggestive of rudimentary photoreceptors, but the majority have only secretory characteristics. Non-myelinated axons are present. There is no doubt that the pineal gland is involved in reproductive function, probably by the action of pineal hormones on the neuronal projections from the hypothalamus to the hypophysis. The pineal gland is strongly influenced by light, via the eyes and brain, and also via the cranial cervical ganglion which supplies axons to the gland. The pineal gland still responds to light even after the cranial cervical ganglion and the eyes have been removed, and it therefore seems likely that light passes through the wall of the enucleated orbit and stimulates the hypothalamus directly, and the latter in turn affects the pineal gland hormonally.

Cerebral hemisphere: external structure

The pointed olfactory bulb, which is relatively small compared to that of mammals, projects from the rostral end of the brain (Fig 15–2a). The median fissure separates the left and right cerebral hemispheres dorsally. The surface of each hemisphere is almost smooth. On the dorsal surface, however, a groove called the vallecula (strictly, the telencephalic vallecula) arises near the rostral end of the median fissure and passes caudally diverging slightly from the midline. Medial to the vallecula, and hidden between the vallecula and the median fissure, is a smooth bolsterlike ridge running parallel with the midline, the *sagittal eminence* ('Wülst'). The *hippocampus* lies on the medial aspect of the hemisphere, i.e. hidden within the median fissure (Figs 15–5 and 15–6). Caudally the cerebral hemisphere overlaps the optic lobe (Fig

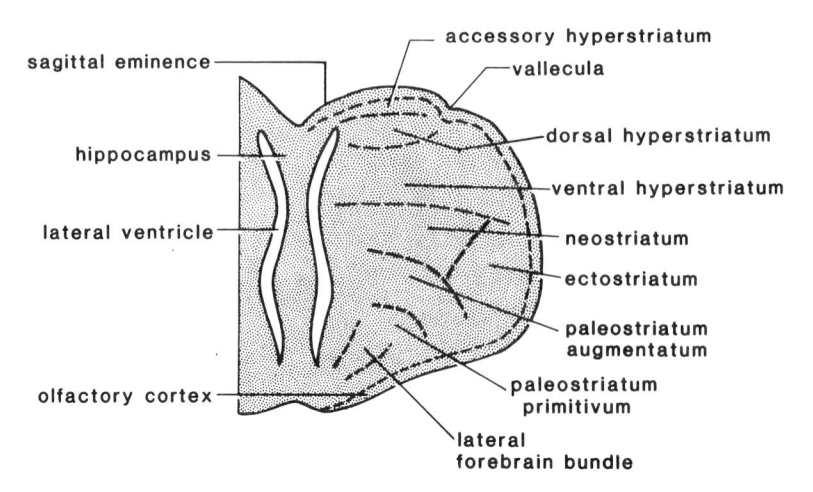

Fig 15–6 Diagrammatic transverse section of the cerebral hemisphere of the domestic fowl.

15–2a). In the domestic fowl the *lateral ventricle* is medial in the rostral part of the hemisphere (Fig 15–6), but in the caudal part it extends medially, dorsally and laterally over the hemisphere (Fig 15–5); in these areas it lies very near the dorsolateral surface of the hemisphere, being covered by a thin layer of cortex which is only about 1 mm deep or less. A choroid plexus is present in the caudomedial part of the ventricle.

Cerebral hemisphere: internal structure

Cortex. The avian cerebral hemisphere is covered by a superficial mantle of 'general cortex' (Fig 15–5), forming the parahippocampal area. The avian general cortex is entirely primordial in character, consisting of only one or two layers. It is divided into three regions, namely the limbic cortex, parahippocampal area and the true olfactory cortex. The *limbic cortex* (rhinencephalon) includes the hippocampus and adjacent cortical regions which together form a substantial part of the cortex covering the dorsomedial region of the hemisphere. The limbic system also includes non-cortical structures (Fig 15–5) such as the caudal one-third and the most medial part of the archistriatum (the archistriatum caudale and mediale); these regions are a major forebrain complex which is connected with the hypothalamus (as in vertebrates generally) and may represent the mammalian amygdala. The *parahippocampal area* is a variable expanse of cortical mantle situated dorsolaterally on the hemisphere. It separates the limbic cortex from the *olfactory cortex* (Fig 15–5), which covers the more ventral regions of the hemisphere (it includes the piriform and prepiriform cortex). As already stated, the avian general cortex is thin in the histological sense since it consists of only a few layers; it is also thin in the macroscopic sense being simply the shell of tissue, often less than 1 mm thick in the domestic fowl, which forms the outer wall of the lateral ventricle.

The most striking structural difference which distinguishes the mammalian from the avian brain is the replacement (in the mammal) of the general cortex by a large expanse of multi-layered cortex with five or six clearly recognizable layers, generally called the *neocortex*. There is no structure in the avian cortex which has essentially the same organization as the mammalian neocortex: indeed, birds do not have a neocortex.

Below the surface of the avian cerebral hemisphere there are large areas of grey matter which form the bulk of the hemisphere. These have long been regarded as homologous to the striatal complex of mammals. For this reason they were commonly known in birds as the *corpus striatum*, although it is now becoming apparent that, at the most, only a small part of these grey areas is homologous to the mammalian striatal complex. The term corpus striatum itself is therefore no longer valid for the avian brain. However, the suffix 'striatum' does occur in many of the terms used for the main subdivisions of these great areas of grey matter, as will shortly be apparent.

The deeper, or more precisely the more ventral, elements of this great complex of grey matter within the avian cerebral hemisphere consists of a medial region, the *paleostriatum primitivum* (Fig 15-6), and a larger and somewhat more dorsolateral area, the *paleostriatum augmentatum* (Fig 15-6). The paleostriatum primitivum appears to be the homologue of the mammalian globus pallidus; the paleostriatum augmentatum is believed to be the homologue of the mammalian caudoputamen.

The more superficial (dorsal) elements of the complex are sharply subdivided into at least four components, i.e. the ectostriatum, archistriatum, neostriatum and hyperstriatum. These areas lie dorsally on the deeper parts, and greatly exceed them in thickness. The *neostriatum* forms the massive central region (Figs 15-5 and 15-6). The *ectostriatum* (Fig 15-6), and more caudally the *archistriatum* (Fig 15-5), form the lateral regions. The *hyperstriatum*, divided into dorsal, ventral and accessory areas (Fig 15-6), constitutes the most dorsal region. Together, these more superficial (dorsal) elements account for the majority of the avian telencephalon. They are entirely unrepresented in the mammalian brain. Unfortunately, there is no name to cover their complete assemblage. In view of the abundant use of the suffix 'striatum' among their components it is tempting to name them the avian striatal complex. However, this falsely suggests homology to the well-established mammalian striatal complex. In the final paragraphs below about the brain they will be called the avian *telencephalic complex*.

The *sagittal eminence* consists of a thin outer coat of general cortex and an inner region formed by the accessory and dorsal hyperstriatum (Fig 15-6); it is the latter two areas which receive the optic projections from the principal optic nucleus of the thalamus.

The lateral forebrain bundle. This great fibre tract (Fig 15-6), the fasciculus lateralis prosencephali, is the main efferent and afferent roadway connecting the hemisphere with the more caudal parts of the brain, and is analogous to the massive internal capsule of mammals. It is subdivided into several tracts (e.g. thalamofrontal, striomesencephalic, striothalamic and quintofrontal tracts).

Relationship between the neocortex of mammals and the telencephalic complex of birds. The avian telencephalic complex has much in common, both functionally and anatomically, with the mammalian neocortex. Firstly, they both receive tectothalamic optic projections; in birds, these have their final relay station in the rotund nucleus (p. 252). Secondly, they both receive direct retinal projections, which in birds are formed via the principal optic nucleus of the thalamus (p. 252). Thirdly, they both receive auditory projections, the ovoidal nucleus being the avian thalamic relay station for this pathway (p. 252). Finally, they both project motor pathways to the hindbrain and spinal cord (p. 241).

In view of these broad similarities it has now been suggested that neurons homologous to those of the avian telencephalic complex are represented in the mammalian neocortex. This means that the neocortex is not simply an entirely new development in mammals, as hitherto supposed. On the contrary, it implies the existence of a neuronal population which occupies either the telencephalic complex in birds or the neocortex in mammals. At present, a missing link in this hypothesis is the difficulty in finding in the avian telencephalic complex a somatic sensory projection area comparable to the somatic sensory areas of the mammalian neocortex. As stated above (p. 252), the medial lemniscal system of birds scarcely reaches the thalamus. Nevertheless, the concept of the interchangeability of the telencephalic complex and neocortex is a valuable unifying factor in comparative neuroanatomy.

CRANIAL NERVES

There are twelve pairs of cranial nerves in birds, as in mammals. Their approximate positions in relation to the skull of the domestic fowl are indicated in Fig 15–7. The more caudal nerves, particularly IX, X, XI and XII, are complicated by numerous anastomoses (Figs 15–2b and 15–7) and the sources of the fibres in their branches are therefore uncertain.

I. Olfactory nerve

The olfactory nerve is entirely sensory. It joins the olfactory bulb by means of 10 to 30 fine rootlets. Its nerve fibres have their cell bodies in the olfactory epithelium of the nasal cavity (p. 310). If the rootlets from the olfactory bulb are traced rostrally, they combine into a single main trunk which leaves the cranial cavity through the olfactory foramen and enters the bony orbit; thus, in contrast to mammals, there is no sievelike cribriform plate. It then runs rostrally along the dorsal border of the interorbital septum (Fig 15–7). At the nasal bone it divides into two branches, which distribute themselves to the olfactory epithelium covering the dorsal and ventral surfaces of the caudal concha, the adjoining roof of the nasal cavity, and the dorsal region of the nasal septum adjacent to the concha (Fig 7–1C and –1D). The varying degree of development of the olfactory bulb is discussed on page 311.

NERVOUS SYSTEM 257

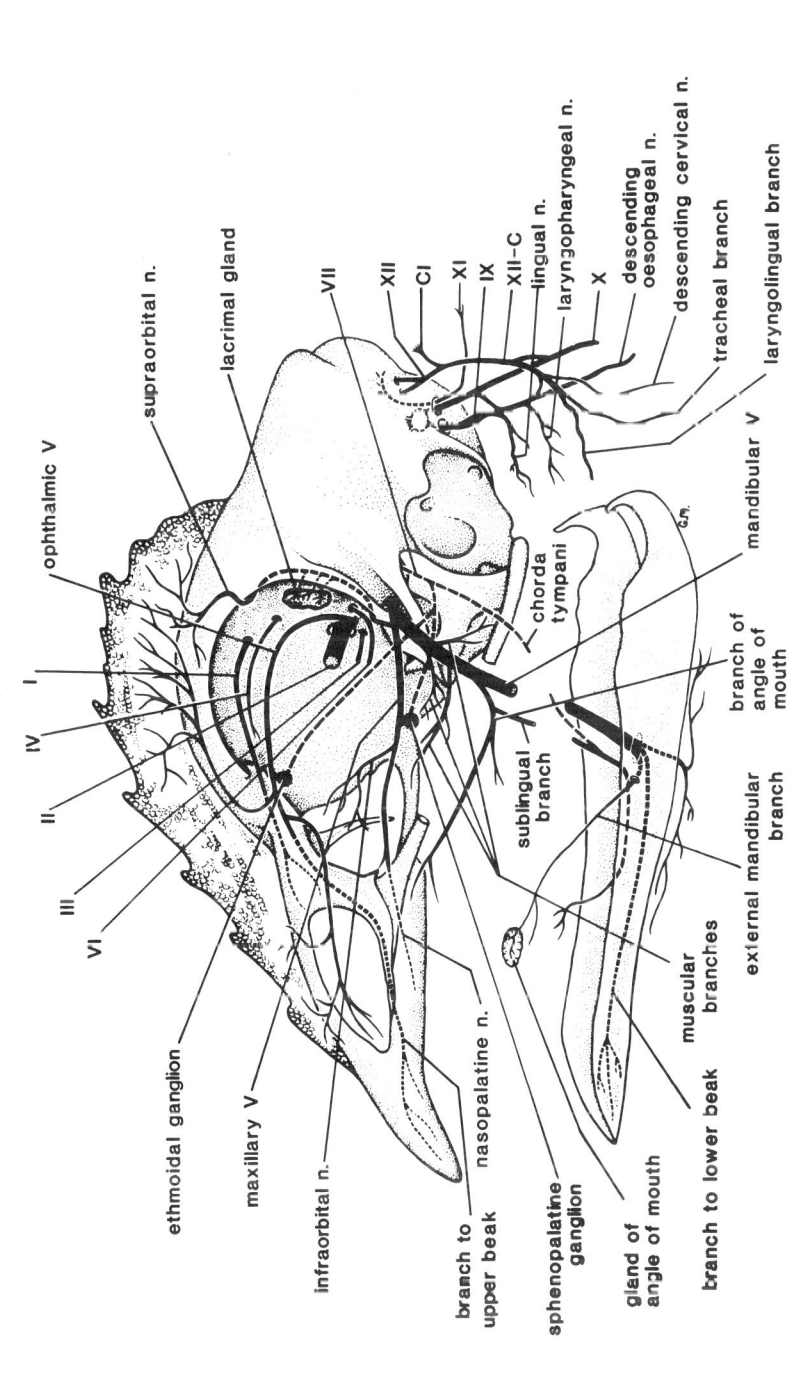

Fig 15–7 Diagram of the cranial nerves of the domestic fowl. n. = nerve; Roman numerals = cranial nerves; C1 = first cervical spinal nerve. The mid-region of the left jugal arch has been removed. The facial nerve connects to the ethmoidal ganglion and sphenopalatine ganglion by the dorsal and ventral branches (dotted lines) of the palatine nerve of VII. The lateral branch of ophthalmic V is shown as a continuous line and the medial branch as a broken line. The proximal ganglia of IX and X fuse to form a common ganglion embedded in the skull and this is shown by dotted lines. The label IX points to the distal ganglion of IX.

II. Optic nerve

The optic nerve, which is sensory from the retina, is much larger than all the other cranial nerves. If the left and right optic nerves are added together their total cross-sectional area exceeds that of the cervical spinal cord, thus reflecting the great development of vision in birds. The optic nerve is particularly huge in falconiform and corvid species and relatively small in nocturnal birds. The cell bodies of its fibres are the ganglion cells of the retina (p. 291). These fibres are unmyelinated as they traverse the retina, but become myelinated on penetrating the sclera at the optic disc beneath the pecten (Fig 16–7B). The optic nerve enters the cranial cavity via the optic foramen (Figs 4–1 and 15–7) and then decussates almost completely at the *optic chiasma*. From the chiasma it continues as the *optic tract* and projects to the principal optic nucleus of the thalamus.

III. Oculomotor nerve

The oculomotor arises from the midbrain and enters the orbit either with the optic nerve through the optic foramen or independently through its own oculomotor foramen as in Fig 15–7. It then divides into a dorsal and ventral branch (Fig 15–9). The dorsal branch supplies the dorsal rectus muscle and the levator of the upper eyelid. The ventral branch innervates the ventral rectus, medial rectus and ventral oblique muscles. The ventral branch carries the *ciliary ganglion* (Fig 15–9) which in turn gives off the *iridociliary nerve* (previously the long ciliary nerve). The latter receives a branch from ophthalmic V (Fig 15–9), and then penetrates the sclera to distribute parasympathetic efferent fibres to the ciliary body and iris.

IV. Trochlear nerve

As in mammals this nerve arises from the dorsal surface of the midbrain (Fig 15–2b), its axons having decussated dorsally over the cerebral aqueduct. It emerges through the trochlear foramen into the orbit (Fig 15–7), where it supplies somatic motor fibres to the dorsal oblique muscle (Fig 15–9).

V. Trigeminal nerve

The trigeminal nerve arises as a single massive trunk from the brainstem at the level of the caudal border of the optic lobe, immediately enlarges into the trigeminal ganglion and then divides into two great trunks, the ophthalmic nerve and the combined maxillary and mandibular nerves (Fig 15–2b); the latter two nerves (maxillary and mandibular V) become separated as they emerge from the skull or immediately after they have emerged.

Ophthalmic nerve. The ophthalmic nerve is the main sensory nerve of the nasal cavity and the wall of the eyeball. It usually enters the orbit either

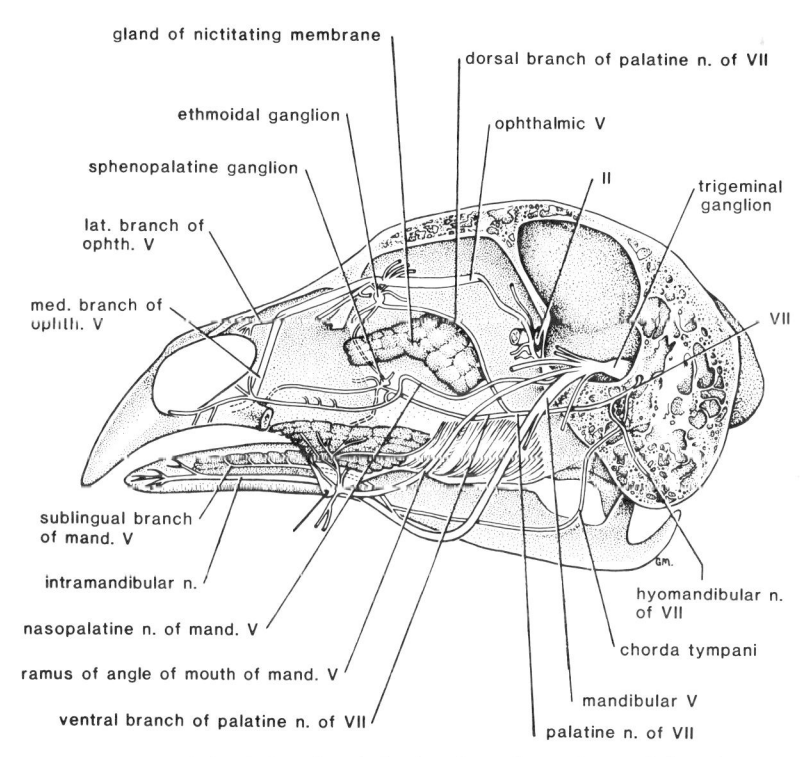

Fig 15–8 Diagram of the left side of the head to show the peripheral course of the trigeminal and facial nerves in the domestic fowl. The caudal half of the left mandible and the caudal half of the left jugal arch have been removed, together with part of the left wall of the cranium. The rostral half of the left mandible still remains but has been partly removed to show the intramandibular nerve. Based on Watanabe and Yasuda (1970), with kind permission of the editor of the *Japanese Journal of Veterinary Science*.

through its own ophthalmic foramen as in Fig 15–7, or with the oculomotor nerve. As it approaches the optic nerve (Fig 15–9) it forms a branch to the *iridociliary branch* of the oculomotor nerve; the iridociliary branch supplies the wall of the eyeball. The ophthalmic nerve then curves dorsally and rostrally across the back of the eyeball (Fig 15–9). At the rostral end of the orbit it connects with the *ethmoidal ganglion* (p. 261) and finally divides into a lateral and a medial branch (Figs 15–7 and 15–8). The lateral branch is sensory to the upper eyelid, the skin of the forehead and comb, and the rostral part of the nasal cavity. The larger medial branch (broken line in Fig 15–7) runs on the nasal septum (Fig 7–1B, –1C and –1D) and supplies the nasal mucosa, the palate, the edge of the upper beak and the extremely important dermal receptor complex in the tip of the upper beak (the bill tip organ in the maxillary nail of ducks and geese, see p. 25). The ophthalmic nerve itself is presumably purely afferent as in mammals, its fibres being somatic afferent.

Maxillary nerve. The maxillary nerve usually emerges from the skull with the mandibular nerve (the two being combined into a single trunk), through the maxillomandibular foramen (Fig 15–7) and forms three main branches. Its *supraorbital nerve* (Fig 15–7) innervates the conjunctiva and skin of the upper eyelid, and also the comb. The *infraorbital nerve* (Fig 15–7), which is very small compared to that of mammals, supplies the lower eyelid and its conjunctiva, and also the skin of the rictus. The *nasopalatine nerve* (Figs 15–7 and 15–8) connects with the *sphenopalatine ganglion*, thus receiving postganglionic parasympathetic fibres of the facial nerve which go to the lacrimal gland and glands of the nasal mucosa. The nasopalatine nerve terminates by supplying receptor endings in the lateral margin (tomium) of the upperbill. The maxillary nerve, too, is presumably purely afferent as in mammals, its fibres again being somatic sensory.

Mandibular nerve. The typical emergence of this nerve from the skull in combination with the maxillary nerve has just been described. Alternatively, the mandibular nerve escapes from the cranial cavity through the mandibular foramen. It gives motor branches to the muscles of mastication, and by means of the branch of the angle of the mouth (Figs 15–7 and 15–8) supplies the skin and mucosa at the rictus. The main continuation of the nerve is the *intramandibular nerve* (Fig 15–8) which passes through the length of the mandible in the mandibular canal, giving off branches to the skin associated with the lower beak. It also innervates the horn including the functionally important dermal receptor complex in the tip of the lower beak (the bill tip organ in the mandibular nail of ducks and geese). The sublingual branch, which innervates the floor of the oral cavity though not the tongue, also travels initially in this canal. The mandibular nerve therefore contains both somatic afferent and special visceral efferent fibres.

VI. Abducent nerve

The abducent nerve arises from the rostral end of the medulla oblongata near the ventral midline (Fig 15–2b). After a long intracranial course it leaves the cranial cavity through the abducent foramen and runs over the back of the eyeball (Fig 15–9). It innervates the lateral rectus muscle and the two striated muscles which move the third eyelid (the quadratus and pyramidalis muscles). This is an exclusively somatic motor nerve.

VII. Facial nerve

The facial nerve arises from the ventrolateral aspect of the medulla oblongata (Figs 15–2a and –2b), enters the internal acoustic meatus with the VIIIth nerve, and then traverses the facial canal. After emerging from the foramen of the facial nerve it divides into two main branches, the palatine nerve and the hyomandibular nerve. Just before it divides, the main trunk carries the very small *geniculate ganglion* on its dorsal aspect. The *palatine nerve* itself divides into a dorsal branch to the ethmoidal ganglion, and a ventral branch to the sphenopalatine ganglion (Fig 15–8). The palatine nerve appears to be the homologue of the greater superficial petrosal nerve of mammals. The *hyo-*

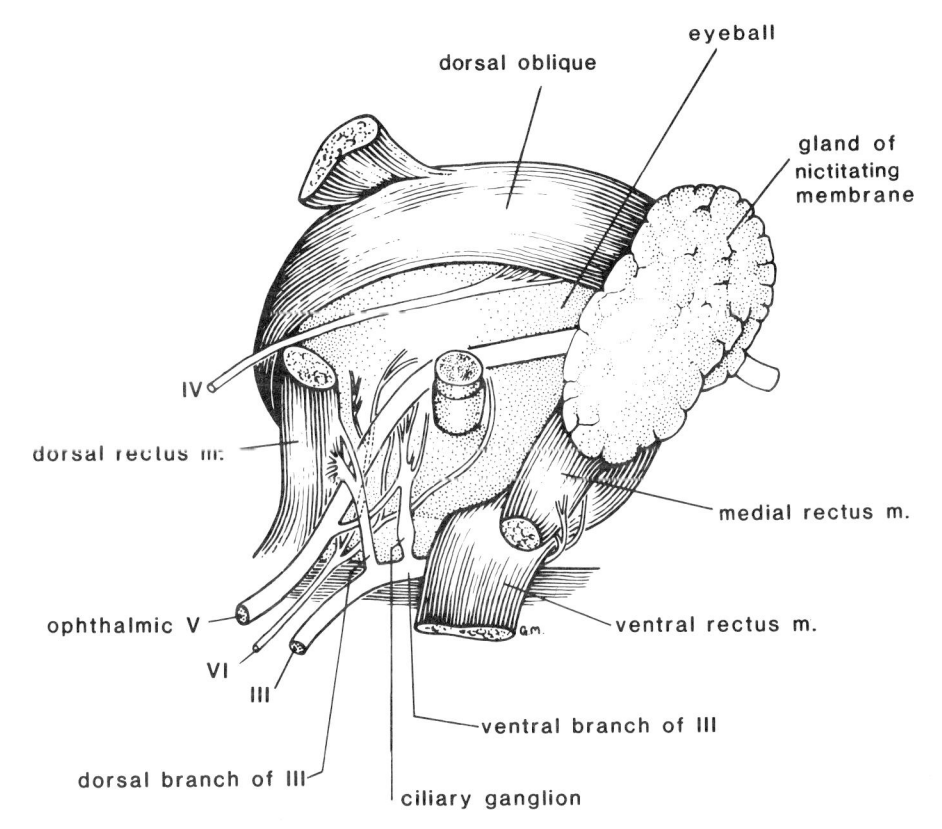

Fig 15-9 Caudal view of the left eyeball of the domestic fowl to show the extraocular muscles and their associated nerves. The ventral branch of the IIIrd nerve gives off a dorsal ramus which expands into the ciliary ganglion; this ramus continues as the iridociliary nerve (unlabelled) towards the sclera, being joined by a branch from ophthalmic V. The transected stump of the optic nerve (unlabelled) lies in the centre of the diagram; the main trunk of ophthalmic V passes just dorsal to it. The middle part of the dorsal rectus muscle has been removed in order to expose the IVth nerve, the part of the dorsal rectus muscle (unlabelled) that inserts on the eyeball is shown as a transected stump at the top of the diagram. Roman numerals = cranial nerves; m. = muscle. Based on Cords (1904).

mandibular nerve innervates the depressor muscle of the mandible and the stylohyoid muscle.

The *ethmoidal ganglion* contains the cell bodies of postganglionic neurons of the facial nerve. These innervate the gland of the nictitating membrane, the nasal gland, and glands of the nasal and the rostral palatine mucosa. The *sphenopalatine ganglion* probably distributes parasympathetic postganglionic fibres of the VIIth nerve to glands of the nasal mucosa and of the caudal part of the palate.

The *chorda tympani* usually arises from the main trunk of VII just before it divides (Fig 15-8), i.e. directly from the geniculate ganglion. It then passes through the tympanic cavity and escapes from the skull near the quadratosquamosal joint. Having entered the mandibular canal it joins the sublingual branch of mandibular V (Fig 15–8). Its preganglionic parasympathetic fibres synapse in the *mandibular ganglion*, and the postganglionic axons innervate the rostral mandibular salivary glands. The mandibular ganglion in birds consists of two or more small aggregations of nerve cell bodies within the mandibular canal. Some axons of VII from *taste buds* in the upper and lower bill may travel in branches of the trigeminal nerve. The afferent component of the avian facial nerve is small compared to the mammalian facial nerve.

VIII. Vestibulocochlear nerve

This nerve arises from the medulla oblongata close to the facial nerve (Figs 15–2a and –2b).

The *vestibular part* consists of two components, the rostral and caudal vestibular nerves. Each of these carries a part of the vestibular ganglion and each distributes three branches to ampullary cristae and to the maculae of the utricle and saccule as shown in Fig 16–16. The *cochlear part* includes the cochlear nerve which consists of the central processes of nerve cells of the cochlear ganglion lying along the edge of the basilar membrane; it also includes the lagenar nerve which arises from the lagenar ganglion innervating the lagenar macula (Fig 16–16).

IX. Glossopharyngeal nerve

The IXth, Xth and XIth nerves arise as a group of continuous fine rootlets from the ventrolateral border of the medulla oblongata (Figs 15–2a and –2b). The proximal ganglion of the glossopharyngeal nerve fuses with the proximal ganglion of the vagus to form one shared ganglion. In the domestic fowl this ganglion is entirely enclosed by bone within a common foramen, but is large enough to be visible macroscopically. The IXth nerve leaves the common ganglion and emerges from the skull through its own glossopharyngeal foramen (Fig 15–7). It then passes in close contact with the lateral surface of the cranial cervical sympathetic ganglion. A few millimetres further on it expands slightly into its *distal ganglion* (also known as the petrosal ganglion) (Fig 15–7), which can scarcely be seen macroscopically. Immediately peripheral to this ganglion the IXth nerve makes a substantial anastomosis with the vagus (Fig 15–7), and then forms its three terminal branches. The first of these, the *lingual nerve*, innervates the epithelium of the whole of the tongue (thus apparently replacing the lingual branch of the Vth nerve of mammals); the branch to the tongue carries all the taste fibres from the lingual taste buds and also sends parasympathetic motor fibres to the lingual salivary glands. The second terminal branch is the *laryngopharyngeal nerve*. This gives two or three small rami to the pharynx and also forms branches innervating the muscles and mucosa of the larynx. These laryngeal axons are probably vagal in origin, perhaps corresponding to the cranial laryngeal nerve of mammals. The third terminal branch, the *descending oesophageal nerve* (Fig 15–7), is the direct continuation of the main trunk. It descends the neck in contact with the jugular vein and distributes fibres to the cervical oesophagus and trachea. Near the syrinx it anastomoses with the recurrent nerve of the vagus and their combined fibres supply the crop. The descending oesophageal nerve also anastomoses with the hyoglossocervical nerve. In birds the glossopharyngeal nerve does *not* innervate the carotid body. The so-called *precarotid* or 'subcarotid' trunk, which was supposed to carry glossopharyngeal fibres to the avian carotid body and was considered to be the homologue of the mammalian carotid body nerve, has not been found by recent investigators.

In its own right the glossopharyngeal nerve probably supplies afferent fibres to the tongue and pharynx. However, the extent to which it really supplies afferent and efferent fibres to the larynx, oesophagus, crop and trachea is uncertain, owing to its anastomoses with the vagus and hypoglossocervical nerves.

X. Vagus nerve

The vagus arises from the medulla oblongata from a series of fine rootlets continuous with those of nerves IX and XI (Figs 15–2a and –2b), which in the domestic fowl combine to form the proximal ganglion common to nerves IX and X (see under **Glossopharyngeal nerve**). As in the mammal the vagal component of this ganglion is believed to consist of somatic afferent neurons. The vagus leaves this ganglion and emerges from the skull through its own separate peripheral opening (Fig 15–7), the foramen of the vagus nerve. On leaving its foramen the vagus connects with the cranial cervical ganglion. For about 5 mm after its emergence from the skull the vagus is enclosed in a common sheath with the XIth nerve. Just below the cranial cervical ganglion it anastomoses with the IXth nerve. By this anastomosis it is believed to distribute fibres to the larynx and pharynx, by means of the laryngopharyngeal branch of IX. In some species, however, this anastomosis is absent. The vagus gives no direct branches to the larynx and pharynx.

The main trunk of the vagus nerve descends the neck in contact with the jugular vein, forming no branches visible to the naked eye. At the thoracic inlet it expands into the slender *distal vagal ganglion* (nodose ganglion) immediately caudal to the thyroid gland (see Figs 12–2 and 12–3). The nerve cell bodies which form this ganglion belong exclusively to afferent neurons, presumably visceral afferent in function.

The thoracic vagus forms the following branches: (1) the *nerve to the carotid body* arises from the caudal end of the distal vagal ganglion and consists almost entirely of afferent axons forming synapses with the type 1 cells of the carotid body; (2) fine *glandular filaments* are distributed to the thymus and to the thyroid, parathyroid and ultimobranchial glands; (3) in some species an *aortic nerve* arises from the distal vagal ganglion or from the cranial cardiac nerve, and carries the baroreceptor axons which have their endings in the media of the root of the aorta and pulmonary trunk, and probably also chemoreceptor axons from the aortic and pulmonary bodies; (4) the *cranial cardiac nerve* joins the vagus immediately caudal to the distal vagal ganglion, and appears to consist of afferent fibres from the heart; (5) the *recurrent nerve*

(Fig 12–2) anastomoses with the IXth nerve and the hypoglossocervical nerve and together they distribute fibres to the oesophagus and crop, and to the tracheal and syringeal muscles (the precise sources of these fibres being uncertain); (6) the recurrent nerve forms the *pulmoesophageal nerve* which sends motor fibres to the oesophagus and carries afferent fibres from the lung (including those that are CO_2 sensitive); (7) about six fine *pulmonary rami* arise from the vagus at the level of aorta and pulmonary trunk and are distributed to the lung; and (8) two or three *caudal cardiac nerves* distribute parasympathetic motor fibres to the heart.

The abdominal vagus is formed by the partial or complete union of the left and right vagi into a single trunk on the ventral surface of the proventriculus; this common trunk supplies the two stomachs, the duodenum and the liver. Distal to this level it has not been possible to establish the abdominal distribution of the vagus nerves since they become involved in the sympathetic plexuses that are associated with the great visceral branches of the aorta, including notably the coeliac plexus.

XI. Accessory nerve

The rootlets of origin of the *spinal component* of the XIth nerve from the avian spinal cord arise only from the levels of the first two cervical segments. The spinal component continues rostrally through the foramen magnum, and is then augmented by the cranial component which arises as three or four rootlets from the medulla oblongata (Figs 15–2a and –2b). The main trunk of the accessory nerve becomes enclosed with the vagus in a common sheath within the vagal foramen of the skull. A few millimetres after the vagus emerges from its foramen a group of accessory fibres apparently departs from the vagus as the *external ramus* of the vagus nerve, and innervates the cucullaris capitis muscle which may be homologous to the trapezius muscle of mammals. Other fibres of the accessory nerve may well remain in the vagus and be distributed elsewhere within vagal branches.

XII. Hypoglossal nerve

The XIIth nerve arises from a series of rootlets on the ventral aspect of the medulla oblongata (Figs 15–2a and –2b). These form two trunks which leave the skull by two hypoglossal foramina (Fig 15–7). After uniting, the combined trunks absorb a large part of the first cervical nerve (and perhaps some of the second cervical nerve) thus forming the *hypoglossocervical nerve*. This nerve passes obliquely over the Xth and IXth nerves, anastomosing with them at this level (Fig 15–7).

Just distal to its anastomosis with the vagus, the hypoglossocervical nerve gives off the slender *descending cervical nerve* (Fig 15–7), which supplies the tracheal muscles. Near the larynx the hypoglossocervical nerve divides into its two terminal branches, the *laryngolingual ramus* to the tongue muscles and the *tracheal ramus* which innervates the tracheal muscles and the intrinsic muscles of the syrinx; the tracheal ramus is (almost certainly) the main or only true motor pathway to the syringeal muscles (but see also the recurrent nerve of the vagus). The tracheal branch descends alongside the trachea and has been interpreted as the homologue of the ramus descendens of the mammalian hypoglossal nerve.

SPINAL NERVES

General principles

The first spinal nerve emerges between the atlas and the occipital bone, and in most birds the last spinal nerve escapes between the last free caudal vertebra and the pygostyle. The spinal nerves are numbered by simply counting the paired nerves in sequence. In the domestic fowl there are typically 41 pairs of spinal nerves. Each nerve escapes through a single intervertebral foramen, except in the region of the synsacrum where there is one opening for the dorsal root and another for the ventral root of each spinal nerve. Because of the great variation in the number of vertebrae in the different species, the total number of pairs of spinal nerves varies correspondingly, being for example only 38 in the pigeon and 51 in the Ostrich. As mentioned on page 53, the first thoracic vertebra is identified as the cranialmost vertebra carrying a rib with both a vertebral and a sternal part, the sternal part articulating directly with the sternum. This enables the last cervical spinal nerve to be identified as the nerve arising immediately cranial to the first thoracic vertebra. On this basis, the typical number of cervical spinal nerves is 15 or 16, with a maximum of 26 in swans. In the domestic fowl there are usually 17 cervical spinal nerves and sometimes 18. The smallest number probably occurs in parakeets in which the number is apparently as low as 12. It is not possible to distinguish between the last thoracic, the lumbar, sacral and the first few caudal spinal nerves; all of these emerge from the synsacrum. This difficulty is overcome by giving the term 'synsacral' to all the spinal nerves which emerge from the synsacrum.

Typically, each spinal nerve arises by a *dorsal root* and a *ventral root*. In general, afferent fibres are carried by the dorsal root and efferent fibres by the ventral root. The afferent fibres in the dorsal root have their cell bodies in the dorsal root ganglion. However, it has been claimed that some afferent fibres occur in the ventral roots and some efferent fibres in the dorsal roots of birds. Typically, each dorsal root possesses a *dorsal root ganglion*, but the first cervical spinal nerve lacks such a ganglion, and the second possesses only a vestigial dorsal root ganglion. The dorsal root ganglia of the cervical and thoracic spinal nerves usually lie within the intervertebral foramen, whereas those of the synsacral and caudal nerves are placed just peripheral to the intervertebral foramen. In the domestic fowl the dorsal root ganglia of the thoracic and the more cranial synsacral segments fuse with the corresponding paravertebral ganglia of the sympathetic chain. The dorsal and ventral roots consist of very fine fanlike filaments which are difficult to see with the naked eye. In each segment the dorsal root combines with the ventral root to form a mixed spinal nerve. The number of nerve fibres, and hence the actual size of both the dorsal and the ventral roots, vary with the functional importance of the segments of the body which they innervate. In strong flying birds the number of fibres and the size of the roots, and hence also the size of the spinal nerves themselves, is much greater in the region of the brachial plexus (Fig 15–10) than elsewhere. In fast running birds like the Ostrich, the roots and spinal nerves are greatly enlarged in the region of the lumbosacral plexus.

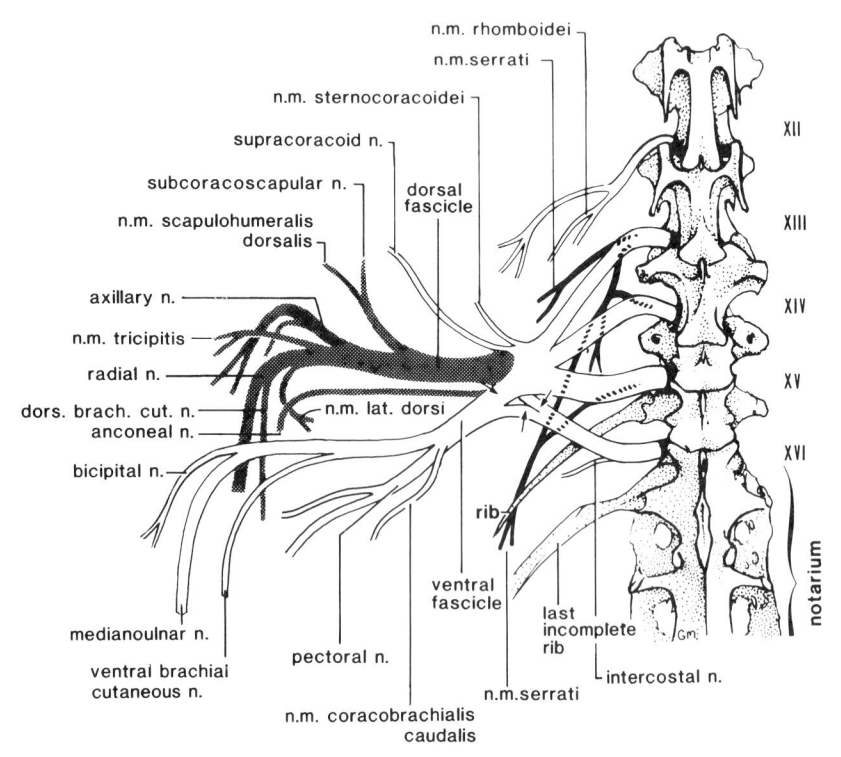

Fig 15–10 A ventral view of the right brachial plexus of the domestic fowl. The accessory plexus is shown in black. XII to XVI = levels of the ventral rami of the spinal nerves forming the roots of the plexus. The four main roots (XIII to XVI inclusive) unite to form three trunks, indicated by arrows, in which there is an exchange of fibres. These trunks resolve themselves into a dorsal fascicle (shaded) and a ventral fascicle (unshaded). The nerves supplying individual muscles are designated by the abbreviations n.m. Thus n.m. tricipitis is a nerve to the triceps muscle. Other abbreviations: dors. brach. cut. n. = dorsal brachial cutaneous nerve; n.m. lat. dorsi = nerve to the latissimus dorsi muscle. From Baumel (1975), with kind permission of the publisher.

Having emerged from its intervertebral foramen, each spinal nerve divides into three main branches. The first of these is the relatively small *meningeal branch*, which re-enters the vertebral canal to innervate the spinal meninges. The second is the *dorsal ramus*, which supplies the epaxial aspect of the body, i.e. the musculature and skin dorsal to the vertebral column. Because the vertebrae of birds tend to fuse together the epaxial muscles are weakly developed and therefore the dorsal rami are relatively much smaller than those of mammals. An exception, however, occurs in the neck which is highly mobile and therefore requires well-developed epaxial (and hypaxial) muscles. The third main branch of a spinal nerve is the *ventral ramus*. This supplies the hypaxial musculature and skin, and is generally much larger than the dorsal ramus. It is the ventral rami that form the great nerve plexuses of the wing and hindlimb, i.e. the brachial and lumbosacral plexuses.

The spinal nerves of the neck

As already stated the number of cervical spinal nerves depends on the number of cervical vertebrae and ranges from 12 to 26. Like those of mammals, the cervical spinal nerves of birds leave the vertebral canal *cranial* to the vertebra of the same segment, except for the *last* cervical nerve which emerges caudal to the last cervical vertebra. Hence the number of cervical spinal nerves will exceed the number of cervical vertebrae by one; thus if there are 14 cervical vertebrae in a certain species there will be 15 pairs of cervical spinal nerves. The first two cervical nerves are atypical in being mainly motor, either lacking a dorsal root ganglion (C1) or possessing only a vestigial ganglion (C2). Most of the first cervical spinal nerve, and perhaps some of the second, are absorbed by the hypoglossal nerve to form the *hypoglossocervical nerve*. The third and each of the subsequent cervical spinal nerves form, at each segment of the neck, a dorsal cervical cutaneous nerve from the dorsal ramus and a ventral cervical cutaneous nerve from the ventral ramus. These pairs of dorsal and ventral cervical cutaneous nerves form a succession of ringlike sensory cutaneous fields along the neck. In the neck as in all other parts of the body, the cutaneous nerves innervate the feather follicles and their papillae, and also the bands of smooth muscle which move the feathers. Complex laminated sensory nerve endings (Herbst corpuscles) are found throughout most of the skin, not only in the feathered areas but also in the apteria and the scaly integument of the feet.

The spinal nerves of the trunk

The number of 'thoracic' spinal nerves in birds is uncertain since there is no reliable way of identifying the last thoracic vertebra (p. 51) and hence the last thoracic spinal nerve. The number is generally estimated to be about five to ten pairs in the various species. Unlike the cervical spinal nerves, each thoracic spinal nerve leaves the vertebral canal *caudal* to its related vertebra, and hence caudal to its related rib. The ventral rami of thoracic nerves form intercostal nerves which run in the intercostal spaces. These lie immediately under the parietal pleura, along the caudal border of the corresponding rib. An intercostal nerve innervates not only the intercostal muscles but also the musculature of the thoracic wall and the abdominal muscles. The costoseptal muscle (Fig 5–1), which in the past has been erroneously referred to as the 'diaphragm' (p. 136), is innervated by intercostal nerves. The first two synsacral nerves also supply many branches to the skin and muscles of the abdominal region.

The nerves of the wing

The nerves of the wing arise from the *brachial plexus*. In most birds the brachial plexus itself is formed from the ventral rami of four or five spinal nerves. In the domestic fowl (Fig 15–10) these are typically nerves XIII to XVI and sometimes also XVII. In the pigeon the plexus lies unusually far cranially, since it arises from nerves X to XV inclusive. In many passerine species the origins are from nerves XII to XV inclusive. The cranial–caudal extremes occur in the Common Swift (X to XIV) and in swans (XXII to XXVI). The ventral ramus of each of these spinal nerves constitutes a *root* of the plexus. The roots combine with each other to form two or three short *trunks* of the plexus (arrows in Fig 15–10). Each trunk tends to divide into a dorsal and ventral division. The dorsal divisions then unite to form a dorsal fascicle and the ventral divisions combine to form a ventral fascicle (Fig 15–10). There is also, in many birds, an *accessory brachial plexus*. This is a network of small nerves which arise from two to four of the more cranial roots of the main plexus. In Fig 15–10 the accessory plexus is shaded black. It distributes branches to the rhomboidalis and serratus muscles, which arise on the axial skeleton and insert on the scapula thereby stabilizing the scapula and contributing to the locomotory movements of the wing. The dorsal fascicle forms peripheral nerves which supply the musculature and integument of the dorsal or extensor surface of the wing. In the shoulder region the ventral fascicle forms nerves which innervate the ventral or flexor aspect of the wing. The very large *pectoral nerves* control the all-important muscles of the downstroke of the wing. These nerves arise typically from the ventral fascicle itself (Fig 15–10). The *supracoracoid nerve* which is also important in flight because it activates the upstroke of the wing by the supracoracoid muscle, again arises from the ventral fascicle (Fig 15–10). A number of other nerves which control the muscles of the shoulder joint are shown in Fig 15–10, including nerves to the sternocoracoid, coracobrachialis, subcoracoscapularis, subscapularis and scapulohumeralis muscles.

In the brachial region the *ventral fascicle* innervates the flexor muscles of the wing by means of branches of the *medianoulnar nerve*, which is the direct continuation of the ventral fascicle (Fig 15–11A). The *bicipital nerve* innervates the biceps brachii, the main flexor of the elbow joint. The *ventral propatagial nerve* (Fig 15–11A) supplies the ventral part of the leading edge of the propatagium. A short distance above the elbow joint the medianoulnar nerve divides into the ulnar and median nerves (Fig 15–11A).

The *ulnar nerve* and its branches supply some of the flexor muscles of the carpus and some of the muscles of the manus. It also supplies sensory fibres to the ventral aspect of the skin in the region of the primary and secondary flight feathers and to the joints of the elbow, carpus and hand. At the elbow the nerve divides into its relatively small cranial ramus and the larger caudal ramus (Fig 15–11A). The cranial ramus ends in metacarpal rami to the skin and

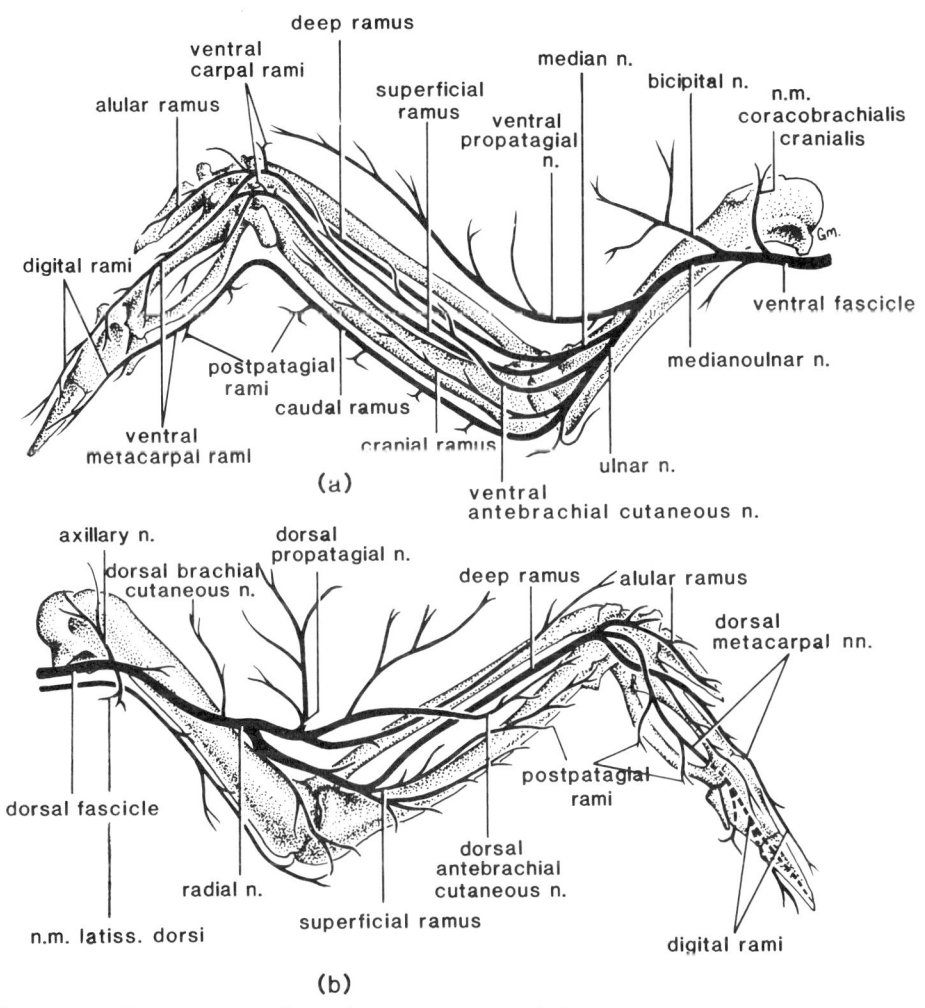

Fig 15–11 Diagrams to show the innervation of the right wing of the pigeon. (a) Nerves on the ventral aspect of the wing; (b) nerves on the dorsal aspect of the wing. latiss. = latissimus; m. = muscle; n. = nerve; nn. = nerves. From Baumel (in Breazile and Yasuda, 1979), with kind permission of the publisher.

joints of the manus (Fig 15–11A). The caudal ramus forms postpatagial rami, and ventral metacarpal and digital rami supplying the skin and joints of the manus (Fig 15–11A). The ulnar nerve innervates the flexor carpi ulnaris, flexor digiti minoris and interosseus ventralis muscles.

The *median nerve* innervates the brachialis muscle, most of the flexor muscles of the carpus and digits, and several of the muscles of the manus. It also distributes afferent fibres to the skin of the ventral surface of the forearm and manus including the propatagium, and to the ventral aspects of the elbow, carpal and digital joints. At the elbow joint it divides into its superficial and deep rami (rs and rp in Fig 15–11A). The superficial ramus supplies the superficial and deep digital flexor muscles. The deep ramus supplies the ulnometacarpalis ventralis muscle and then enters the manus where it innervates the abductor digiti majoris and flexor alulae muscles. Finally, it forms an alular ramus, ventral carpal rami, ventral metacarpal rami, postpatagial rami and digital rami (Fig 15–11A) which innervate the joints and skin of the manus.

The *dorsal fascicle* gives rise to motor nerves of the brachial region which elevate the humerus. Thus it forms the *axillary nerve* (Fig 15–11B) which innervates the major and minor deltoid muscles and supplies afferent fibres to the shoulder joint and the skin in the dorsal region of the shoulder, arm and propatagium. The dorsal fascicle then distributes motor fibres to the triceps brachii muscles (Fig 15–10) and to the smooth muscle of the expansor secundariorum. The dorsal fascicle also supplies a branch to the latissimus dorsi muscle (Fig 15–11B). Cutaneous branches from the dorsal fascicle are distributed to the caudodorsal skin of the distal region of the arm and elbow and to the follicles of the proximal secondary flight feathers.

The *radial nerve* is the direct continuation of the dorsal fascicle. It supplies motor branches to the extensor muscles of the carpal and digital joints, and cutaneous branches to the skin of the dorsal aspect of the propatagium, forearm and manus, including the dorsal aspect of the follicles of the primary and secondary flight feathers. One of its first branches is the dorsal cutaneous antebrachial nerve to the caudodorsal region of the forearm. Then follows the large dorsal propatagial nerve (Fig 15–11B) which innervates the dorsal surface of the propatagium. Just proximal to the elbow the radial nerve sends motor fibres to the extensor metacarpi radialis muscle, and sensory fibres to the dorsal aspect of the elbow joint. Distal to the elbow it divides into its superficial and deep rami (Fig 15–11B). The superficial ramus supplies, for example, the extensor digitorum communis and extensor metacarpi ulnaris and ends in postpatagial rami (Fig 15–11B) to skin and follicles related to the secondary flight feathers. The deep ramus innervates several extensor muscles including the ulnometacarpalis dorsalis, interosseus dorsalis and adductor alulae muscles. It ends by forming dorsal metacarpal nerves with postpatagial and digital rami (Fig 15–11B); these supply the skin and follicles of the primary flight feathers, and also the digital joints.

The nerves of the pelvic limb

The pelvic limb is innervated by branches of the lumbosacral plexus. In most species of birds the lumbosacral plexus is formed by the ventral rami of eight spinal nerves, as in Fig 15–12. It is divided into a lumbar and a sacral plexus. The *lumbar plexus* typically has three roots derived from three synsacral nerves as in Fig 15–12, i.e. from the second synsacral nerve, which arises between the second and third synsacral vertebrae, and from the next two successive synsacral nerves. The roots of the lumbar plexus typically form two trunks (the uppermost two arrows in Fig 15–12). The *sacral plexus* (Fig 15–12) usually has six roots formed by the ventral rami of six spinal nerves. Of these, the most cranial (synsacral nerve no. 4) also constitutes the most caudal root of the lumbar plexus. Since this nerve contributes one root to the lumbar plexus

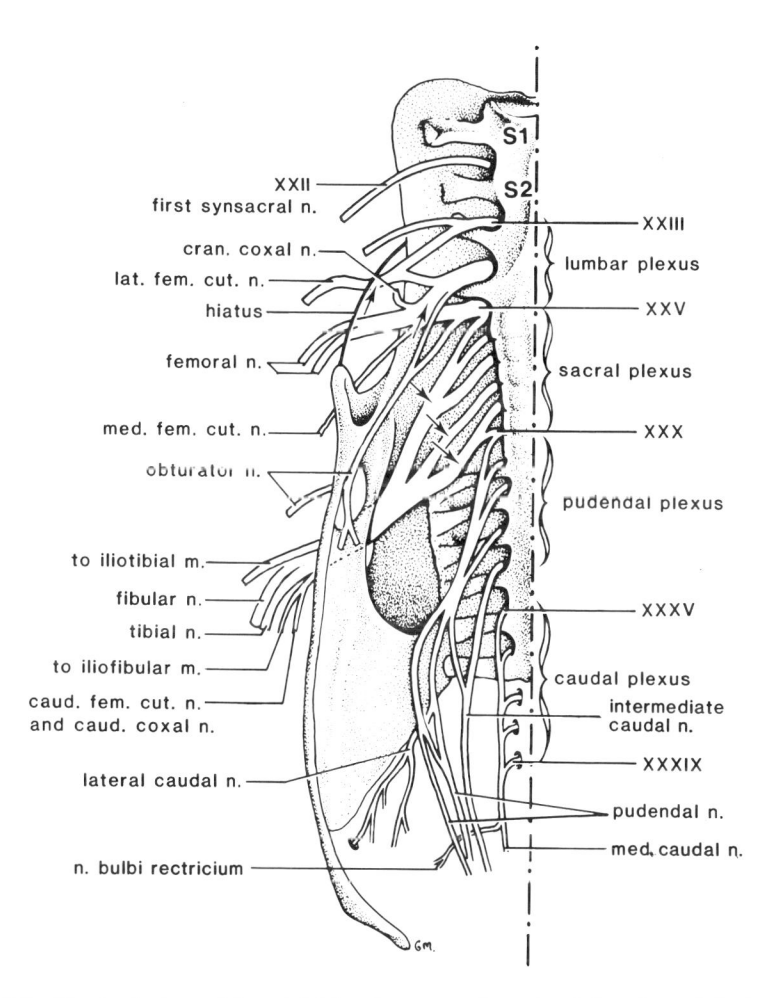

Fig 15–12 Ventral view of the right lumbar, sacral, pudendal and caudal plexuses of the domestic fowl. XXIII–XXXIX = ventral rami of spinal nerves; the two arrows pointing cranially = the two trunks of the lumbar plexus; the three arrows pointing caudally = the three trunks of the sacral plexus; S1 and S2 = first and second synsacral vertebrae (these were vertebrae XXI and XXII in this specimen); hiatus = aperture between the lateral border of the ilium and the abdominal muscles, through which the peripheral nerves arising from the lumbar plexus escape from the pelvis. The lumbosacral plexus of this species is typical of birds generally. It has eight roots. The lumbar plexus has three roots and the sacral plexus six roots. Nerve XXV (the fourth synsacral nerve) contributes a root to both the lumbar and the sacral plexus, and is known as the 'furcal' nerve. The pudendal plexus has five roots (nerves XXX–XXIV inclusive). This is one more root than is typical of birds generally; the extra root is derived from nerve XXX, which divides to give a root to both the sacral and the pudendal plexus and is consequently known as the 'bigeminal' nerve. The caudal plexus is formed by five roots (nerves XXXV–XXXIX inclusive). caud. = caudal; cran. = cranial; cut. = cutaneous; fem. = femoral; med. = medial; n. = nerve; m. = muscle. From Baumel (1975), with kind permission of the publisher.

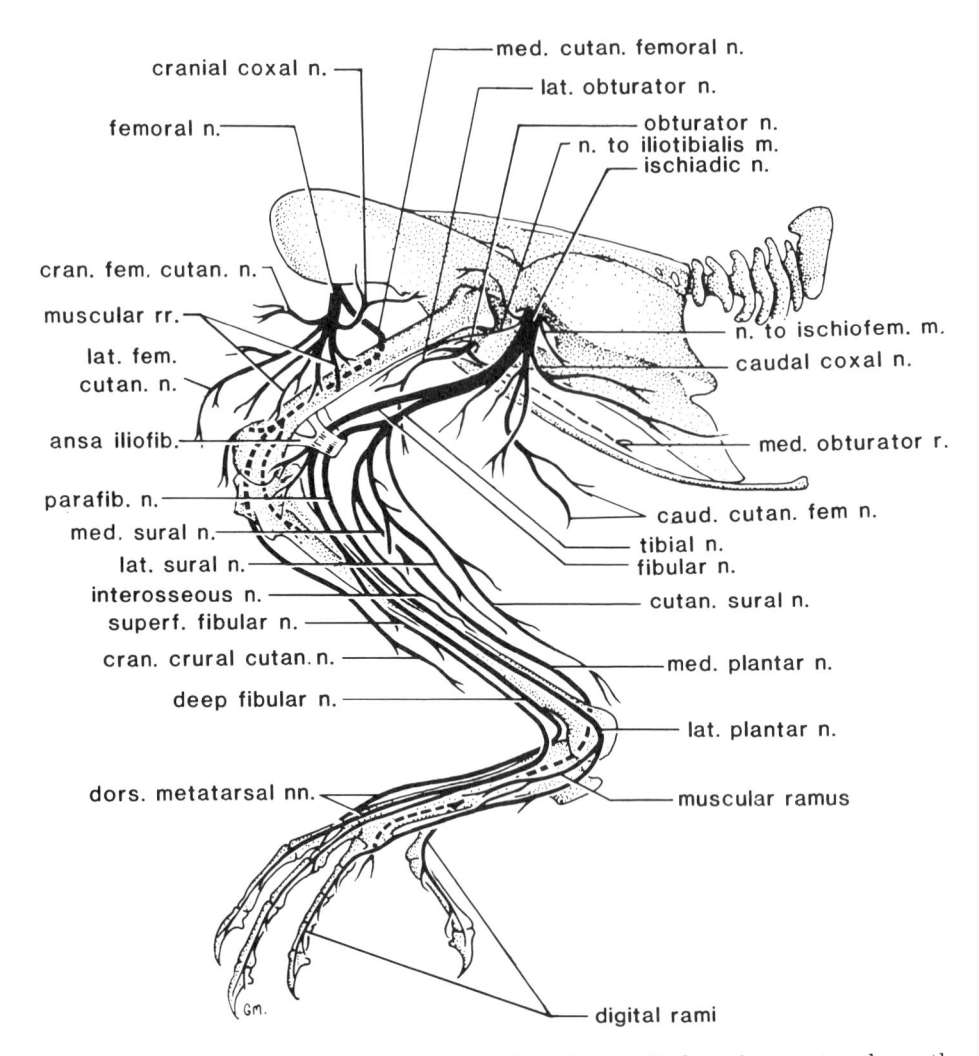

Fig 15–13 Lateral view of the left pelvic limb of the pigeon, to show the principal nerves. cran. = cranial; cutan. = cutaneous; dors. = dorsal; fem. = femoral; iliofib. = iliofibular; ischiofem. = ischiofemoral; lat. = lateral; m. = muscle; med. = medial; n. and nn. = nerve and nerves; parafib. = parafibular; r. and rr. = ramus and rami; superf. = superficial. From Breazile and Yasuda (1979), with kind permission of J.J. Baumel and the publisher.

and one to the sacral plexus it has been termed the 'furcal' nerve. The last root of the sacral plexus is synsacral nerve no. 9 (Fig 15–12). The roots of the sacral plexus tend to form three trunks (the lowermost three arrows in Fig 15–12).

The *lateral femoral cutaneous nerve* (Figs 15–12 and 15–13) supplies the skin of the craniolateral region of the thigh, the lateral region of the knee and the proximal part of the leg. The *medial femoral cutaneous nerve* (Figs 15–12 and 15–13) innervates the iliofemoralis internus muscle, distributes cutaneous fibres to the medial aspect of the knee and the medial proximal part of the leg, and then continues as the cranial crural cutaneous nerve (Fig 15–13) to supply skin along the cranial aspect of the leg. The *femoral nerve* forms the cranial coxal nerve (Figs 15–12 and 15–13) which supplies the muscle mass on the dorsal surface of the wing of the ilium, cranial to the acetabulum. Another cutaneous branch of the femoral nerve is the cranial femoral cutaneous nerve (Fig 15–13) which innervates a large area of skin on the cranial aspect of the thigh. The femoral nerve supplies motor fibres to the extensors of the knee joint; these include the femorotibialis and iliotibialis muscles. It also innervates the ambiens muscle. The femoral nerve supplies some articular fibres to the knee joint. The *obturator nerve* (Figs 15–12 and 15–13) divides into a medial obturator ramus which innervates the medial obturator muscle, and a lateral obturator ramus (Fig 15–13) to the lateral obturator and the pubo-ischio-femoralis muscles. The lumbar plexus forms one or two long nerves to the iliotrochanteric muscles. These muscles have been considered (not correctly) to be homologues of the gluteal musculature of mammals.

Of the nerves arising from the sacral plexus, the caudal coxal nerve supplies the caudo-ilio-femoralis muscle and the lateral and medial flexor cruris muscles (Fig 15–13). The caudal cutaneous femoral nerve is distributed to the skin of the caudal region of the thigh (Fig 15–13). There are muscular rami to the iliotibial, iliofibular, iliotrochanteric and ischiofemoral muscles. The *ischiadic nerve* consists of the tibial and fibular nerves which are united together within a common epineurial sheath. The ischiadic nerve is the largest nerve in the avian body. The *tibial nerve* is the larger of the two terminal branches of the ischiadic nerve. It divides (Fig 15–13) into the medial and lateral sural nerves. The *medial sural nerve* supplies the extensors of the intertarsal joint (hock or ankle joint), of which the gastrocnemius is the most important. It also supplies the deep flexors of the digits, which are important in the action of grasping. The medial sural nerve continues into the medial plantar nerve (Fig 15–13) which supplies the skin on the medial aspect of the hock joint and metatarsal region. The *lateral sural nerve* also innervates the gastrocnemius and completes the innervation of the flexors of the digits. The parafibular nerve (Fig 15–13) is also a branch of the tibial nerve. It continues directly into the lateral plantar nerve, which supplies motor fibres to some of the small intrinsic abductor, adductor and flexor muscles of the digits. It also innervates the skin on the lateral aspect of the distal part of the leg, the hock and the tarsometatarsus, and supplies the metatarsal pad of the foot. It ends by forming digital rami.

The *fibular nerve* at first lies within a common epineurial sheath with the parafibular nerve. Half-way down the leg it divides into the superficial and deep fibular nerves (Fig 15–13). It innervates the flexors of the hock joint and the extensors of the digits, including the tibialis cranialis and extensor digitorum longus muscles. The *superficial fibular nerve* continues into metatarsal and digital nerves (Fig 15–13). It innervates the small intrinsic extensors of the digits and supplies sensory fibres to the knee joint and hock joint. It also supplies the skin of the tarsometatarsus and digits. The *deep fibular nerve* continues into dorsal metatarsal and dorsal digital nerves. It contributes to the innervation of the small intrinsic extensor and abductor muscles of the digits. Its sensory fibres include articular fibres to the hock joint and small branches to the toes.

The nerves of the tail

The pudendal and caudal plexuses are the sources of the nerves which supply the tail. In the domestic fowl the *pudendal plexus* arises from the ventral rami of four or five spinal nerves, numbers XXX to XXXIII or XXXIV inclusive; these are synsacral nerves 9 to 12 or 13. In some specimens there is an additional root from the last root of the sacral plexus; since this supplies both the pudendal plexus and the sacral plexus, it is then known as the 'bigeminal' nerve (nerve XXX in Fig 15–12). In the pigeon there are five roots from spinal nerves XXVII to XXXI. The nerves arising from the pudendal plexus innervate the striated muscles of the tail and cloaca, and the skin in the region of the tail and vent. The *lateral caudal nerve* (Fig 15–12) supplies the muscle of the rectrical bulb (m. bulbi rectricium) and the lateralis caudae, depressor caudae, pubocaudalis internus and transversus cloacae muscles. Cutaneous branches of this nerve innervate the skin on the ventrolateral aspect of the tail, and the skin in the ventral abdominal region, and the vent itself. The *intermediate caudal nerve* (Fig 15–12) supplies motor fibres to the pubocaudalis externus and sphincter cloacae muscles, and supplies cutaneous branches to the skin of the dorsal and ventral lips of the vent and the ventral surface of the tail. The *pudendal nerve* (Fig 15–12) runs with the ureter to the dorsolateral aspect of the cloaca. Here it forms the *cloacal plexus*, a network of fine nerves and ganglia on the cloacal wall to which the intestinal nerve also contributes. Visceral (autonomic) branches of the pudendal nerve emerge from the cloacal plexus as ureteral, oviductal and deferential branches which supply the terminal regions of the urogenital ducts. Other visceral filaments of the pudendal nerve emerge from the cloacal plexus as *cloacal rami* and innervate the phallus, paracloacal vascular body, and the coprodeum, urodeum and proctodeum. Yet other visceral fibres of the pudendal nerve are believed to traverse the cloacal plexus and enter the intestinal nerve. Somatic branches of the pudendal nerve run in the cloacal rami to the striated muscles of the cloaca and to the skin round the vent and on the ventral aspect of the tail. A number of direct branches from the pudendal plexus supply the feather follicles of the rectrices and their smooth muscles and also innervate the striated adductor rectricium muscle.

In the domestic fowl (Fig 15–12) and the pigeon the *caudal plexus* is formed by the ventral rami of about five spinal nerves, i.e. nerves XXXV to XXXIX inclusive in Gallus and XXXII to XXXVI inclusive in the pigeon. These roots form the medial caudal nerve (Fig 15–12) which supplies the depressor caudae muscle and the muscle of the rectrical bulb.

The roots of the lumbar plexus are in contact with the dorsal surface of the cranial division of the kidney. The sacral plexus is totally embedded within the middle division of the kidney, and the beginning of the ischiadic nerve runs through kidney tissue. Some of the roots of the pudendal plexus are embedded in the caudal division of the kidney.

AUTONOMIC NERVOUS SYSTEM

As in mammals the autonomic nervous system can be divided into two systems: (1) the craniosacral (parasympathetic) system, which has a craniosacral efferent outflow from the brain and sacral spinal cord and functions to conserve bodily reserves; and (2) the thoracolumbar (sympathetic) system, which has an efferent outflow from thoracolumbar segments of the spinal cord and functions to induce physiological responses appropriate to flight and fight. Both systems are constructed on a basis of preganglionic and postganglionic efferent pathways; the preganglionic endings are cholinergic in both systems, but the postganglionic endings are mainly cholinergic in the craniosacral system and noradrenergic in the thoracolumbar system. The parasympathetic and sympathetic systems cooperate to preserve homeostasis of the internal environment, typically by means of a dual innervation for each organ. The autonomic nervous system is often arbitrarily regarded as purely efferent; however, this concept should be avoided since the autonomic nervous system operates by reflex arcs like all other parts of the nervous system, and virtually all the nerves which form it contain both afferent and efferent fibres. Indeed some major autonomic nerves are predominantly afferent. For example, the cervical vagus in the domestic fowl is at least 60 per cent afferent, and the afferent information which it transmits to the brain about arterial pressure, blood gases, and the mechanical and chemical events in the heart, lungs and alimentary tract are indispensable to life.

The classical concept of the pre- and postganglionic efferent pathways as outlined above is beginning to crumble as more knowledge of the structure and function of the autonomic nervous system becomes available. Recent morphological and pharmacological observations on mammalian and particularly on avian tissues have made it clear that the organization of the autonomic nervous system is in fact much more complicated than these simple generalizations suggest.

Craniosacral system

The cranial division of this system is formed by the IIIrd, VIIth, IXth and Xth cranial nerves. These distribute parasympathetic efferent pathways to the ciliary body and iris (IIIrd nerve), the glands of the orbit and nasal cavity (VIIth nerve), the salivary glands (VIIth and IXth nerves), and the heart, lungs and alimentary tract at least as far as and probably beyond the duodenum (Xth nerve). The caudal cardiac nerves of the vagus tonically inhibit the heart rate, and the pulmonary rami cause contraction of the atrial muscles of the parabronchial wall. The gastrointestinal fibres are presumably excitatory to the glands and smooth muscle of the gut.

The 'sacral' division presumably supplies the rest of the small intestine and all of the large intestine, as well as the urogenital and cloacal organs. The sacral parasympathetic outflow arises from the pudendal plexus from spinal nerves XXX to XXXIII inclusive in the domestic fowl (Fig 15–12). These form the *pudendal nerve* (also known as the pelvic nerve), which accompanies the

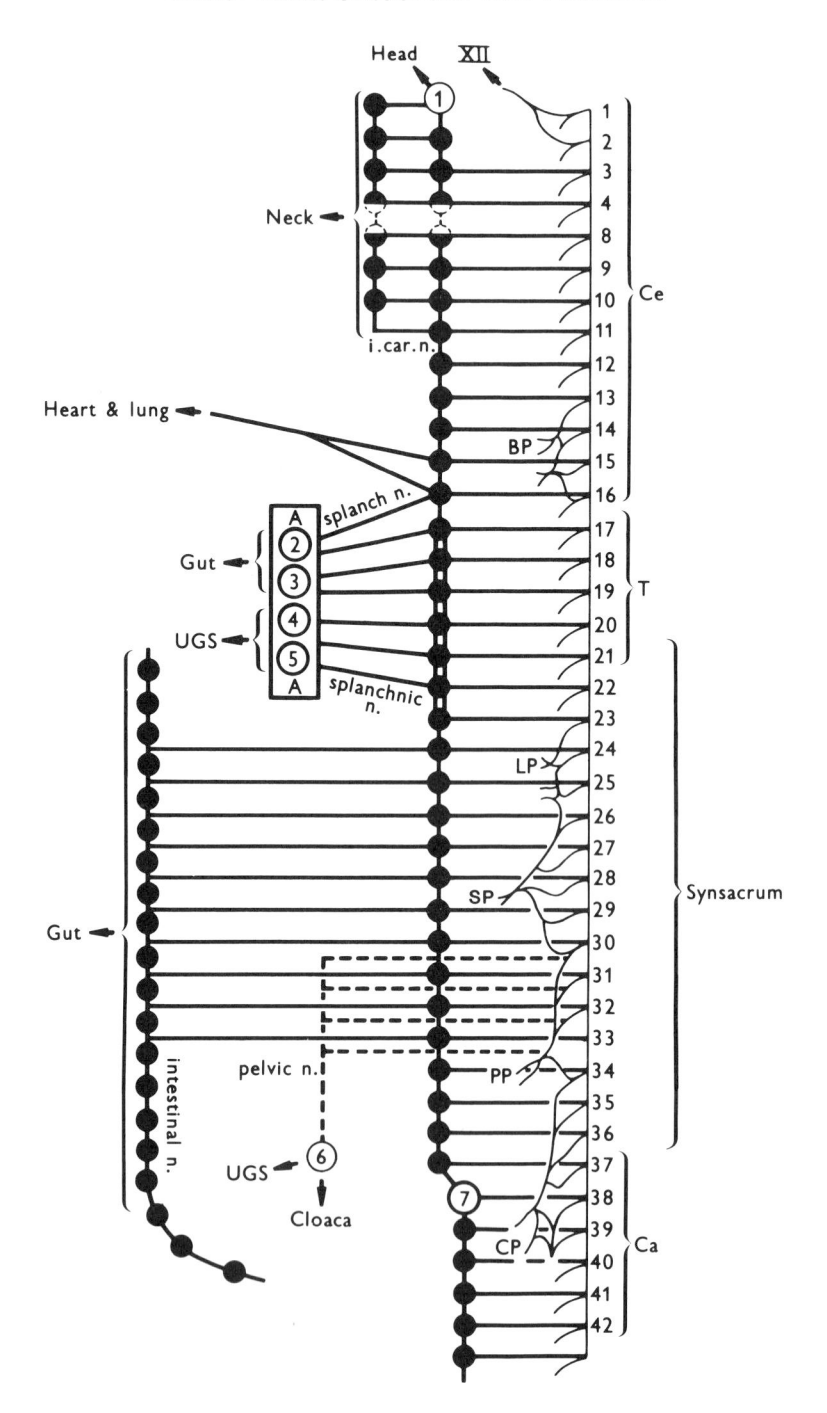

Fig 15-14 Ventral view diagram of the right side of the thoracolumbar division, and of the sacral part of the craniosacral division, of the visceral (autonomic) part of the peripheral nervous system of the domestic fowl. 1 to 42, segmental spinal nerves (curved) and rami communicantes (straight lines). Ce. = cervical; T = thoracic;

ureter to the dorsal wall of the cloaca. There it contributes to the network of nerves and cloacal ganglia known as the cloacal plexus. The intestinal nerve (see **Thoracolumbar system**) also contributes to the cloacal plexus, which is therefore a combination of parasympathetic and sympathetic elements. The cloacal plexus supplies branches to the oviduct, ductus deferens, ureter and cloaca.

Thoracolumbar system

The thoracolumbar (sympathetic) system is based on the chain of paravertebral ganglia segmentally arranged, and a series of subvertebral ganglia associated with the coeliac artery and other great branches of the aorta (Fig 15–14). The chain of paravertebral ganglia (the sympathetic chain) extends from the base of the skull to the pygostyle. The cranial end of the chain is formed by the *cranial cervical ganglion*. This is the largest sympathetic ganglion in the bird. It lies close to the medial aspects of the IXth and Xth nerves as they leave the skull. It distributes the equivalent of grey rami communicantes to the cranial nerves, and also gives rise to fibres which follow the great arteries of the head. By these pathways the thoracolumbar system distributes sympathetic fibres to the iris and ciliary body, and to all the glands and many of the blood vessels of the head.

The *paravertebral trunk* in the neck travels in the transverse foramina of the cervical vertebrae with the vertebral artery, like the mammalian vertebral nerve. In contrast to the mammal, where the segmental ganglia in the embryonic neck eventually fuse into only three cervical vertebral ganglia, the bird has a segmental paravertebral ganglion at each cervical segment. Only the last two or three of these cervical ganglia have macroscopically visible *rami communicantes*; all the other cervical ganglia lie so close to their spinal nerves that the rami communicantes are concealed. Accompanying the internal carotid artery throughout the length of the neck is a smaller sympathetic trunk, also with segmental ganglia, the *cervical carotid nerve*. This receives a communicating filament from each cervical paravertebral ganglion (Fig 15–14).

Ca. = caudal; BP = brachial plexus; LP = lumbar plexus; SP = sacral plexus; PP = pudendal plexus; CP = caudal plexus. XII = XIIth nerve. The nerves of the thoracolumbar system are shown in solid lines. The circles are ganglia of the thoracolumbar system; solid black circles on the right side of the diagram represent the paravertebral ganglia of the paravertebral trunk; i.car.n., which in the diagram passes vertically in the neck region, represents the cervical carotid nerve accompanying the internal carotid artery; splanch n. = splanchnic nerves connecting the paravertebral ganglia of segments 16 to 22 to A, the aortic plexus. Circles containing the numbers 2, 3, 4 and 5, represent the following subvertebral ganglia: 2 = coeliac ganglion; 3 = cranial mesenteric ganglion; 4 and 5 = adrenal ganglia. ① represents the cranial cervical ganglion. The intestinal nerve and its ganglia supplying the gut, are shown on the extreme left of the diagram, together with their connections to the paravertebral ganglia. The sacral part of the craniosacral (parasympathetic) system is shown in broken lines arising from spinal nerves 30 to 34; pelvic n. = the pudendal nerve; UGS = urogenital system; ⑥ = a representative of the cloacal ganglia.

The paravertebral ganglia of the thoracic and cranial synsacral segments are fused to the dorsal root ganglia, and therefore rami communicantes are absent in these regions. The connecting links of the thoracic part of the chain divide (Fig 15–14), passing dorsal and ventral to the heads of the ribs.

The *cardiac nerve* (commonly known as the cardiac sympathetic nerve) arises from the first thoracic paravertebral ganglion in the domestic fowl, and from the last cervical and first three thoracic ganglia in the pigeon. The cardiac nerve carries cardioaccelerator fibres to the heart; these fibres end mainly on the sinuatrial node, right atrial myocardium and atrioventricular node, and less extensively on the myocardium of the left atrium and the two ventricles. It has been claimed that some sympathetic fibres join the vagus and ascend the neck as a vagosympathetic trunk, but this is doubtful.

The *splanchnic nerves*, about seven altogether, are all essentially segmental nerves arising from the thoracic and synsacral paravertebral ganglia of (about) segments 16 to 22 inclusive (Fig 15–14); because of the reduction of the hypaxial musculature these nerves are easily visible on the bodies of the thoracic vertebrae and synsacrum. The thoracic splanchnic nerves project to a cluster of subvertebral ganglia which surround the roots of the coeliac and cranial mesenteric arteries forming the *coeliac* and *cranial mesenteric ganglia* (Fig 15–14). These ganglia send postganglionic fibres to the alimentary tract via the branches of their related arteries. The synsacral splanchnic nerves project to the many other large and small subvertebral ganglia (the adrenal ganglia) which lie on the capsule of the adrenal glands (Fig 15–14) and supply the ovary extensively as well as the oviduct and kidneys. The many fibres passing to and from all these ganglia produce a more or less continuous aortic plexus along the ventral aspect of the aorta. Attempts to divide this plexus into renal, adrenal, aortic and ovarian plexuses cannot really be justified, so closely blended are the nerves and ganglia. Caudal to the gonads the paravertebral ganglia and rami communicantes of the more caudal synsacral segments become relatively more obvious, the paravertebral ganglia no longer being fused to the dorsal root ganglia. The left and right sympathetic (paravertebral) trunks penetrate the kidney and converge on a median ganglion impar (Fig 15–14) at the level of the free caudal vertebrae. In some species there may be several unpaired ganglia imparia. Caudal to this point the left and right paravertebral trunks continue as a single midline chain of ganglia. The ganglia of the caudal segments have distinct rami communicantes.

The *intestinal nerve* (Remak's nerve) is a large midline ganglionated nerve arising from the plexuses of sympathetic fibres associated with the cranial mesenteric artery, aorta and caudal mesenteric artery. It travels in the mesentery, parallel with and close to the mesenteric suspension of the gut from the duodenum to the cloaca, and innervates the intestines. The intestinal nerve receives parasympathetic and sympathetic fibres at its cranial and caudal ends, and sympathetic fibres along its whole length. Nerve cell bodies are scattered along the nerve, forming recognizable ganglia at 30 to 50 sites. Probably these ganglia contain both sympathetic and parasympathetic cell bodies. Caudally the intestinal nerve ends in the *cloacal plexus*, where it is believed to acquire parasympathetic fibres from the *pudendal nerve*.

Postmortem examination of autonomic nerves

The diagnosis of the neural form of Marek's disease is commonly based on the postmortem examination of the brachial, lumbar and sacral plexuses, intercostal nerves, ischiadic nerve and parts of the paravertebral trunks (sympathetic chains). Less attention tends to be paid to the autonomic nerves because of their small size. However, it is possible to identify nearly all neural cases of Marek's disease by examining only the aortic plexus, intestinal nerve, splanchnic nerves, brachial plexus and the sacral plexus, in that order. Anatomically, this involves a minimum of dissection and is quicker and more reliable than the more conventional procedure.

ORIENTATION AND NAVIGATION

Birds undoubtedly have a remarkable capacity for orientation and navigation. These abilities depend extensively on the special senses, but there are other factors which appear to operate beyond the range of sensory phenomena familiar to man. Moreover, the many cues which appear to be involved in avian orientation and navigation presumably require integration within the central nervous system. For these reasons, this brief discussion is placed at the end of the chapter on the nervous system rather than the special sense organs.

Celestial cues

It is certain that the sun and stars are dominant orientational cues for birds. However, these factors can only provide compass information. To gain orientational information from the sun's direction, the bird must also know the time so that it can compensate for the changing position of the sun during the day; there is, indeed, evidence that birds are able to couple their circadian rhythm with the sun's azimuth (direction from the observer). On the other hand using the patterns of the stars does not require compensation for time, and it has been shown that some species can use a star compass to determine directions without needing time compensation. Thus the sun compass and star compass are major cues for orientation in birds. Nevertheless, there is good evidence that neither of these celestial factors is *essential* for accurate orientation. For example, experienced homing pigeons can orientate themselves accurately towards home from release points which are far distant and unfamiliar even when there is total cloud cover, and so also can nocturnal migrants when the stars are invisible through heavy overcast. It is therefore apparent that birds must use 'back-up' cues. Furthermore, they need additional information to tell them when they have actually *arrived* at their destination. Common sense seems to suggest that birds could achieve this by learning to recognize *visually* the main geographical landmarks in a particular vicinity. How else could swallows return to the same individual building year after year? However, evidence suggests that physical landmarks are *not* used much in migratory orientation; in fact pigeons that have flown daily over their home base totally fail to recognize it if they are released only half a mile away after they have been shifted six hours out of phase with true sun time. Moreover, pigeons wearing lenses that restrict vision to 2 or 3 m can home successfully from a distance.

Polarized light

Because of the geometric relationships between the position of the observer, the plane of polarization of sunlight and the position of the sun, the position of the sun can be determined if the polarization can be detected. There is no direct evidence that orientating birds do use polarized light in this way. However, if they can do so they could, like honeybees, use a derivative of the sun compass when the sun itself is overcast but blue sky remains.

Ultraviolet light

Experiments have indicated that homing pigeons have some capacity to detect ultraviolet light. Honeybees analyze polarization in the ultraviolet wavelengths, and the possibility that pigeons can also do this is under investigation.

Olfaction

It has been suggested that young pigeons learn to correlate particular odours with winds coming from certain directions. It seems likely that odours are one of the many cues that birds may use when navigating, but the extent to which they use such information remains controversial.

Detection of infrasound

It has been shown that homing pigeons are sensitive to sound frequencies below 1 Hz, much lower than the generally assumed lower limit of 100 Hz. At this level of sensitivity it would be possible for a bird to detect effective signals from the background noise of the environment. Such low infrasonic frequencies are capable of travelling hundreds or perhaps thousands of kilometres with little loss of energy. Therefore birds could determine their approximate position by detecting the infrasound of waves breaking on the shore or wind whistling through mountain tops. Comparisons between the two ears to determine the direction from which such sounds are coming would not be possible for such very long wavelengths which exceed 3 km at 0.1 Hz. However, the Doppler shift when a bird flies away from or towards a source of infrasound would be well within the range of the bird's capacity for detection. Therefore a bird may be able to detect direction from infrasounds while it is flying, even if it cannot when stationary.

Magnetic detection

The results of attaching bar magnets or Helmholtz coils to pigeons suggest that the earth's magnetic field does provide compact information which experienced birds use mainly when the sun compass is not available. There is evidence that, like the honeybee, birds are extremely sensitive to magnetic changes of less than 10^{-3} G, and do utilize them when they are orientating themselves despite the weakness of the earth's magnetic field (about 0.5 G). Ferromagnetic material, probably magnetite, has been found in a small piece of tissue about 2 mm^2 in area sited between the dura mater and the skull of pigeons. This tissue consists of connective tissue and nerve fibres, together with numerous clusters of minute electron-opaque structures which are rich in iron. Such structures could attract or repel each other in a way which could indicate the direction of the earth's magnetic field. Whatever the magnetic mechanism may be, it is apparently the *alignment* of the magnetic vector and not its polarity that influences the birds' behaviour. Thus certain species of bird orientate themselves northward in the spring whether the magnetic field vector points north and down, as is normal, or south and up; they reverse their orientation to southward if the magnetic vector points north and up or south and down.

Gravity

Gravity undergoes a *regular* variation in a north–south gradient because the earth is not a perfect sphere. It also changes *irregularly* because of the varying density of the material in the earth's crust at different sites. Hence gravity cues could aid navigation both by indicating the north–south axis and by providing a topographical map. Observations indicate a significant correlation between the orientation capacity of pigeons and the changing relative positions of the earth, sun and moon. However, there is still no direct evidence that birds can detect the extremely minute differences in gravity (less than 10 gal) which would be required as gravitational cues in long-distance navigation.

Barometric pressure

There is evidence that birds about to migrate do have a remarkable capacity for detecting changes in barometric pressure. For example, autumnal migrations in eastern North America tend to take place on the east side of a high pressure zone after the passage of a cold front, whereas spring flights tend to take place on the west side of a high pressure zone in front of an advancing low pressure area. It has been shown that homing pigeons are sufficiently sensitive to barometric pressure to recognize a change in altitude of only 10 m, or even less. The ability to use barometric pressure as an altimeter would be useful to a bird flying in cloud. Furthermore, such an ability is essential if a bird is to utilize changes in gravity, since gravity decreases with increasing altitude.

Conclusions

It is evident that a whole variety of different cues can be exploited depending on the species, age and experience of the bird, on the weather, on the season of the year, and on the geographical location. No single cue so far discovered appears to be essential and no one system of cues can explain all aspects of avian orientation. The problem now is to discover how the bird integrates all the different cues which it has the capacity to utilize. As yet, not enough is known to reveal any complex or complexes of orientational and navigational systems which can fully account for what the birds themselves do actually achieve.

Further reading

Akester, A.R. (1979) The autonomic nervous system. In *Form and Function in Birds* (Ed.) King, A.S. & McLelland, J. Vol. 1. London and New York: Academic Press.

Akker, L.M. van den (1970) *An Anatomical Outline of the Spinal Cord of the Pigeon*. Van Gorcum: Arren.

Baptista, L.F. & Abs, M. (1983) Vocalizations. In *Physiology and Behaviour of the Pigeon* (Ed.) Abs, M. London and New York: Academic Press.

Baumel, J.J. (1958) Variation in the brachial plexus of *Progne subis*. Acta anat., 34, 1–34.

Baumel, J.J. (1975) Aves nervous system. In *Sisson and Grossman's The Anatomy of the Domestic Animals* (Ed.) Getty, R. Vol. 2. Philadelphia, London and Toronto: Saunders.

Bennet, T. (1974) Peripheral and autonomic nervous system. In *Avian Biology* (Ed.) Farner, D.S. & King, J.R. Vol. 4. New York and London: Academic Press.

Breazile, J.E. (1979) Systema nervosum centrale. In *Nomina Anatomica Avium* (Ed.) Baumel, J.J., King, A.S., Lucas, A.M., Breazile, J.E. & Evans, H.E. London and New York: Academic Press.

Breazile, J.E. & Yasuda, M. (1979) Systema nervosum peripheriale. In *Nomina Anatomica Avium* (Ed.) Baumel, J.J., King, A.S., Lucas, A.M., Breazile, J.E. & Evans, H.E. London and New York: Academic Press.

Bubien-Waluszewska, A. (1981) The cranial nerves. In *Form and Function in Birds* (Ed.) King, A.S. & McLelland, J. Vol. 3. London and New York: Academic Press.

Bubien-Waluszewska, A. (1984) Somatic peripheral nerves in birds. In *Form and Function in Birds* (Ed.) King, A.S. & McLelland, J. Vol. 4. London and New York: Academic Press.

Cords, E. (1904) Beiträge zur Lehre von Kopfnerven der Vögel. *Anatomische Hefte*, 26, 50–97.

Gilbert, A.B. (1969) The innervation of the ovary of the domestic fowl. *Q. Jl exp. Physiol.*, 54, 404–411.

Hsieh, T.M. (1951) The sympathetic and parasympathetic nervous systems of the fowl. PhD Thesis, University of Edinburgh.

Jungherr, E.L. (1969) The neuroanatomy of the domestic fowl. *Avian Dis.*, Special Issue.

Keeton, W.T. (1979) Avian orientation and navigation: a brief overview. *Br. Birds*, 72, 451–470.

McLelland, J. & Abdulla, A.B. (1972) The gross anatomy of the nerve supply to the lungs of *Gallus domesticus*. *Anat. Anz.*, 131, 448–453.

Malinovsky, L. (1962) Contribution to the anatomy of the vegetative nervous system in the neck and thorax of the domestic pigeon. *Acta anat.*, 50, 326–347.

Nauta, W.J.H. & Karten, H.J. (1970) A general profile of the vertebrate brain, with sidelights on the ancestry of cerebral cortex. In *The Neurosciences* (Ed.) Schmitt, F.O. New York: Rockefeller University Press.

Pearson, R. (1972) *The Avian Brain*. London and New York: Academic Press.
Schrader, E. (1970) Die Topographie der Kopfnerven von Huhn. Inaugural Dissertation, Free University of Berlin.
Wakley, G.K. & Bower, A.J. (1981) The distal vagal ganglion of the hen (*Gallus domesticus*), a histological and physiological study. *J. Anat.*, 132, 95–105.
Watanabe, T. (1960) On peripheral course of the vagus nerve in the fowl. *Jap. J. vet. Sci.*, 22, 145–154.
Watanabe, T. (1964) Peripheral courses of the hypoglossal, accessory and glossopharyngeal nerves. *Jap. J. vet. Sci.*, 26, 249–258.
Watanabe, T. (1972) Sympathetic nervous system of the fowl. Part 2. Nervus intestinalis. *Jap. J. vet. Sci.*, 34, 303–313.
Watanabe, T., Isomura, G. & Yasuda, M. (1967) Distribution of nerves in the oculomotor and ciliary muscles. *Jap. J. vet. Sci.*, 29, 151–158.
Watanabe, T. & Yasuda, M. (1970) Peripheral course of the trigeminal nerve in the fowl. *Jap. J. vet. Sci.*, 32, 43–57.
Yasuda, M. (1960a) The distribution of the nerves to the trunk. *Jap. J. vet. Sci.*, 22, 434.
Yasuda, M. (1960b) On the nervous supply of the thoracic limb in the fowl. *Jap. J. vet. Sci.*, 22, 89–101.
Yasuda, M. (1960c) On the nervous supply of the coccygeal region. *Jap. J. vet. Sci.*, 22, 503.
Yasuda, M. (1961) On the nervous supply of the hind-limb. *Jap. J. vet. Sci.*, 23, 145–155.
Yasuda, M. (1964) Distribution of the cutaneous nerves of the fowl. *Jap. J. vet. Sci.*, 26, 241–254.

Chapter 16

SPECIAL SENSE ORGANS

EYE

Nearly all birds are intensely visual animals. A simple indication of this fact is the *size* of the eye which is extremely large in relation to the rest of the head. Indeed some owls and hawks have eyes which are absolutely as large or even larger than those of a man. The head accounts for about 10 per cent of the total body weight in both a man and a starling, but the weight of the eyes forms less than 1 per cent of the weight of the human head; in the starling it forms about 15 per cent. The eye of the Ostrich is about 50 mm in diameter and is absolutely the largest among the contemporary terrestrial vertebrates. In many birds the two eyes together outweigh the brain, but the weight ratio between the eyes and the brain in the domestic fowl is about 1:1. The relatively large eye in birds as a group permits a correspondingly large image to be projected on the retina, thus contributing to the acuity of avian vision.

In species with narrow heads such as pigeons the *position* of the eyes is lateral in the skull, while birds with broader heads such as falconiforms have eyes which are directed more frontally. Thus in pigeons the angle between the right and left *bulbar axes* is about 145° (the bulbar axis being a line passing through the centre of the cornea and lens to the retina). In predators like the Common Kestrel the angle may be reduced to 90° or less. Species with laterally placed eyes have a larger *visual field* (300° in pigeons) than those with frontally directed eyes (150° in the Barn Owl). On the other hand, the *binocular field* of vision in pigeons (24°) is correspondingly less than that of owls (60–70°). In binocular vision both eyes are focused on the same object, and the movement of the two eyes becomes coordinated. Monocular vision occurs when only one eye is focused on one object at any particular moment. Most birds can use binocular vision, but some (e.g. penguins) cannot.

The eyeball consists of a small anterior region covered by the cornea, a much larger and almost hemispherical posterior region covered on the outside by the sclera, and a variably shaped intermediate region based on the scleral ossicles and uniting the other two regions. In the *'flat'* eyeball (Fig 16–1A) found in the majority of diurnal birds with narrow heads like the domestic fowl, the intermediate region is a flat disc almost parallel with the surface of the body and the bulbar axis is relatively short. Because of the shortness of the axis in this type of eye the image thrown on the retina is relatively small, so visual acuity is correspondingly low in these species. In diurnal birds with wider heads, like passeriforms and birds of prey, the intermediate region is cone-shaped producing a *'globular'* eye (Fig 16–1B) and relatively greater visual acuity. Nocturnal birds of prey, e.g. owls, have a *'tubular'* eye (Fig 16–1C), in which the intermediate region is relatively elongated. Fig 16–1 shows that in all these different forms of avian eye the *shape of the retina* resembles a parabolic reflector (like the dish aerial of a radio telescope). Consequently, the

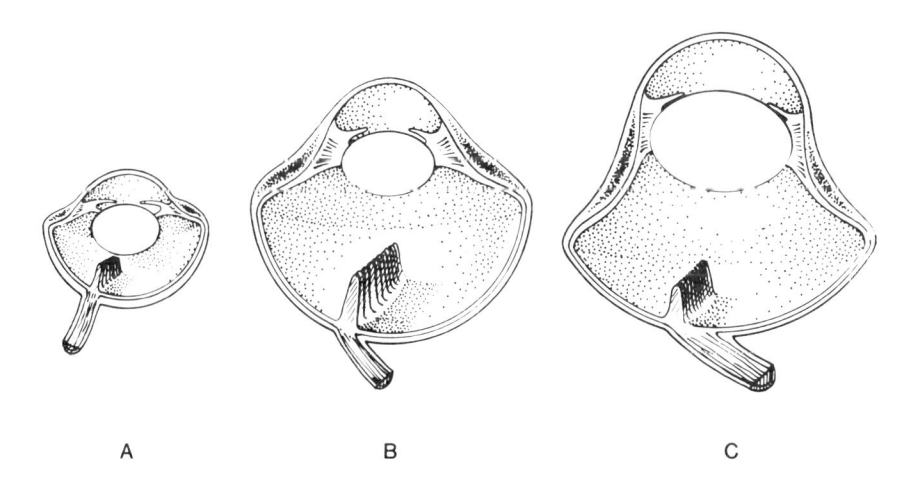

Fig 16–1 The ventral half of the left eyeball of **A** a 'flat' eye as in swans, **B** a 'globular' eye as in eagles, and **C** a 'tubular' eye as in owls. In all three forms the eyeball consists of a relatively small corneal region, a variable intermediate region supported by the scleral ossicles (dark shading), and a more or less hemispherical 'posterior' region. In all birds the eyeball is somewhat asymmetrical, in that the bulbar axis begins slightly towards the nasal side of the midline (to the right in these diagrams). From Walls (1942), with kind permission of the Cranbrook Institute of Science, Michigan.

avian retina lies fairly near the point of focus for all directions of incident light (Fig 16–2a): in the more spherical mammalian eye, the retina lies anterior to the focus of any light which enters the eye obliquely. Thus the mammalian retina has only one region of acute vision and relatively poor visual acuity elsewhere, whereas the avian retina has all round visual acuity. In all birds, even those with laterally placed eyes, the eyeball is somewhat asymmetrical. Thus the intermediate region is slightly shorter on the nasal side than on the temporal side (Fig 16–1). Consequently, the bulbar axis when projected forwards tends to be directed somewhat towards the midline, and presumably this favours binocular vision.

The *wall* of the eyeball consists of the same general layers as in mammals, i.e. an outer fibrous tunic comprising the cornea and sclera, a middle vascular tunic, and an inner nervous layer or retina (Fig 16–3).

Fibrous tunic

The fibrous tunic is a tough layer which maintains the shape of the eyeball and protects the deeper layers within it. The *cornea* in most birds has a relatively small area compared to the rest of the eyeball. Its area is particularly small in underwater swimmers, but it tends to be relatively more extensive and more strongly curved in species such as eagles and owls with globular or tubular eyes (Fig 16–1B and –1C). As in mammals it consists of an anterior (outer) stratified squamous epithelium, an anterior (outer) limiting lamina (Bowman's membrane), a substantia propria which forms the great bulk of the corneal wall and consists of bundles of collagen fibres, a posterior (inner)

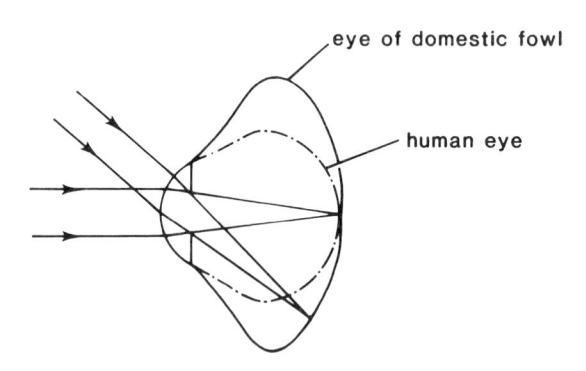

Fig 16–2a Diagram of the eye of a bird and a man superimposed, the scales being adjusted so that the polar diameters are equal. The avian retina lies near the position of focus for all directions of incident light, but the retina of the human eye lies anterior to the focal surface except at the bulbar axis. From King-Smith (1971), with kind permission of the publisher.

limiting lamina (Descemet's membrane), and a posterior (inner) layer of simple cuboidal epithelium. In the domestic fowl all these layers are present, but in some species the anterior limiting lamina is not differentiated. The thickness of the cornea is about 450 μm in the domestic fowl, which is about midway in the range of thickness in the various species. The difference between the refractive indices of the cornea and of air is relatively great, and therefore the cornea is very important in refracting light in air. Under water, however, the cornea has no power to act like a lens since its refractive index is practically the same as that of water.

In birds generally, as in many reptiles, the whole of the *sclera* is reinforced by a continuous layer of hyaline cartilage, except in the region of the scleral ossicles (Fig 16–3). In the zone which is nearest to the cornea the wall of the sclera is modified into a ring of small, roughly quadrilateral, overlapping bones, the *scleral ossicles*, which strengthen the eyeball and provide attachments for the ciliary muscles. The number of ossicles varies from 10 to 18, but in most species including the domestic fowl there are usually 14 or 15. In many species, including falconiforms, hummingbirds, woodpeckers and passeriforms, the scleral cartilage round the optic nerve is ossified forming a U-shaped bone, the os nervi optici. The *scleral venous sinus* (canal of Schlemm) is quite conspicuous in some species but in others is small or almost invisible. It lies at the *limbus*, i.e. at the junction between the cornea and sclera (Fig 16–3). In this region a wide meshed plexus of connective tissue fibres, the *trabecular reticulum* (pectinate ligament), joins the limbus to the iris and to the ciliary body. The spaces between these fibres form the *spaces of the*

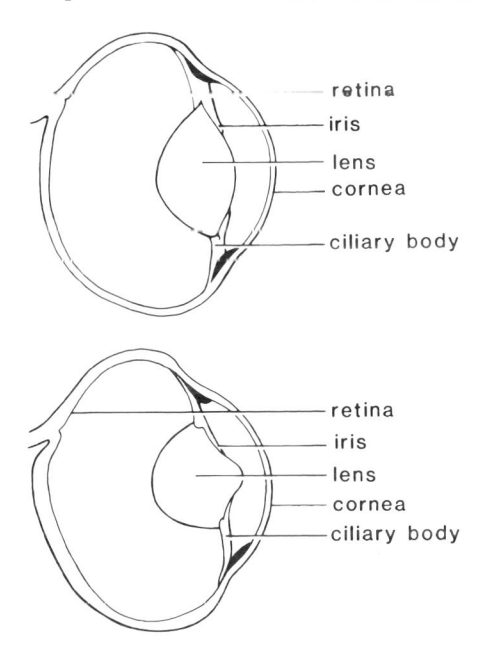

Fig 16–2b Diagrammatic sagittal sections of the eyes of the Hooded Merganser with accommodation relaxed (top) and with accommodation induced (bottom). Coordinated action of the sclerocorneal muscles and the sphincter muscle of the iris have forced the lens against the iris so that the central part of the lens bulges through the pupil. From Sivak (1980), with kind permission of the editor of *Trends in Neurosciences*.

irido-corneal angle (spaces of Fontana) through which the aqueous humour drains into the scleral venous sinus.

Vascular tunic

The vascular tunic consists of the choroid, the ciliary body and the iris (Fig 16–3). The *choroid* is the thick, highly vascular and darkly pigmented layer which coats the retina and contributes nutrition to the tissues of the eyeball. It is particularly vascular in many divers. A *tapetum lucidum*, the brilliantly coloured area which reflects light in many nocturnal vertebrates, has been found in only a few birds (e.g. the nocturnal goatsuckers). The choroid is continued by the ciliary body and iris. The *ciliary body* suspends the lens by the zonular fibres, and also forms small folds, the ciliary processes, which

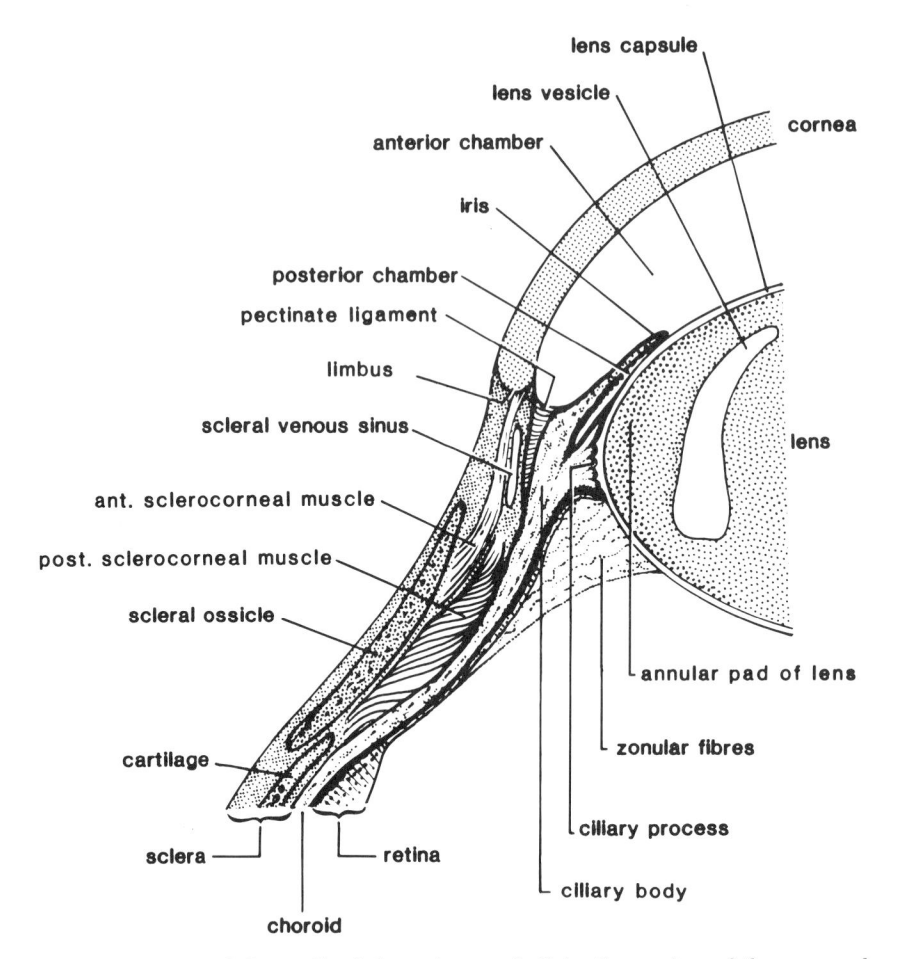

Fig 16–3 Diagram of the wall of the avian eyeball in the region of the cornoscleral junction and ciliary body. The scleral wall is supported by cartilage and, near the corneoscleral junction (the limbus), by the scleral ossicles. The thick choroid continues into the ciliary body, which suspends the lens by the zonular fibres. The ciliary processes form aqueous humour, which eventually drains away by percolating through the pectinate ligament into the scleral venous sinus. During accommodation the posterior sclerocorneal muscle moves the ciliary body towards the bulbar axis, thus pressing the ciliary body against the annular pad of the lens; the anterior sclerocorneal muscle may pull the cornea posteriorly, thus increasing the curvature of the cornea. The diagram is composite from numerous sources including Slonaker (1918), Walls (1942), Pumphrey (1961) and Evans (1979).

produce the aqueous humour. The ciliary processes are firmly pressed against the rim of the lens by the ciliary muscles (the muscles of accommodation), and are directly attached to the lens capsule (Fig 16–3). There are two ciliary muscles (sclerocorneal muscles) in birds (Fig 16–3), i.e. the anterior sclerocorneal muscle (Crampton's muscle) and the posterior sclerocorneal muscle (Brücke's muscle). These muscles are striated in birds, in contrast to the smooth muscle of the ciliary muscle in mammals. Two other possible striated ciliary muscles, Muller's muscle and the temporal ciliary muscle, are believed to be simply subdivisions of the posterior sclerocorneal muscle. These various subdivisions of the ciliary musculature may be largely artificial separations caused mainly by branches of the ciliary nerve. In general, the greatest development and greatest subdivision of the ciliary muscles occur in hawks where a relatively high degree of accommodation is required for focusing on moving prey; the least development and subdivision are found in ground-feeding and seed-feeding birds such as the House Sparrow and Eastern Meadowlark.

The mechanism of *accommodation* is entirely different in mammals and birds. In mammals the ciliary muscles act by reducing the tension on the zonular fibres and thus allowing the elastic lens passively to assume a more spherical shape. In many birds the posterior sclerocorneal muscle brings about accommodation directly by forcing the ciliary body against the lens so that the curvature of the lens is increased. In nocturnal birds including owls, but also in hawks, this muscle is weak, but accommodation in this group is achieved by the anterior sclerocorneal muscle. The attachments of this muscle from the cornea to the sclera (Fig 16-3) enable it to distort the cornea by pulling the corneoscleral junction posteriorly, thus increasing the curvature of the cornea at its centre. In diving birds this mechanism would be ineffective while the bird is under water, since the cornea is then no longer a significant refracting surface; as would therefore be expected, the anterior muscle is very much reduced in diving birds. Diving creates problems in focusing, because of the sudden need for an additional 20 diopters of refraction to compensate for the loss of corneal refraction. Some diving birds such as terns seem to lack the means to overcome this difficulty and are consequently long-sighted (hypermetropic) while in the water; they detect their fish from above, but often miss them in the water. Penguins, on the other hand, can see accurately in the water but are somewhat short-sighted (myopic) on land. However, many other diving birds, including cormorants, diving ducks and dippers, have truly amphibious vision. In these birds the lens is particularly soft and the two sclerocorneal muscles are enormously powerful. Even so, additional accommodation is probably produced by compression of the anterior part of the lens by the strongly developed sphincter muscle of the iris. Furthermore, it has been suggested that the coordinated action of the sclerocorneal muscles and the iris sphincter could force the soft lens against the iris and hence cause the centre of the lens to bulge through the pupil (Fig 16-2b) thus greatly increasing its curvature at the bulbar axis. The power of accommodation of the cormorant has been shown to be four to five times greater than that of the young human adult. As mentioned below under **Retina**, kingfishers seem to have adopted yet another device, i.e. a second fovea, to obtain sharp vision under water. To sum up, three mechanisms of accommodation occur in birds: (1) compression of the whole lens by the posterior sclerocorneal muscle, in many diurnal birds; (2) distortion of the centre of the cornea by the anterior sclerocorneal muscle, in nocturnal birds and hawks; and (3) compression of the front of the lens by the combined action of the sclerocorneal muscles and the sphincter muscle of the iris perhaps forcing the lens to bulge through the pupil, in many diving birds. To these can be added the possible utilization of a fovea for underwater vision, in kingfishers.

The *iris* in most birds is dark in colour, ranging from brown to black, but in some species it is highly coloured (e.g. yellow in most owls, green in cormorants, red in night-herons and pale blue in the Gannet). Refractive cells (iridiocytes) form a *tapetum lucidum of the iris* in several columbiform species, and are responsible for the rapid changes in colour of the iris which occur in these species when excited. The shape of the pupil is round in almost all birds, but it forms a horizontal oval when dilated in some species (e.g. thick-knees); in skimmers, but no other known species of bird, it is a vertical slit. Measurements of the area of the skimmer's pupil when constricted and dilated suggest that the vertical shape in bright light achieves a greater reduction in pupillary area than would a constricted circular pupil; on the other hand when the bird is in almost complete darkness the pupil is circular and relatively wide in relation to the diameter of the eye. These characteristics may enhance the skimmer's ability to feed even during the darkest nights, and yet protect the retina when the bird is active in brilliant sunlight. In the domestic fowl the iris can become mis-shaped in ocular leucosis. In contrast to mammals the *sphincter* and *dilator muscles of the pupil* are again striated. The movements of the avian pupil can be very extensive and much faster than in mammals, and yet the pupil seems unexpectedly unresponsive to light possibly because of inhibition by the brainstem. In birds (and also in teleost fish and frogs, but not in mammals) the amount of light reaching the visual cells of the retina can also be regulated by photomechanical changes in the cells of the pigment epithelium of the retina. The processes of the pigment cells extend between the photosensitive tips of the receptor cells (Fig 16–5). In light adaptation the pigment migrates inwards within these processes, shielding the outer segments of the receptor cells.

Lens

The lens of birds is much softer than that of mammals. A lens vesicle, filled with fluid, lies between the annular pad and the body of the lens (Fig 16–3), contributing to the general softness of the lens. This softness facilitates the rapid accommodation which typifies the avian eye. The anterior surface of the lens is generally much flatter in diurnal species than in nocturnal and aquatic birds. In all birds the lens includes the annular pad (Ringwülst) round its equator, adjacent to the ciliary processes. The pad is particularly well developed in diurnal predators, but is reduced in nocturnal species, diving birds and flightless birds. In primates the lens acts as a yellow filter which cuts off light of wavelengths below 400 nm and therefore renders ultraviolet radiation invisible: the cornea and lens of diurnal birds are optically clear and appear to transmit wavelengths down to about 350 nm, thus rendering near ultraviolet radiation visible and absorbing only those ultraviolet wavelengths which are not physiologically destructive.

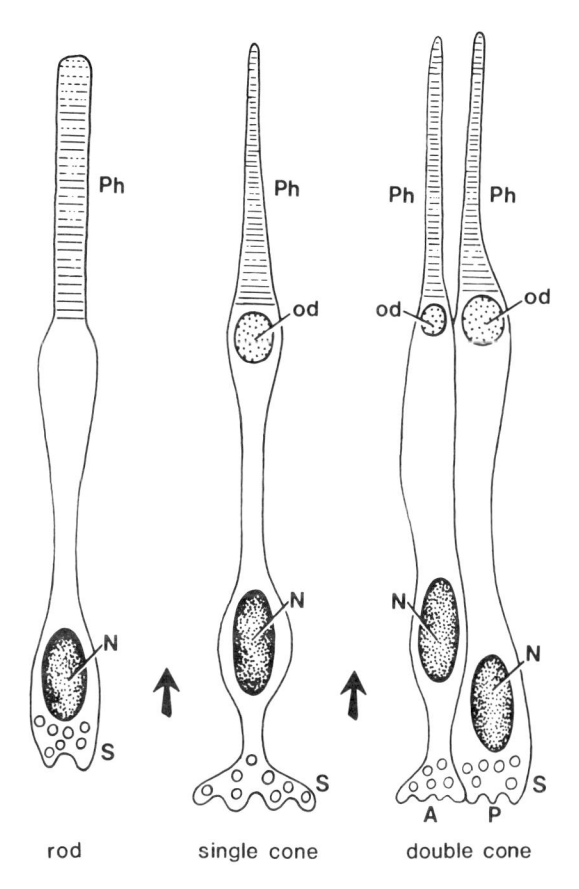

Fig 16–4 Cones are of two types: single and double. Double cones consist of a principal cone (P) and an accessory cone (A). They are widespread among vertebrates, except for placental mammals. Oil droplets (od) are present in single cones and in the principal cone of double cones, and may or may not be present in the accessory cone also depending on the species. They are never present in rods. S = synaptic zone; Ph = photosensitive laminae; N = nucleus; arrows = direction of light. Based on King-Smith (1971) and Sillman (1973), with kind permission of the publisher.

Retina

In contrast to the retina of mammals, that of birds is relatively thick and has no blood vessels. It resembles the mammalian retina, however, in consisting of a non-nervous pigment epithelium and a nervous layer formed essentially from rods and cones, bipolar cells and ganglion cells (Fig 16–5). The ganglion cells form the axons of the optic nerve. Cones (Fig 16–4) are responsible for visual acuity and colour vision. Diurnal birds have far more cones than rods over the entire retina, the cone density being greater than in man. In diurnal predators and passerines the few rods are confined to the periphery. Rods (Fig 16–4) are sensitive to the intensity of light; nocturnal birds such as owls have some cones but mostly rods. Although it has not been directly demonstrated, it

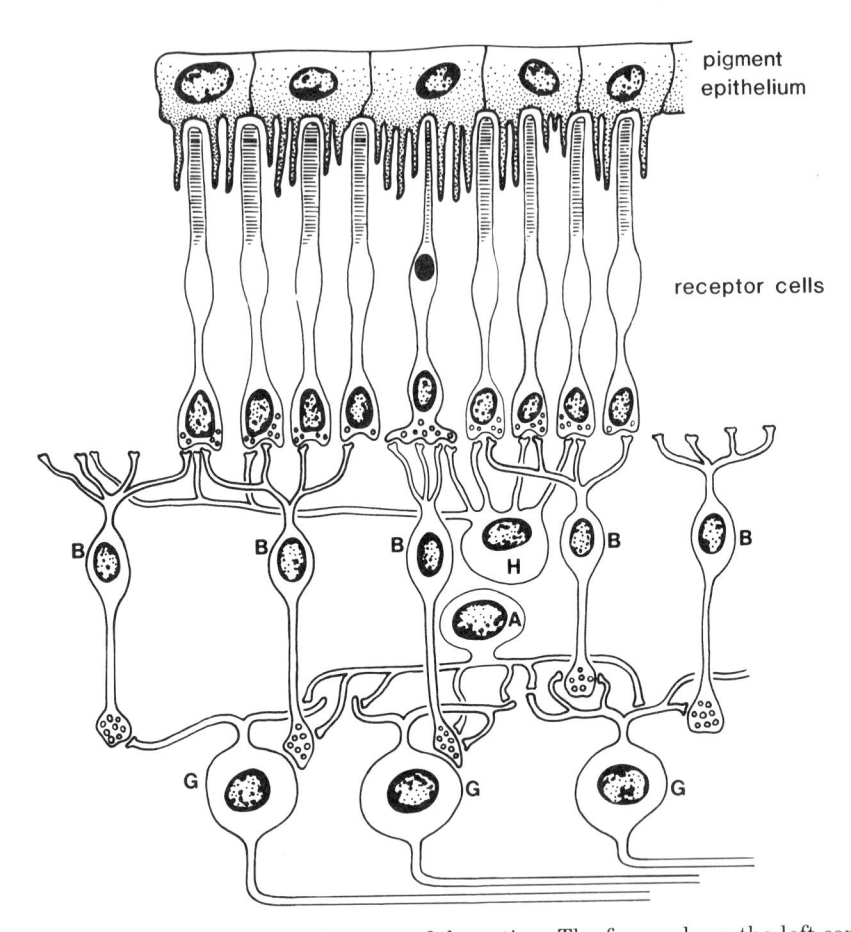

Fig 16–5 Simplified 'wiring' diagram of the retina. The four rods on the left converge onto one bipolar neuron (B). So also do the four rods on the right. The ganglion cells (G) on the right and left receive synapses from two or more rod bipolar cells. Thus many rods *converge* on one ganglion cell. In contrast the single cone in the middle (identified by its oil droplet, see Fig 16–4) may project on a single bipolar cell, which in turn may project onto one ganglion cell, thus creating a 'private line' for the cone. The horizontal cell (H) and the amacrine cell (A) cause *divergence* of neural activity from side to side.

is generally believed that in diurnal birds one cone synapses with a single bipolar cell, which in turn synapses with a single ganglion cell (Fig 16–5). Such a one-to-one projection of the cones to the brain would give each cone a 'private line' to the brain, and thus improve visual acuity. In contrast, it is widely accepted that several rods synapse with a single bipolar cell, and several of these bipolar cells synapse with a single ganglion cell (Fig 16–5). This convergence of many rods on one ganglion cell would give the eye of a nocturnal bird good visual sensitivity to small amounts of light. At the back of the eyeball the axons of the ganglion cells collect together at the *optic disc* where they pass through the wall to form the optic nerve. Since there are no rods or cones at the optic disc it is also known as the *blind spot*.

As in mammals, a part of the retina of birds is thickened into a mound where there is a relatively high concentration of cones and other nervous elements but no rods thus forming a special region of maximum optical resolution known as the *central area* (Fig 16–6). Some birds have a laterally-placed special region, which is then known as the *temporal area*. The central area may have a *fovea*, which in principle resembles that of primates in being a shallow bowl with its concavity facing the vitreous body (Fig 16–6); at the bottom of the fovea the cones are extremely closely packed, and since the non-receptor elements are displaced peripherally the light can pass almost directly to the receptors in this part of the retina. Compared to the fovea of primates, however, that of birds is much deeper. It has been suggested that the deep

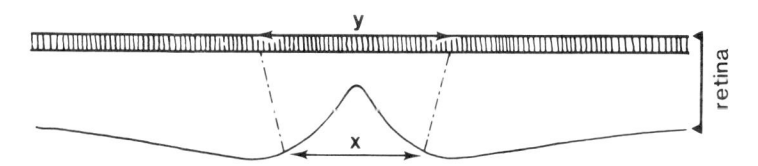

Fig 16–6 Diagram of an avian central area with fovea. The central area is a slight mound caused by packing in a relatively large number of cones at the expense of rods. The fovea is the depression in the centre of the mound. Many of the non-receptor elements, especially bipolar neurons, are displaced from the fovea thus causing the mound. The diagram shows how the convex lips of the fovea might magnify the image. Refraction of light as it passes through the retina may magnify the image at x, increasing it to y at the photoreceptive zone of the retina (hatched). Based on Walls (1942), with kind permission of the Cranbrook Institute of Science, Michigan, and Sillman (1973), with kind permission of the publisher.

fovea of birds may increase their visual acuity. The two steeply convex lips of the fovea could act like the edges of two adjacent convex lenses, refracting the rays of light so that they diverge and thereby magnify the area of the image by as much as 30 per cent (Fig 16–6). It has also been proposed that the fovea of the round central area fixes the eye on an object by refraction from its steep sides and increases the sensitivity to movement of the object. The temporal area of maximum optical resolution may also have a fovea.

There are three main types of arrangement of the fovea and area (or areas) of maximum optical resolution. (a) The great majority of species have a single *round central area* (in each eye) close to the optic axis, with a fovea. The domestic fowl and a few other species have a round central area but no fovea. (b) Two foveate areas, comprising a *central area* close to the optic axis and a laterally situated *temporal area*, occur in each eye in a number of species. Examples are terns, swallows and falconiforms, all of which pursue fast-moving prey or feed on the wing and therefore require a very accurate perception of distance and relative speed. The temporal fovea is so positioned that the image of the object is formed on the temporal fovea of both the left and the right eye simultaneously, indicating that they function in stereoscopic binocular vision. Kingfishers also have two foveae in each eye. These birds see well under water despite having no power of accommodation. It is suggested that in kingfishers one fovea is adapted for aerial and the other for aquatic vision. (c) Owls are unique among birds in possessing a single foveate temporal area in each eye. (d) Water birds and birds which live in open plains have a *horizontal central area* which is expanded into a ribbonlike band. This type of area can occur regardless of whether the retina is afoveate, monofoveate or bifoveate. The eye of these birds is held in such a position that the long axis of the horizontal central area lies close to the horizontal plane. The ribbonlike area could then fix the horizon accurately as a reference point.

There has been much discussion of the popular idea that diurnal birds have far greater *visual acuity* than mammals. Visual acuity is the sharpness with which detail is perceived, i.e. the resolving power of the eye. Factors which influence visual acuity include (1) the relative size of the eye, (2) the accuracy of focus on the various regions of the retina, (3) the possible magnifying capacity of the fovea, (4) the absence of blood vessels in the retina, (5) the fineness of 'grain' of the retina, (6) the degree of convergence of receptor cells on the ganglion cells, and (7) the amount of contrast between an object and its background. Of these factors, 1, 2, 3 and 6 have already been discussed, and all of them favour greater visual acuity in birds; 7 is considered below under colour vision. Factor 4 again indicates greater acuity in birds. In mammals the floor of the fovea is devoid of blood vessels and can be used for acute vision, but the rest of the retina suffers from optical interference by the retinal vessels. The complete absence of retinal vessels in birds means that the whole of the retina is free of such interference. Factor 5, retinal grain, is important for the same reason as the fineness of grain in photographic film is a factor in photographic detail. In diurnal birds the retina is heavily dominated by cones. In the fovea of the hawk there are about 300 000 cones per mm^2, whereas the human fovea contains only about 147 000 cones per mm^2. Furthermore, nearly every cone may be individually represented in the optic nerve of a diurnal bird, whereas in the human eye there are 6 to 7 million cones but only a million axons in the entire optic nerve. Moreover, diurnal birds seem to be able to detect a rapidly flickering stimulus better than man; thus the pigeon's retina can distinguish between individual flashes at a rate of 140 per second, while the maximum rate for man is about 70 per second. Measurements including counts of cones in the fovea suggest, nevertheless, that the resolving power of passerine and faconiform birds is in fact only about twice to three times greater than that of man or monkey. On the other hand, the anatomical characteristics of the eye make it likely that a diurnal bird could see a whole panorama as accurately as a mammal could see a single detail. Thus both a bird and a man could see a mouse from a height of 250 feet, but the man could only do so if his attention were accurately directed to it: the bird should be able to see it without looking directly at it. Moreover, the bird should be able to see in a single glance all the mice in a field, whereas the man could only do this by scanning the area laboriously with his central area and fovea. In other words the bird should be able to assimilate detail very much faster than a mammal. Furthermore, because it can follow a faster flickering stimulus, the bird should be better able to detect and follow movement.

Very few birds are active at night. Those that are, amount to less than 3 per cent of all the avian species, and the 146 species of owl account for well over half of these. The owls are the only nocturnal birds which are thought to rely at all extensively on vision for feeding. The *sensitivity* of the owl's eye to small amounts of light has therefore been much debated. The receptor cells of the owl's eye consist predominantly of rods but the ganglion cells are very numerous, indeed more so even than in the human fovea, so there appears to be relatively little convergence. Cones are present, and there is a temporal fovea placed for binocular vision. Measurements suggest that the owl's eye is in fact, only about two-and-a-half times more sensitive than that of a man. This superiority is not a factor of the sensitivity of the rods themselves; it has been suggested that the individual human rod can respond to the absorption of a single quantum of light, thus reaching the theoretical limit of light detection. The difference in absolute sensitivity appears to be due, instead, to the better optics of the owl's eye, which enable it to gather more light than the human eye; thus it loses fewer quanta from absorption, scatter and reflection in the optic media. The f-number of the owl's eye is about 1.30, whereas that of man is about 2.10. This produces a retinal image in the owl which is about two-and-a-half times brighter than that in a man. This optical advantage is, however, too small to explain the extraordinary ability of owls to fly between obstacles and to hunt successfully in what looks to us like pitch darkness. The reason lies much more in the sense of hearing. Owls can locate sounds with sufficient accuracy to catch prey in the total absence of light. However, the Tawny Owl can also see perfectly well in daylight, with colour vision. When food requirements are strong this highly nocturnal species will hunt by day. Its visual acuity is, indeed, equal to that of a pigeon and only slightly inferior to that of a man. The main factor accounting for the visual acuity of the owl is the exceptionally large size of its eye, which therefore produces a very large retinal image. Owls, like man, have a dual purpose eye that is well-adapted for both nocturnal and diurnal vision.

The magnificent plumage of many avian species suggests that birds have *colour vision*, and this is confirmed by behavioural and electrophysiological observations. Spectrophotometry has demonstrated three visual pigments in birds, and there is perhaps a fourth which is sensitive to wavelengths near the ultraviolet. Thus birds are at least trichromatic, and perhaps tetrachromatic. A complication in the avian cone (and in the cones of many amphibia and reptiles) is the presence of a brightly coloured oil droplet at the base of the photosensitive outer segment (Fig 16–4) and therefore interposed between the incident light and the visual pigment; the rods usually lack the oil droplet. The oil droplets are orange, red, yellow or clear, but five types have been identified by their absorbency spectra. One interpretation of this is that avian cones have only one visual pigment and that colour vision depends entirely on the oil droplets. The more widely held view is that colour vision is a function of several (three or four) visual pigments in the cones, and that the oil droplets simply enhance contrast by acting as intraocular light filters thereby intensifying similar colours by reducing the discrimination of others. Thus the yellow droplets would remove much of the blue from the background, hence increasing the contrast between an object and the blue sky. Similarly, the red

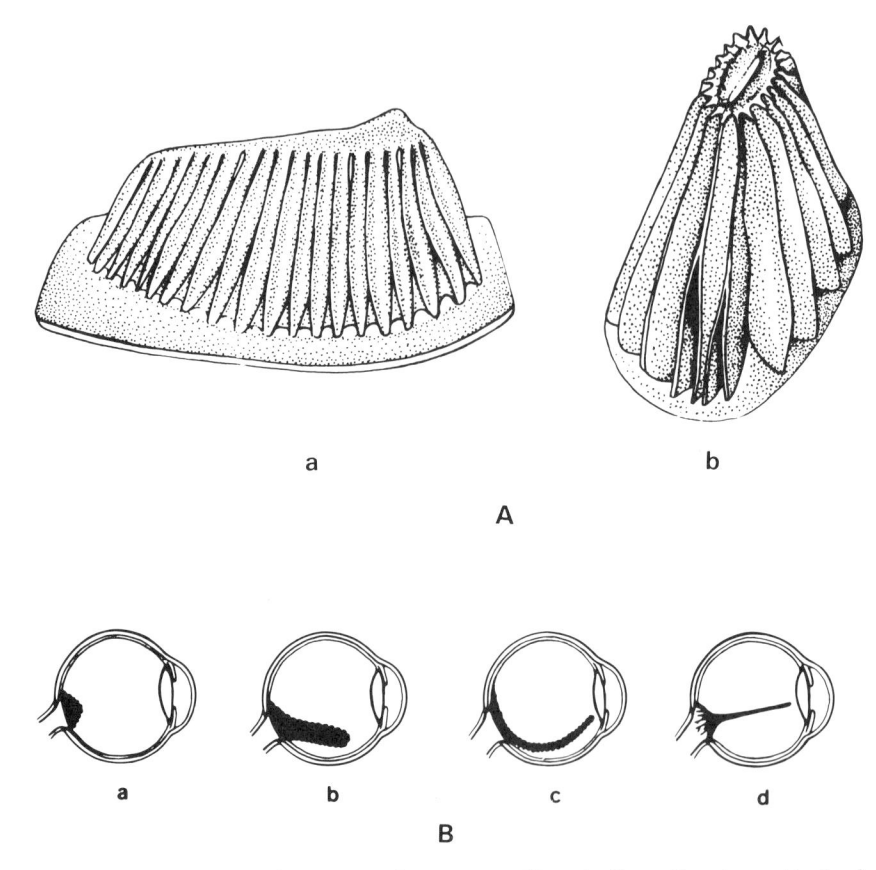

Fig 16–7 **A,** The two main types of pecten oculi. **a** is from the domestic fowl and represents the pleated type which consists of thick vertical pleats and typifies the carinate species. Along the top, the tips of the pleats blend into a longitudinal bridge which adheres to the vitreous body. **b** is from the Ostrich, being of the vaned type which carries 25 to 30 thin vertical vanes and is characteristic of many ratites. From Walls (1942) with kind permission of the Cranbrook Institute of Science, Michigan. **B,** Diagrammatic sagittal sections through the eyeball to show variations in the folds, length and curvature of the pecten. **a,** Common Kestrel; **b,** Barn Swallow; **c,** Blue Jay; **d,** kiwi. From Wood, C.A. (1917) *The Fundus Oculi of Birds Especially as Viewed by the Ophthalmoscope.* Chicago: Lakeside Press.

droplets would remove much of the green from the background, thus improving the contrast between an object and trees. Enhanced contrast would considerably increase visual acuity.

The *pecten* is a thin black structure projecting from the retina into the vitreous body, towards the lens. Its base is always planted on the optic disc, which is functionally the best place for it to be since the optic disc is the blind spot and the pecten is non-sensory. In general, it is small and simple in nocturnal birds, and large and elaborate in diurnal birds, although there are many species variations. Basically, however, there are two main types of pecten (Fig 16–7A). In the pleated type of pecten, which occurs in carinates, the surface of the ridge is thrown into narrow thick vertical corrugations or pleats (**a** in Fig 16–7A). The tips of the pleats are generally held together distally by a bridge which is strongly adherent to the vitreous body. The number of pleats varies widely between species, the pecten of active diurnal birds tending to be larger and more folded than that of nocturnal species; the pecten of the domestic fowl has 16 to 18 pleats while that of owls has only five to eight pleats. In the vaned type, which occurs in most ratites such as the Ostrich and rheas, the ridge carries 25 to 30 wide thin vertical vanes projecting all round it (**b** in Fig 16–7A). The pecten of kiwis differs from the two main types in being a simple conical structure without any pleats or vanes at all. The curvature and length of the pecten also vary greatly (Fig 16–7B). In structure the pecten is extremely vascular, consisting mainly of modified small blood vessels and melanotic pigment cells. Both the luminal and external surfaces of the endothelial cells form large numbers of microfolds, which greatly increases the surface area. It is widely accepted that the main function of the pecten is to provide nutrients to the avascular retina by diffusion through the vitreous body. A similar but smaller and simpler structure, the conus papillaris, occurs at the same site in reptiles which like birds do not have blood vessels within the retina. Numerous other functions have been suggested but not conclusively substantiated. They include a role in accommodation by varying the pressure in the vitreous body; maintaining a high ocular temperature at high altitude where the ambient temperature is extremely low; acting as a dark mirror by decreasing glare and reflecting an image onto the retina of an object approaching from the direction of the sun; protecting the central area from the sun; and providing a fixed point as a navigational aid to fix the position of the sun during migration.

The chambers of the eye and vitreous body

The *anterior chamber* is bounded by the cornea and the iris and the *posterior chamber* by the iris and the lens (Fig 16–3), the anatomy of these chambers being essentially the same as in mammals. They contain the *aqueous humour*, which is formed by the ciliary body and is responsible for the intraocular pressure, thus maintaining the global shape of the eye. The aqueous humour percolates through the spaces of the iridocorneal angle and then drains into the venous circulation via the scleral venous sinus. The *vitreous body*, as in mammals, is a clear translucent gel which fills the eyeball between the lens and retina.

Eyelids

The cornea is protected by the upper and lower eyelids and the nictitating membrane. In many or most birds the eyelids close only in sleep, and the nictitating membrane alone is responsible for blinking. On the other hand, in the American Dipper there is a regular and frequent blinking of the eyelids. In the domestic fowl the dorsal border of the mobile part of the upper eyelid forms a deep groove housing several species of lice and fleas; the comparable groove on the lower lid is much shallower. The *lower eyelid* is thinner, more extensive and more movable than the *upper eyelid* being mainly responsible for closing the eyes. It is devoid of glands. When the eyelids are open their edges form a circle round the edge of the iris so that very little of the white of the eye (the sclera) is visible. In altricial birds the lids remain closed for a short period after hatching. Rows of small overlapping bristle feathers take the place of eyelashes. At least three striated muscles act on the eyelids. The levator of the upper lid is innervated by the oculomotor nerve, and the depressor of the lower lid by the mandibular nerve. Two other muscles have been named as eyelid muscles, the m. tensor periorbitae and the m. orbicularis palpebrarum,

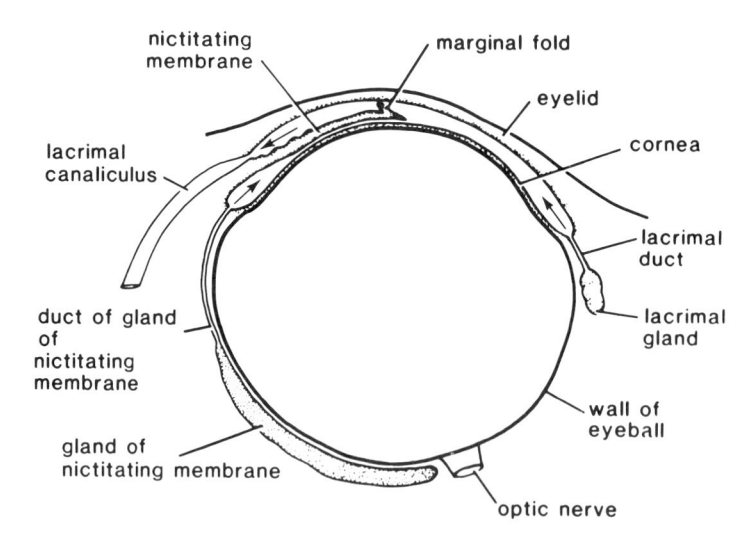

Fig 16–8 Diagrammatic horizontal section through the eye to show the positions of the gland of the nictitating membrane and lacrimal gland, and the direction taken by their secretions (arrows). From Slonaker (1918).

but it is not clear whether or not these are one and the same muscle. However, it seems certain that sphincter-like striated muscle fibres encircle the eyelids and are innervated by the mandibular nerve. It is also probable that smooth muscle fibres are present in the avian, as in the mammalian, eyelid.

The *nictitating membrane* (third eyelid) lies beneath the eyelids on the nasal side of the orbit, and can be freely moved transversely across the front of the eye (Fig 16–8). It darts across the eye about 30 to 35 times a minute in the domestic fowl, and also moves in response to objects approaching the eye suddenly or striking the comb, wattles or ear lobes. The free edge of the membrane is stiffened by a connective tissue band, and on its outer surface it has an anteriorly-directed scooplike projection, the marginal fold (Fig 16–8). In contrast to mammals, two striated muscles, namely the quadratus and pyramidalis muscles (Fig 16–9), are responsible for movements of the nictitating membrane. These muscles are under the control of the sixth cranial nerve. At its insertion on the back of the eyeball the quadrate muscle forms a fibrous pulleylike sheath just dorsal to the optic nerve; the tendon of the pyramidalis muscle passes mediolaterally through the pulley and inserts on the ventral part of the free edge of the membrane (Figs 16–9 and 16–10). Since the dorsal part of the free edge of the membrane is fixed to the eyeball, the membrane makes a pendulumlike movement as it passes to and fro across the eye. As the nictitating membrane travels across the eye from the nasal to the temporal side it sweeps the surface of the cornea distributing the secretion of the gland

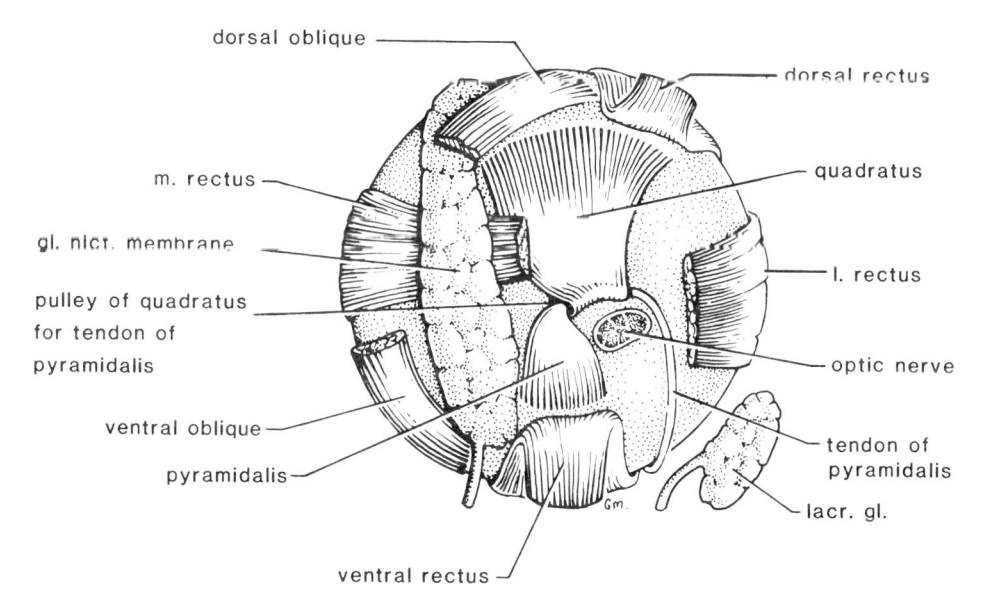

Fig 16–9 View of the posterior surface of the right eyeball showing the extraocular muscles, and the pyramidalis and quadratus muscles. The extraocular muscles have been cut, leaving only their insertions on the eyeball. The quadratus muscle arises from the dorsal aspect of the eyeball and curves ventrally towards the optic nerve. Its ventral border makes a sling through which the tendon of the pyramidalis muscle passes on its way to the nictitating membrane (see Fig 16–10). The pyramidalis muscle acts *directly* on the nictitating membrane, whereas the quadratus muscle acts *indirectly* on the membrane, via the pyramidalis tendon. Most of the relatively large gland of the nictitating membrane (gl. nict. membrane) lies on the ventral and caudomedial aspect of the eyeball; the much smaller lacrimal gland (lacr. gl.) lies on the caudolateral aspect (see Fig 16–10). Both glands are superficial to the extraocular muscles, and each opens into the conjunctival sac by a single duct. l. = lateral; m. = medial. Based on Gadow and Selenka (1891) and Slonaker (1918).

of the nictitating membrane over the cornea, but the marginal fold is flattened so that excess fluid can flow onto the anterior surface of the membrane (Fig 16–8). As the membrane makes its return journey medially across the eye the marginal fold swings outwards to become scooplike, and thus sweeps the excess of fluid into the nasal commissure of the eyelids where it drains into the lacrimal apparatus (Fig 16–8). In a few species, including the American Robin, American Dipper, and owls, the membrane is cloudy. In most birds, however, vision is not severely impaired when the eye is covered by the nictitating membrane because the membrane is usually transparent in diurnal species. Indeed it has even been suggested that some birds may fly with the membrane covering the cornea, thus protecting it from desiccation. In a number of diving birds, including auks, divers and some diving ducks, the central part of the membrane has a transparent window. In other diving birds, including cormorants and certain diving ducks such as pochards, mergansers and goldeneyes, the membrane is transparent all over. It has long been held that in diving birds the transparent region of the membrane is highly refractive and bends light under water thus compensating for the loss of corneal refraction when

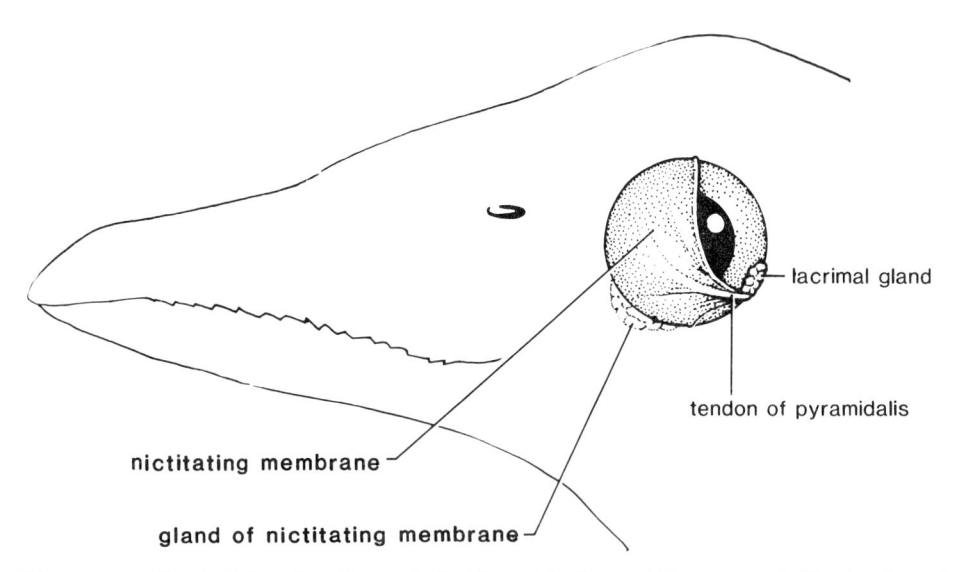

Fig 16-10 Head of Greylag Goose, left side, with the eyelids removed. The tendon of the pyramidalis muscle has pulled the nictitating membrane partly across the anterior surface of the eyeball. The two glands are included schematically to show their approximate position in relation to the eyeball. From Gadow and Selenka (1891).

submerged. It has now been shown, however, that the refractive index and curvature of the membrane are virtually the same as those of the cornea, so the membrane does not after all possess a refractive function under water.

Lacrimal apparatus

The gland of the nictitating membrane (Harderian gland). In birds generally, this relatively large compound tubular or tubuloalveolar gland is tongue-shaped and has a pink or yellow colour which is very similar to that of the muscles of the eyeball. In the domestic fowl and the Rook it is about $18 \times 8 \times 2$ mm in size. It lies on the ventral and caudomedial surface of the eyeball (Figs 16-9 and 16-10), but often remains in the orbit when the eyeball is removed. Its position in the orbit is much deeper than that of the lacrimal gland. The mucoid secretion of the gland discharges through a single duct into the conjunctival pouch between the nictitating membrane and the eyeball, and cleans and moistens the cornea. In birds generally, the gland becomes infiltrated by plasma cells which are derived from the cloacal bursa. These cells produce specific antibody in response primarily to local antigenic stimulation of the eye, thereby protecting the eye against microbial invasion.

The lacrimal gland. In birds generally, the lacrimal gland is much smaller ($7 \times 2 \times 1.5$ mm in the domestic fowl) than the gland of the nictitating membrane, this relationship being the reverse of that in mammals. It is usually reddish-brown in colour. It lies in the region of the temporal (caudal or lateral) commissure of the eyelids and is firmly attached to the orbital rim (Figs 16–9 and 16–10). In contrast to the gland in mammals, it drains by a single duct which opens into the conjunctival sac on the bulbar surface of the lower lid. The gland is absent in a few species including penguins.

The drainage system. The lacrimal secretions drain by the lacrimal ostia of the upper and lower eyelids. The ostium of the upper lid is a large opening about 3 mm in diameter in the domestic fowl, that of the lower lid being only about one-third of this diameter. In this species the two ostia are close together within about 1–2 mm of the medial commissure of the eyelids. Each ostium leads into a lacrimal canaliculus (Fig 16–8). After a few millimetres the two canaliculi join to form the spacious nasolacrimal duct (Fig 7–1B) which passes through the dorsal and medial wall of the infraorbital sinus and opens into the nasal cavity through an elongated slit. This opening lies dorsal to the rostral end of the choanal opening and ventral to the middle nasal concha (Fig 7–1B).

Extraocular muscles

Eye movements in birds are generally limited since the eyeball almost completely fills the orbit. The eyes of many species, however, are capable of forward convergence towards the tip of the beak, this being demonstrated in an extreme form by the Eurasian Bittern which can direct its gaze forwards and horizontally over the surface of the ground even while its beak is pointing vertically upwards in the typical cryptic posture. There is relatively great mobility of the eyes in toucans and hornbills. The small eye movements of the majority of birds, however, are compensated by the great mobility of the head and neck. In contrast to mammals, movements of both eyes are quite independent. They are controlled by the dorsal and ventral oblique muscles and by the dorsal, medial, ventral and lateral rectus muscles (Figs 15–9 and 16–9) resembling those of mammals. (See Chapter 15 for their innervation by the oculomotor, trochlear and abducent nerves.) However, unlike many mammals and reptiles the retractor bulbi muscle is absent. Although the extraocular muscles in owls are fairly well developed, the enormous tubular eyes in these species are only capable of small movements.

EAR

External ear

The external ear is a relatively short canal extending ventrally and caudally from the external acoustic meatus to the tympanic membrane. The external acoustic meatus is a small aperture, nearly always circular, which opens externally on the side of the head. In the domestic fowl it is 4–5 mm in diameter. In most birds the meatus is covered by specialized contour feathers, the ear coverts (Fig 14–2), and only rarely is it naked as in vultures and the Ostrich. The coverts lying on the rostral aspect of the meatus reduce the drag caused by turbulence in flight and thus diminish the masking of sound by noise generated from turbulence in the external ear; since the barbs of these ear

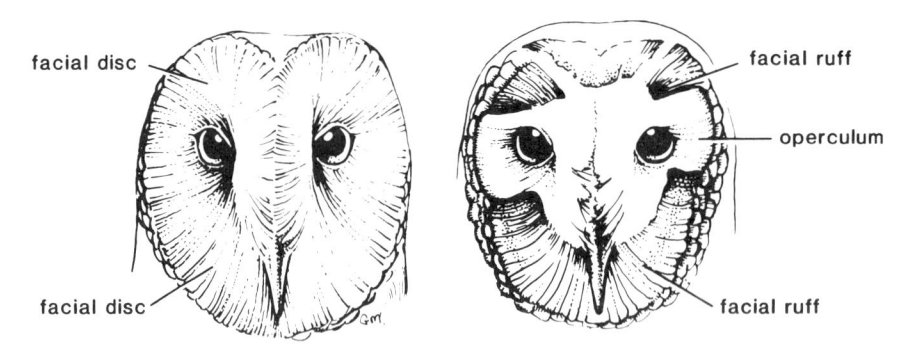

Fig 16–11 Facial disc, facial ruff and operculum of the Barn Owl. In the drawing on the left the facial disc is intact. In the drawing on the right the facial disc has been removed to show the facial ruff and the operculum on each side. The left and right opercula are asymmetrical. The external acoustic meatus lies between the operculum and the facial ruff. Redrawn from Knudsen and Konishi (1979), with kind permission of the editor of *Journal of Comparative Physiology*.

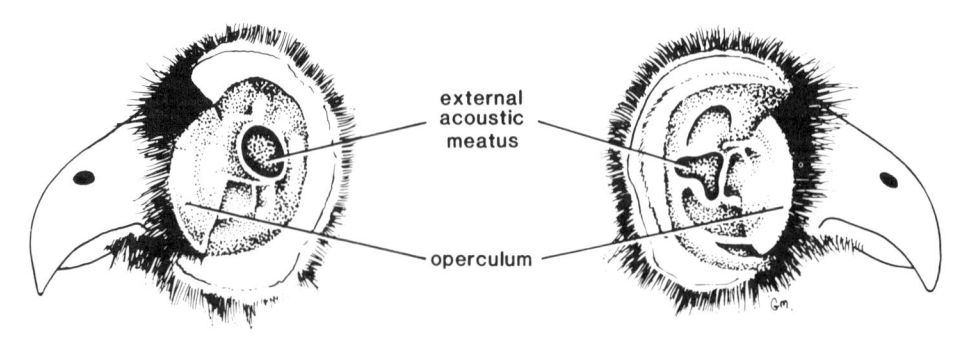

Fig 16–12 Left and right lateral views of the external aperture of the ear in the Long-eared Owl. The operculum (also known as the pre-aural flap) is a flap of feathered skin which can be moved to aid localization of sound sources. In both drawings it has been moved rostrally to expose the external acoustic meatus. The meatus is asymmetrical in shape in this species of owl. Also the skull bones are asymmetrical, the meatus being more dorsal on the left side. Redrawn from Pycraft (1910).

coverts lack barbules the sound waves are not obstructed. On the caudal aspect of the meatus the specialized feathers combine into a tight funnel which is particularly enlarged in songbirds, parrots and falconiform species such as the Osprey. In many species of owl the rostral border of the external meatus has a vertical skin flap, the operculum (Fig 16–11), bearing a row of feathers along and at right angles to its edge. The operculum (also commonly known as the concha) can be erected by striated muscle to assist in locating sounds. In the Barn Owl the facial ruff, which consists of small, curved, stiff feathers supporting the lower feathers of the facial disc, is an effective sound reflector, similar to a man cupping his hands behind his ears. If the operculum swings rostrally (like a door on a hinge) its feathers form, in conjunction with the peripheral feathers of the disc, a vertical slitlike aperture resembling the semi-tubular concha of many mammals. Acoustic location is apparently improved still further in owls by asymmetry of the external ear (Figs 16–11

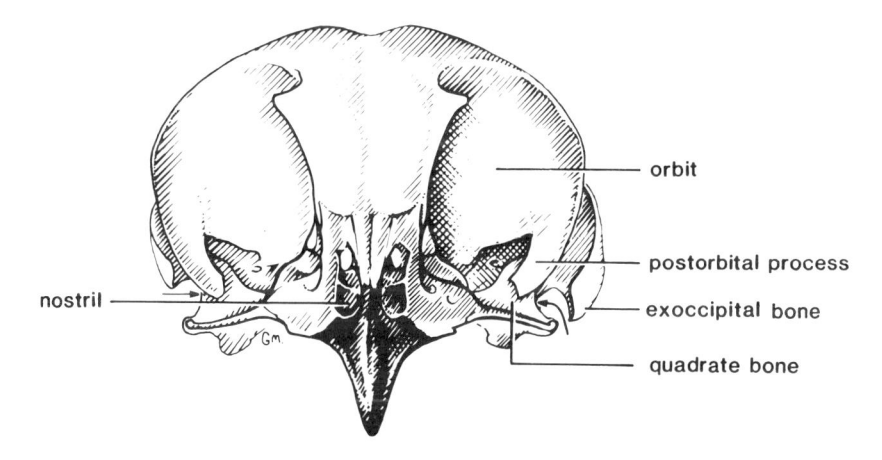

Fig 16–13 Skull of the Boreal Owl showing its asymmetry. The site of the middle ear cavity (arrows) lies in the general region caudal to the postorbital process, and immediately caudal to the quadrate bone. From Sillman (1973), with kind permission of the publisher.

and 16–12) which is sometimes even accompanied by asymmetry of the skull itself (Fig 16–13). In diving birds the external aperture is protected by stronger feathers and by reduction of its diameter. Deep diving species can close the external ear canal altogether.

Middle ear

The middle ear (Fig 16–14) is the air-filled cavity between the *tympanic membrane* and the inner ear. Unlike that of mammals, the tympanic membrane projects outwards rather than inwards. In owls the membrane is exceptionally large in diameter. The tension in the tympanic membrane is altered by the columellar muscle, which attaches to the extracolumellar cartilage and to the tympanic membrane itself. Since this muscle arises from the second embryonic pharyngeal arch and has a motor nerve supply from the facial nerve, it is presumably homologous to the mammalian stapedius muscle, although its action resembles that of the mammalian tensor tympani muscle. Vibrations of the tympanic membrane are carried to the perilymph of the inner ear by the extracolumellar cartilage which is in contact with the tympanic membrane, and the rodlike bony *columella* which is implanted medially in the

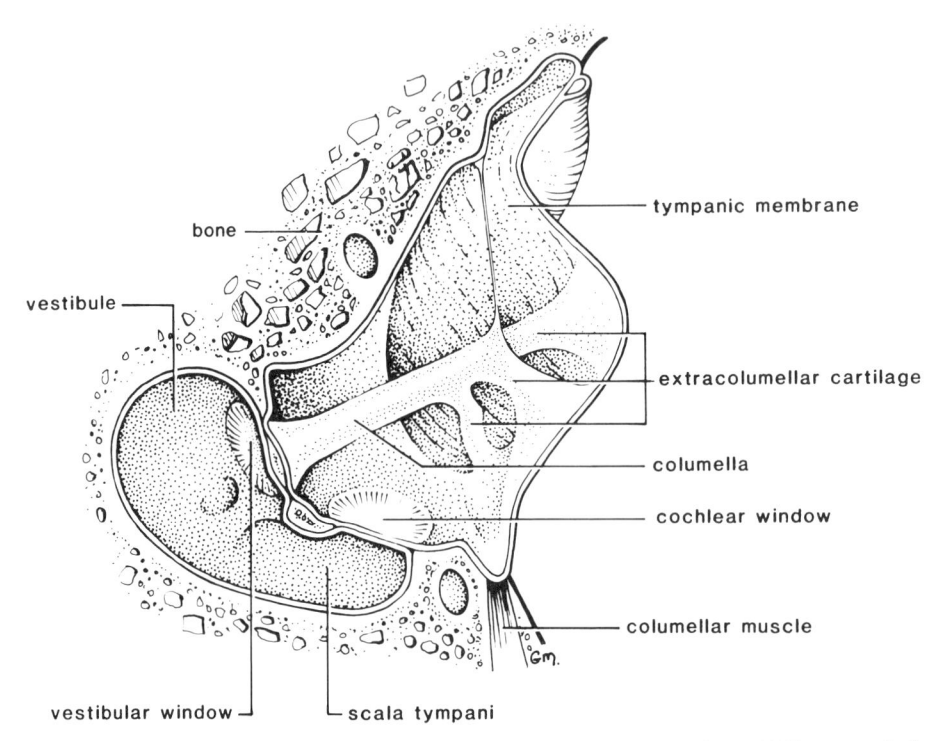

Fig 16–14 Diagrammatic transverse section through the right middle ear of the domestic fowl. In the diagram, dorsal is upwards and lateral is to the right. Ossicular conduction from the tympanic membrane to the vestibular window is achieved by the columellar complex, consisting of the extracolumellar cartilage laterally and the bony columella medially. The columellar muscle attaches to the tympanic membrane rather than to the columella, but is innervated by the facial nerve. The vestibular and cochlear windows have both been partly transected by the plane of the section. Redrawn from Pohlman (1921).

vestibular window. The columella is homologous to the mammalian stapes; the mammalian incus and malleus are homologous to the avian quadrate and articular bones respectively. The columella has taxonomic significance. The typical avian columella is reptilian with a flat footplate and a straight bony shaft arising from the centre of the footplate. In a few groups, for example storks, the columella is tubular and its shaft has many perforations, especially near the footplate. Medially the base (footplate) of the columella is attached to the margin of the vestibular window, and its medial surface is therefore in direct contact with the perilymph of the vestibule of the inner ear (Fig 16–14);

probably it is also in direct contact with perilymph in the cistern of the vestigial scala vestibuli (Fig 16–15). In most birds the footplate of the columella is flat and moves in and out of the window like a piston, but in owls the footplate is hemispherical and the columella is oblique to the window so that rocking movements must occur (the mammalian stapes also undergoes rocking movements). The cochlear (round) window lies near the vestibular window and is in contact with the scala tympani of the inner ear (Fig 16–15). The membrane which covers the cochlear window is particularly thin and transparent in passerine species and owls. The compression of the perilymph of the inner ear, which occurs when the columella pushes the vestibular window inwards, is accompanied by an outward movement of the membrane of the cochlear window. The *pharyngotympanic tube* (Eustachian tube) connects the cavity of the middle ear to the oropharynx via the infundibular cleft (Fig 6–1), thereby equalizing the pressure on either side of the tympanic membrane.

Inner ear

The inner ear consists of bony and membranous labyrinths. The bony labyrinth, comprising the vestibule, semicircular canals and the cochlea, encloses the membranous labyrinth. The space between the bony and membranous labyrinths is filled with a fluid, the *perilymph*. The cavities of the membranous labyrinth are occupied by *endolymph*. Of the membranous labyrinth, the utricle, saccule and semicircular ducts are concerned with the position and movement of the head in space, whereas the cochlear duct is involved in hearing. The function of the lagena is not clear.

Auditory organ. The avian cochlea differs from the spiral cochlea of most mammals in being a relatively short and only slightly curved tube (about 6 mm in length in the domestic fowl). The cochlea of the owl is relatively long compared to that of other birds. Extending throughout the length of the cochlea is the *cochlear duct* (scala media) which is filled with endolymph, but because of the shortness of the cochlea the avian cochlear duct is only about one-tenth the length of that of a mammal of comparable body size. The scala vestibuli is vestigial, being reduced to a small space adjacent to the vestibular window, namely the cistern of the scala vestibuli (Fig 16–15), and another small space near the lagena known as the fossa of the scala vestibuli (Fig 16–15). The cochlear duct is separated from the vestigial *scala vestibuli* by the thick folded tegmentum vasculosum (corresponding to the vestibular membrane of Reissner in mammals), and from the well-developed *scala tympani* by the basilar membrane (Fig 16–15). The scala tympani and the remnants of the scala vestibuli connect with each other at the apical end of the cochlea via the apical interscalar canal which thus corresponds to the mammalian helicotrema; the two scalae also connect at the base of the cochlea by the basal interscalar canal (ductus brevis) (Fig 16–15). The scala tympani, and presumably also the vestiges of the scala vestibuli, contain perilymph. The blind apex of the cochlear duct is formed by the *lagena* (absent in mammals apart from the egg-laying monotremes) which contains the macula lagenae, a group of sensory cells with otoconia. Afferent nerve fibres from this macula appear to end in auditory centres of the medulla, and presumably have an auditory function.

As in mammals the *basilar membrane* carries the neuroepithelial receptor cells which constitute the organ of hearing. The basilar membrane of birds is considered to be relatively shorter than that of mammals (for instance 3 mm long in the pigeon, 35 mm long in man) because of the relative shortness of the avian cochlea, but it is much wider in birds thus making it possible for a relatively large number of receptor cells to be carried per unit length; indeed,

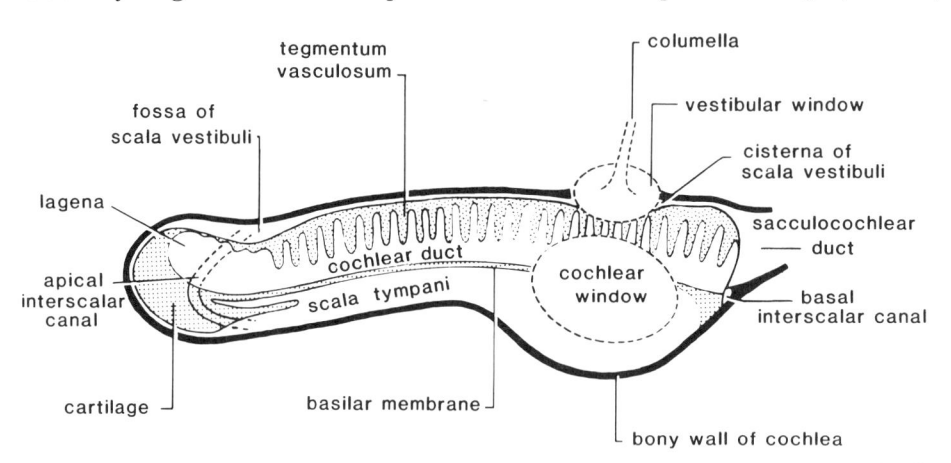

Fig 16–15 The cochlea of a songbird. The cochlear duct is bounded on one side by the basilar membrane, which carries the sensory acoustic epithelium of the papilla basilaris, and on the other side by the thick folded tegmentum vasculosum. The scala vestibuli is vestigial, but remnants of it persist apically as the fossa of the scala vestibuli and basally as the cistern of the scala vestibuli. The scala tympani and the remnants of the scala vestibuli connect with each other via the apical interscalar canal and the basal interscalar canal. Based on Schwartzkopff (1968), with kind permission of the publisher.

the receptor cells are densely packed across almost the whole width. As in mammals the length of the transverse fibres of the basilar membrane, and hence the width of the membrane, increases towards the apex of the cochlear duct. The receptor cells are carried as a continuous ridge, the crista basilaris, along the whole length of the basilar membrane. The crista basilaris is the homologue of the organ of Corti of mammals. In a cross-section it contains about 10 times as many receptor cells as in the organ of Corti, thus compensating for the relative shortness of the cochlear duct. Each *receptor cell* carries a single kinocilium and up to 100 stereocilia, the tips of which are firmly inserted into niches in the *tectorial membrane*. The tectorial membrane spreads like a thick blanket over the crista basilaris, and is much more massive than the thin ribbonlike tectorial membrane of mammals. During movement of the basilar membrane shearing stresses are imposed by the tectorial membrane on the cilia, and these induce the receptor (generator) potential in the receptor cells. The receptor cells make synaptic contacts with axons of the cochlear part of the vestibulocochlear nerve. Experimental evidence suggests that the apex of the cochlear duct responds to low frequency sounds and that the base of the cochlear duct responds to high frequency sounds.

Although the avian cochlear duct is relatively short its greater number of receptor cells per unit length suggests that the total number of receptor cells may be somewhat similar to that of mammals. However, the *auditory performance* of birds differs from that of mammals in several characteristics.

1. The capacity for *frequency analysis* (discrimination of pitch) is good, at least in passerines and parrots where it approaches that of man. However, this capacity is restricted to a narrower band than in mammals, and furthermore within their range of hearing birds are less sensitive to higher and lower tones than man.
2. On the other hand, it seems that the *temporal resolution* of the avian ear is about 10 times faster than that of the human ear. Thus the song of the Chaffinch would have to be slowed down 10 times before the human ear could resolve all the details which are learned by the Chaffinch chick.
3. The capacity for *directional analysis* depends on the recognition of differences in the time of arrival of sounds in each ear, and the perception of different intensities at each ear. Behavioural studies indicate that the power of localization is not particularly remarkable in diurnal birds; the songs of passerines seem to be locatable with more or less equal accuracy by other birds and by man. Localization improves when the sound is repeated, allowing the brain to average the signals; localization is also better when the sound is full of transients as in the songs of passerines, in contrast to a pure tone like the alarm call of hawks which seems to give little or no sense of direction at least to the human ear. Auditory localization is far more accurate in nocturnal birds. As already mentioned in the discussion of visual sensitivity, owls can hunt successfully in total darkness. Indeed, the Barn Owl has an acuity of localization surpassing all terrestrial animals so far tested, including man. In response even to very brief sudden sounds the Barn Owl instantly moves its head to a fixed point which brings the prey into a relative position of maximum acuity, both auditory and visual. The aerial attack that follows can be guided visually or entirely acoustically. Experiments suggest that the accuracy of the acoustic mechanism depends on a comparison of the arrival time of the sound in the two ears, and a comparison of the difference in the intensity of the sound at each frequency at the two ears (the interaural spectrum). There would be a small but distinct time difference between the arrival in the external acoustic meatus, of first the primary sound signals, and then the secondary signals which have been reflected by the contours of the facial ruff and by the flaplike operculum of the external acoustic meatus. The very small delay between these signals would generate, at the basilar membrane, a complex time pattern that would vary with the direction of the sound. The interaural spectrum is evidently crucial, since the Barn Owl is much less accurate at localizing a tonal signal than a noise signal with a wide bandwidth. This can be explained by the asymmetrical

positions of the ears, which cause the left ear to be more sensitive to regions below the horizontal plane, and vice versa for the right ear; thus as sound frequency increases, the plane of equal sensitivity for the two ears rotates from vertical to horizontal. A possible alternative, or perhaps additional mechanism for localization in birds, and especially in owls, depends on acoustic coupling of the two ears across the pneumatic cavities of the skull bones. The principle would be the sensing of pressure differences on either side of the tympanic membrane. If sound strikes both surfaces of the membrane, a net cancelling effect would occur when intensity and phase are closely matched. Such a system relies on null detection, as in radio navigation, and is an extremely accurate method for direction finding.

The necessary neuronal equipment for all this elaborate central analysis appears to be present, since the nocturnal hunting owls (e.g. the Barn Owl) which have asymmetrical ears, also possess an extraordinary number of neurons in the auditory nuclei of the medulla oblongata, exceeding by three- or four-fold the number in a passerine bird (e.g. the Common Crow) of about twice the body weight. Midbrain auditory components are also very well developed in owls. On the other hand the Little Owl, which hunts at dawn, shows no special development of the auditory areas of the medulla oblongata.

Although birds as a group do not hear ultrasonic vibrations, a few nocturnal species (the Oilbird and cave swiftlets, which live in caverns) do use echo location. By this means they avoid obstacles in the dark, but cannot catch moving targets as bats do. These nocturnal species produce short pulses at a frequency of about 4–8 kHz, which are fully audible to man (i.e. not ultrasonic): the pulses emitted by bats have frequencies up to 100 kHz (inaudible to man). For birds flying in the dark the silent period between each signal is very short (about 2–3 ms), showing a similar time resolution to that of bats. However, the lower frequency used by birds enables them only to avoid bars over 6 mm in diameter, whereas bats using the highest frequencies avoid wires of 0.6 mm diameter and can catch prey of this diameter. Apparently, penguins are able to hunt in the water by echo location, although once again the signals are not ultrasonic.

Balance organ. The (membranous) semicircular ducts (Fig 16–16) occupy the (bony) semicircular canals; the (membranous) utricle and saccule lie within the (bony) vestibule. The three *semicircular ducts* (and of course the three semicircular canals that enclose them) are in roughly orthogonal planes. In all birds, the horizontal semicircular duct has a constant orientation, being parallel to the geophysical plane when the head is in the alert position. The rostral and caudal ducts are vertical in position, approximately at right angles to each other and at 45° to the saggital plane; thus the rostral duct on one side is parallel to the caudal duct on the other side. The *semicircular canals* of birds generally are long in comparison with those of other vertebrates (averaging about 10 mm in length in the domestic fowl, compared to 13 mm in man). In the strongest fliers such as the falconiform species the canals are long and thin,

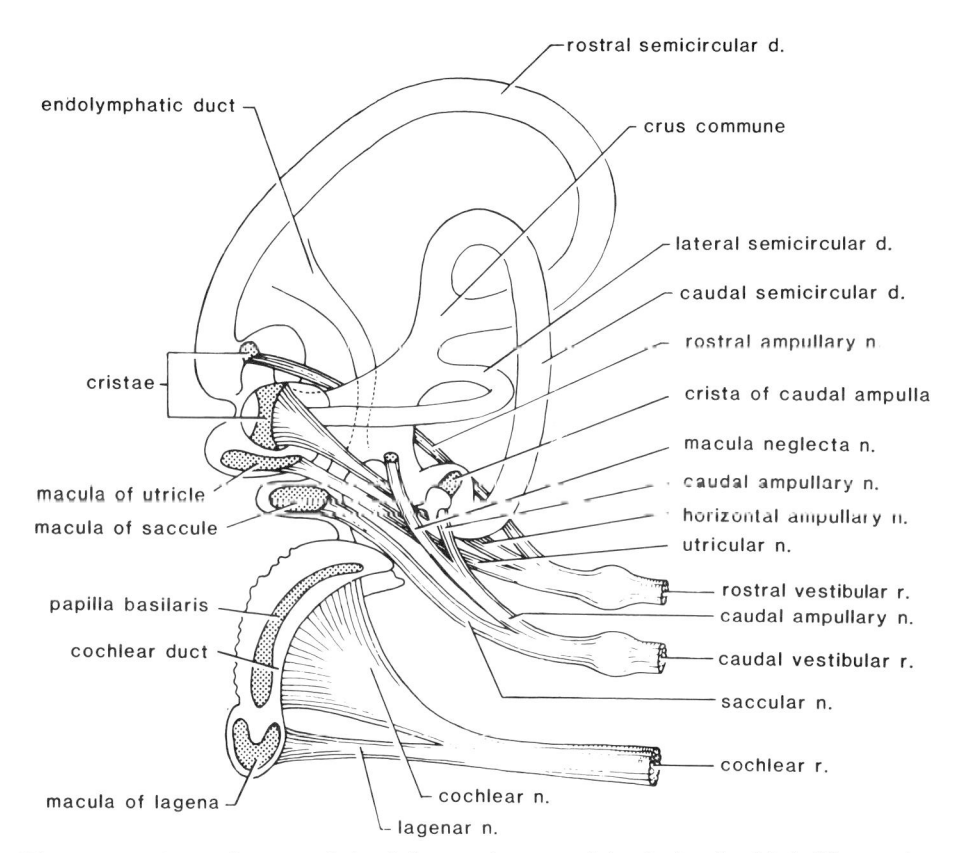

Fig 16–16 Lateral view of the left membranous labyrinth of a bird. The various receptor areas are stippled. d. = duct; n = nerve; r = ramus. The macula neglecta nerve is more correctly named the crista neglecta nerve. The lateral semicircular duct is generally known as the horizontal duct. From Bubien-Waluszewska (1981), with kind permission of the publisher.

and have pronounced osseous ampullae, whereas weaker fliers, including the domestic fowl, have short wide canals. These and other avian elaborations of the canals and their ducts probably reflect their relatively great functional importance to life in three dimensions. The semicircular ducts arise from the saclike *utricle*. At the origin from the utricle each duct has a dilation known as the membranous ampulla (Fig 16–16). The utricle connects by a small opening in its floor with the *saccule*. The saccule in turn connects to the cochlear duct by the *sacculocochlear duct* (ductus reuniens). The *endolymphatic duct* arises from the saccule and ends blindly in the cranial cavity under the dura mater.

Zones of neuroepithelial mechanoreceptor cells are present in the ampulla of each of the three semicircular ducts and in the utricle, saccule and lagena. Those in the ampullae take the form of a ridge, the *ampullary crista* (Fig 16–16). The crista of the horizontal duct is a simple ridge, but the cristae of the rostral and caudal ampullae have an additional transverse ridge, the cruciate septum, which has no receptor cells and divides the sensory epithelium of the crista into two regions. The mechanoreceptor zones of the utricle, saccule and lagena consists of patches of receptor cells, each patch being known as a *macula* (Fig 16–16). The zone in the utricle, the macula of the utricule, is strictly horizontal in its orientation; the macula of the saccule is orientated obliquely, and the macula of the lagena is vertical in position. The utricle has another smaller zone, the crista neglecta, which is a ridge rather than a macula.

The cristae and the maculae have an essentially similar structure. Each *receptor cell* carries a single kinocilium and a cluster of about 50 stereocilia. The cilia of the cristae are inserted into a gelatinous cap, the *cupula*, which is secreted by the supporting cells of the crista. The cupula almost fills the membranous ampulla. The cilia of the maculae are inserted into the gelatinous *statoconial membrane* which covers each macula. The statoconial membrane resembles the cupula, except that it contains numerous crystals of calcium carbonate, the statoconia. The cupulae and statoconial membranes are shifted by appropriate movements of the head, thus applying shearing stresses to the cilia. As in the crista basilaris of the cochlear duct, it is the bending of the cilia which induces the receptor (generator) potential in the receptor cell. Bending the cilia towards the kinocilium excites the receptor cell, but bending them away inhibits the receptor cell. The receptor cells make synaptic contacts with axons of the vestibular part of the vestibulocochlear nerve.

The receptor cells of the maculae tend to be tonically active, responding to the effects of gravity. Their patterns of activity vary, however, with the position of the head, since the shearing stresses in the maculae on each side of the head will vary when the head is tilted. Thus the maculae continuously inform the brain about the *position* of the head in space. The cristae, on the other hand, signal on and off discharges in response to changes in the rate of movement of the head, that is to acceleration or deceleration, but are silent when the head is still. Thus essentially, the cristae inform the brain about *movement* of the head in space.

OLFACTORY ORGANS

The visual life-style of the great majority of birds, coupled with the likelihood that odours would be quickly dispersed high above the surface of the land or water, probably account for the popular assumption that birds have very little sense of smell. Indeed, in both of the two great groups of flying vertebrates, the pterosaurs and birds, the visual region of the midbrain (the mesencephalic tectum, commonly called the optic lobe) became greatly enlarged, and the olfactory lobes were proportionately reduced. Nevertheless there is abundant evidence (anatomical, physiological and behavioural) that birds do perceive olfactory stimuli; what is not yet established, however, is the full role of olfaction in avian life.

Anatomically the olfactory region of the nasal cavity lies in the region of the domelike caudal nasal concha. The outer (nasal) surface of the concha is lined by an olfactory epithelium, which also extends over the adjoining walls of the nasal cavity (Fig 7–1C and –1D). The small septal concha, which is unique to petrels, is also covered with olfactory epithelium. In most birds the caudal concha is a simple dome, but in a few species its surface area is much increased by scroll-formation as in vultures or several very extensive transverse folds as in kiwis. In birds such as swifts in which the caudal concha is absent, the olfactory epithelium covers the roof and lateral walls of the nasal cavity, as well as the dorsal part of the nasal septum. In birds generally the anatomy of the nasal cavity is adapted for effective airflow over the olfactory region.

As in other vertebrates the *olfactory cell* is a bipolar ciliated neuron supported by sustentacular and basal cells. Its terminal cilia and microvilli project from the epithelial surface. The unmyelinated axonal processes of the receptor cells form the olfactory nerve and end in the *olfactory bulb* of the brain. The size of the olfactory bulb relative to the cerebral hemispheres varies greatly. The smallest size occurs in passerines, and also in parrots and woodpeckers. Pigeons, galliforms, falconiforms, waders and gulls form an intermediate group. Larger bulbs have been found in a group of water birds including anseriforms, the Common Loon, American Coot, Horned Grebe and especially in some oceanic procellariiforms (albatrosses, storm petrels); however, there are others with larger olfactory bulbs that are not aquatic, such as the Yellow-billed Cuckoo which lives in trees and the whip-poor-wills which catch insects on the wing. The olfactory bulb reaches its largest relative size in the Brown Kiwi.

Electrical activity comparable to that of macrosmatic reptiles and mammals has been observed in the olfactory pathways of numerous avian species, in response to various odorous stimuli. It is known from behavioural studies that pigeons possess excellent olfactory perception. The nocturnal Brown Kiwi, which is a deplorably noisy sniffer, has no difficulty in finding buried delicacies.

Vultures have been shown to congregate rapidly over odiferous up-drafts, and are therefore presumed to use olfaction for locating the position of carrion. Leach's Storm Petrel and other oceanic procellariiforms have been found to navigate, at least partly, by olfaction, their own nesting material attracting them to their own island in the dark and leading them correctly through a maze to their burrows. However, if their nostrils were plugged or their olfactory nerves were cut the birds failed to return to their burrows within one week. Furthermore, experiments also show that procellariiform species use olfaction to find food at sea. Other examples of specific olfactory functions may occur in African honeyguides which appear to locate beehives by the smell of the beeswax on which they feed, and in pigeons which possibly utilize olfaction in homing. Although the exact contribution of olfaction to the biology of birds remains uncertain it must be assumed that birds do perceive odours to a greater or lesser degree.

TASTE

In the past it has been widely believed that the sense of taste is poorly developed in birds. However, recent work has shown that taste buds are much more numerous than had previously been supposed, at least in the chicken, pigeon and duck. In most of the species which have been investigated, including the domestic fowl, pigeon, swift, falconiforms and several songbirds, the taste buds lie on the base of the tongue. In the Mallard there are no taste buds on the tongue, but groups occur at five other sites as follows: the internal aspect of the tip of the lower bill just caudal to the mandibular nail; the corresponding region at the tip of the upper bill just caudal to the maxillary nail; two sites on the caudal part of the roof of the oropharynx on either side of the midline; and the region of the roof immediately rostral to the choanal opening. The total number of taste buds in the Mallard is less than 500, whereas 10 000 have been reported in man and 17 000 in the rabbit. In the flamingo and oystercatcher areas of taste buds have been found on the floor of the caudal region of the oropharynx just cranial to the laryngeal mound. In parrots they occur on the roof of the oropharynx on either side of the choanal opening, and on the floor of the oropharynx at the rostral end of the laryngeal mound. In general, taste buds are confined to regions where the epithelium is soft, non-cornified and glandular; typically, they have a strong topographical affinity for the ducts of the salivary glands.

Histologically the taste buds of birds are often ovoid with a funnel-shaped outer pore. No distinction can be made between receptor cells and supporting cells. The so-called 'taste hairs' at the apical ends of the receptor cells are apparently artefacts. About 20 to 30 axons enter a typical taste bud. The axons from the taste buds on the tongue and palate travel in the glossopharyngeal nerve; axons from the taste buds in the upper and lower bill of the Mallard travel peripherally in the trigeminal nerve but probably belong to the facial nerve.

Electrophysiological observations show that salts and acids are generally effective stimuli, but sweet substances are not. Strong responses are obtained to distilled water. Behavioural studies confirm that birds can distinguish certain tastes, but in general the acuity of taste is less than that of mammals. There are indications that pigeons have relatively high acuity, and that the Mallard may be intermediate between the pigeon and the domestic fowl. Bitter-tasting substances are rejected fairly uniformly. Salt is also commonly rejected. Sour substances are rejected by several species including the domestic fowl, but quail prefer them. Sweet solutions produce the most unpredictable responses; for example domestic fowl vary individually in their reaction to sugars but consistently reject saccharine.

Although the total number of taste buds is probably much less in birds than in mammals there can no longer be any doubt that many birds do have a sense of taste, but very little is known about the role of taste in the biology of birds. In the Mallard the position of the taste buds does agree, however, with the pathway which the food is believed to take in its journey through the oropharynx. The taste buds at the bill tip enable unpleasant food particles to be rejected immediately; the remaining taste buds are distributed so that the palatability of food can be monitored almost continuously until it is swallowed.

Further reading

Aitken, I.D. & Survashe, B.D. (1976) A procedure for location and removal of the lachrymal and Harderian glands of avian species. *Comp. Biochem. Physiol.*, 53A, 193–195.

Bang, B.G. (1971) Functional anatomy of the olfactory system in 23 orders of birds. *Acta anat.*, 79 (Supplement 58) 1–76.

Berkhoudt, H. (1977) Taste buds in the bill of the Mallard (*Anas platyrhynchos* L.). Their morphology, distribution and functional significance. *Neth. J. Zool.*, 27, 310–331.

Bowmaker, J.K. (1980) Colour vision in birds and the role of oil droplets. *Trends Neurosci.*, 196–199.

Brach, V. (1977) The functional significance of the avian pecten: a review. *Condor*, 79, 321–327.

Bubien-Waluszewska, A. (1981) The cranial nerves. In *Form and Function in Birds* (Ed.) King, A.S. & McLelland, J. Vol. 2. London and New York: Academic Press.

Burns, R. B. (1975) Plasma cells in the avian Harderian gland and the morphology of the gland in the Rook. *Canad. J. Zool.*, 53, 1258–1269.

Correia, M.J., Landolt, J.P. & Young, E.R. (1974) The sensora neglecta in the pigeon: a scanning electron and light microscope study. *J. comp. Neurol.*, 154, 303–316.

Evans, H.E. (1979) Organa sensoria. In *Nomina Anatomica Avium* (Ed.) Baumel, J.J., King, A.S., Lucas, A.M., Breazile, J.E. & Evans, H.E. London and New York: Academic Press.

Fite, K.V. & Rosenfield-Wessels, S. (1975) A comparative study of deep avian foveas. *Brain Behav. Evol.*, 12, 97–115.

Gadow, H. & Selenka, E. (1891) Vögel: I. Anatomischer Theil. In *Bronn's Klassen und Ordnungen des Thier-Reichs*, Bd. 6. Leipzig: Winter.

Gentle, M.J. (1971) The lingual taste buds of *Gallus domesticus* L. *Br. Poult. Sci.*, 12, 245–248.

Goodge, W.R. (1960) Adaptations for amphibious vision in the Dipper (*Cinclus mexicanus*). *J. Morph.*, 107, 79–91.

Hodges, R.D. (1974) *The Histology of the Fowl*. London and New York: Academic Press.

Holden, A.L. (1983) Special senses. In *Physiology and Biochemistry of the Domestic Fowl* (Ed.) Freeman, B.M. Vol. 4. London and New York: Academic Press.

Hutchison, L.V. & Wenzel, B. (1980) Olfactory guidance in foraging by procellariiforms. *Condor*, 82, 314–319.

Jasinski, A. (1973) Fine structure of capillaries in the pecten oculi of the sparrow, *Passer domesticus*. *Z. Zellforsch. Mikrosk. Anat.*, 146, 281–292.

Kare, M.R. & Rogers, J.G. (1976) Sense organs. In *Avian Physiology* (Ed.) Sturkie, P.D. 3rd Edn. New York: Springer-Verlag.

King-Smith, P.E. (1971) Special senses. In *Physiology and Biochemistry of the Domestic Fowl* (Ed.) Bell, D.J. & Freeman, B.M. Vol. 2. London and New York: Academic Press.

Knudsen, E.J., Blasdel, G.G. & Konishi, M. (1979) Sound localization by the Barn Owl (*Tyto alba*) measured with the search coil technique. *J. comp. Physiol. A*, 133, 1–11.

Knudsen, E.J. & Konishi, M. (1979) Mechanisms of sound localization in the Barn Owl (*Tyto alba*). *J. comp. Physiol.*, 133, 13–21.

Krol, C.P.M. & Dubbeldam, J.L. (1979) On the innervation of taste buds by the N. facialis in the Mallard, *Anas platyrhynchos* L. *Neth. J. Zool.*, 29, 267–274.

Lewis, B. & Coles, R. (1980) Sound localization in birds. *Trends Neurosci.*, 102–105.

Lord, R.D. (1956) A comparative study of the eyes of some falconiform and passeriform birds. *Am. Midl. Nat.*, 56, 325–344.

Martin, G. (1978) Through an owl's eye. *New Scient.*, 77, 72–74.

McLelland, J. (1975) Aves sense organs. In *Sisson and Grossman's The Anatomy of the Domestic Animals* (Ed.) Getty, R. Vol. 2, 5th Edn. Philadelphia: Saunders.

Nicol, J.A.C. & Arnott, H.J. (1974) Tapeta lucida in the eyes of goatsuckers (*Caprimulgidae*). *Proc. R. Soc., Lond.*, 187, 349–352.

Pearson, R. (1972) *The Avian Brain*. London and New York: Academic Press.
Pohlman, A.G. (1921) The position and functional interpretation of the elastic ligaments in the middle-ear region of *Gallus*. *J. Morph.*, 35, 229–262.
Portmann, A. (1961) Sensory organs: equilibration. In *Biology and Comparative Physiology of Birds* (Ed.) Marshall, A.J. Vol. II. New York and London: Academic Press.
Pumphrey, R.J. (1961) Sensory organs: vision: hearing. In *Biology and Comparative Physiology of Birds* (Ed.) Marshall, A.J. Vol. II. New York and London: Academic Press.
Pycraft, W.P. (1910) *A History of Birds*. London: Methuen.
Schwartzkopff, J. (1968) Structure and function of the ear and of the auditory brain areas in birds. In *Hearing Mechanisms in Vertebrates* (Ed.) de Reuck, A.V.S. & Knight, J. A Ciba Foundation Symposium. London: Churchill.
Schwartzkopff, J. (1973) Mechanoreception. In *Avian Biology* (Ed.) Farner, D.S. & King, J.R. Vol. III. New York and London: Academic Press.
Sillman, A.J. (1973) Avian vision. In *Avian Biology* (Ed.) Farner, D.S. & King, J.R. Vol. III. New York and London: Academic Press.
Sivak, J.G. (1980) Avian mechanisms for vision in air and water. *Trends Neurosci.*, 314–317.
Sivak, J.G., Bouier, W.R. & Levy, B. (1978) The refractive significance of the nictitating membrane of the bird eye. *J. comp. Physiol.*, 125, 335–339.
Slonaker, J.R. (1918) A physiological study of the anatomy of the eye and its accessory parts in the English Sparrow (*Passer domesticus*). *J. Morph.*, 31, 351–459.
Tanaka, K. & Smith, C.A. (1978) Structure of the chicken's inner ear: SEM and TEM study. *Am. J. Anat.*, 153, 251–272.
Tansley, K. (1965) *Vision in Vertebrates*. London: Chapman and Hall.
Walls, G.L. (1942) *The Vertebrate Eye*. Bulletin no. 19. Michigan: Cranbrook Institute of Science.
Wenzel, B.M. (1973) Chemoreception. In *Avian Biology* (Ed.) Farner, D.S. & King, J.R. Vol. III. New York and London: Academic Press.
Wight, P.A.L., Burns, R.B., Rothwell, B. & McKenzie, G.M. (1971) The Harderian gland of the domestic fowl. 1. Histology with reference to the genesis of plasma cells and Russell bodies. *J. Anat.*, 110, 307–315.
Zusi, R.L. & Bridge, D. (1981) On the slit pupil of the black skimmer (*Rynchops niger*). *J. Field Ornith.*, 52, 338–40.

APPENDIX

A list of the English common names of the birds cited in the text along with their Latin scientific names. The birds are cited alphabetically according to their common names.

Anhinga	*Anhinga anhinga*
Avocet	*Recurvirostra avosetta*
Bananaquit	*Coereba flaveola*
Blackbird	*Turdus merula*
Budgerigar	*Melopsittacus undulatus*
Bustard, Great	*Otis tarda*
Chaffinch	*Fringilla coelebs*
Condor, Andean	*Vultur gryphus*
Coot, American	*Fulica americana*
Cormorant, Great	*Phalacrocorax carbo*
Crossbill, Red	*Loxia curvirostra*
Crow, Common	*Corvus brachyrhynchos*
Cuckoo	*Cuculus canorus*
Cuckoo, Yellow-billed	*Coccyzus americanus*
Dipper, American	*Cinclus mexicanus*
Dove, Rock	*Columba livia*
Duck, Muscovy	*Cairina moschata*
Duck, Musk	*Biziura lobata*
Duck, Ruddy	*Oxyura jamaicensis*
Duck, Tufted	*Aythya fuligula*
Eagle, Steller's Sea	*Haliaeetus pelagicus*
Eagle, White-tailed Sea	*Haliaeetus albicilla*
Emu	*Dromaius novaehollandiae*
Finch, Zebra	*Phoephila guttata*
Flamingo, Greater	*Phoenicopterus ruber*

Fulmar, Northern	*Fulmarus glacialis*
Galah	*Eolophus roseicapillus*
Gannet	*Morus bassanus*
Goldcrest	*Regulus regulus*
Goose, Bar-headed	*Anser indicus*
Goose, Canada	*Branta canadensis*
Goose, Greylag	*Anser anser*
Goose, Hawaiian	*Branta sandvicensis*
Goose, Magpie	*Anseranas semipalmata*
Goose, Spur-winged	*Plectropterus gambensis*
Grebe, Horned	*Podiceps auritus*
Grebe, Little	*Tachybaptus ruficollis*
Grouse, Sage	*Centrocercus urophasianus*
Guineafowl, Helmet	*Numida meleagris*
Gull, Herring	*Larus argentatus*
Gull, Little	*Larus minutus*
Hawfinch	*Coccothraustes coccothraustes*
Heron, Grey	*Ardea cinerea*
Hoatzin	*Opisthocomus hoazin*
Honeyeater, Singing	*Meliphaga virescens*
Hoopoe	*Upupa epops*
Hummingbird, Sword-billed	*Ensifera ensifera*
Hummingbird, Vervain	*Mellisuga minima*
Jackdaw	*Corvus monedula*
Jay, Blue	*Cyanocitta cristata*
Jay, Grey	*Perisoreus canadensis*
Junglefowl, Red	*Gallus gallus*
Kakapo	*Strigops habroptilus*
Kestrel, American	*Falco sparverius*
Kestrel, Common	*Falco tinnunculus*
Kiwi, Brown	*Apteryx australis*
Kookaburra, Laughing	*Dacelo novaeguineae*
Lapwing, Spur-winged	*Vanellus spinosus*
Loon, Common	*Gavia immer*
Macaw, Blue-and-yellow	*Ara ararauna*
Magpie, Common	*Pica pica*
Mallard	*Anas platyrhynchos*
Marabou	*Leptoptilos crumeniferus*
Meadowlark, Eastern	*Sturnella magna*
Merganser, Hooded	*Mergus cucullatus*
Merganser, Red-breasted	*Mergus serrator*

Murre, Common	*Uria aalge*
Nightjar, European	*Caprimulgus europaeus*
Oil-bird	*Steatornis caripensis*
Osprey	*Pandion haliaetus*
Ostrich	*Struthio camelus*
Owl, Barn	*Tyto alba*
Owl, Boreal	*Aegolius funereus*
Owl, Little	*Athene noctua*
Owl, Long-eared	*Asio otus*
Owl, Tawny	*Strix aluco*
Oystercatcher	*Haematopus ostralegus*
Peafowl	*Pavo cristatus*
Pelican, European White	*Pelecanus onocrotalus*
Penguin, Adelie	*Pygoscelis adeliae*
Penguin, Emperor	*Aptenodytes forsteri*
Penguin, Jackass	*Spheniscus demersus*
Penguin, King	*Aptenodytes patagonica*
Plover, American Golden	*Pluvialis dominica*
Pochard, Common	*Aythya ferina*
Prairie Chicken	*Tympanuchus cupido*
Ptarmigan, Rock	*Lagopus mutus*
Ptarmigan, Willow	*Lagopus lagopus*
Quail, Japanese	*Coturnix japonica*
Raven	*Corvus corax*
Razorbill	*Alca torda*
Rhea, Greater	*Rhea americana*
Robin, American	*Turdus migratorius*
Rook	*Corvus frugilegus*
Sandgrouse, Pallas's	*Syrrhaptes paradoxus*
Scoter, White-winged	*Melanitta fusca*
Sea Eagle	see Eagle
Secretary Bird	*Saggittarius serpentarius*
Shearwater, Sooty	*Puffinus griseus*
Shelduck	*Tadorna tadorna*
Skimmer, Black	*Rhynchops niger*
Skylark	*Alauda arvensis*
Snipe, Painted	*Rostratula benghalensis*
Sparrow, House	*Passer domesticus*
Sparrow, Savannah	*Ammodramus sandwichensis*
Sparrow, White-crowned	*Zonotrichia leucophrys*
Sparrowhawk	*Accipiter nisus*

Starling, Common	*Sturnus vulgaris*
Starling, Wattled	*Creatophora cinerea*
Stork, Black	*Ciconia nigra*
Stork, Whale-headed	*Balaeniceps rex*
Storm Petrel, Leach's	*Oceanodroma leucorhoa*
Swallow, Barn	*Hirundo rustica*
Swan, Coscoroba	*Coscoroba coscoroba*
Swan, Mute	*Cygnus olor*
Swan, Whooper	*Cygnus cygnus*
Swift, Common	*Apus apus*
Turkey	*Meleagris gallopavo*
Woodcock, Eurasian	*Scolopax rusticola*
Woodpecker, Green	*Picus viridis*
Woodpecker, Malaysian Grey-breasted	*Hemicircus concretus*
Woodpecker, Orange-backed	*Reinwardtipicus validus*
Woodpecker, White-headed	*Picoides albolarvatus*
Wryneck	*Jynx torquilla*

INDEX

Note: principal page reference is indicated by *italics*

abdomen, 14
abdominal muscles, *57*, 138
abdominal sac, 132
abnormalities of hindlimb, 71
abnormalities of vertebral column, 55
acceleration phase of testis, 170
accessory adrenal glands, 208
accessory brachial plexus, 268
accessory carotid body tissue, 208
accessory genital glands, 171
accessory parathyroid glands, 205
accessory spleens, 234
accommodation, 289
acetabulum, 65
acoustic coupling of both ears, 308
acoustic location, 303
acuity of taste, 312
adaptations of tongue for feeding, 86
adaptive immunity, 225, *234*
adaptive radiation of birds, 5
adenohypophysis, 149, *200*
adrenal ganglia, 278
adrenal gland, 170, *200*, *209*
adrenal portal system, *208*, 222
aerofoil, 75
afferent glomerular arteriole, 181
aggregated lymphoid nodules, *234*
air capillaries, *127*, 138
air cell of egg, *159*, 160
air-cooling in birds, 114
air pathways in lungs and air sacs, 139
air sacculitis, 133
air sac, histology, 135
air sacs, *131*, 139, 140
albumen formation, 153, *156*
albumen ligaments, 159
albumen of egg, 158
aldosterone, 209
alula, 17, *60*, 76
alular patagium, 17
ambiens muscle, 70
ampulla of semicircular duct, 309
amygdala, 254
amylase, 89, 105
anastomotic line of lung, 126

androgens, 151, *169*, 170, 203
aneurism in turkeys, 219
angle of attack, 76
angle of mouth, 10
anisodactyl foot, 19
ankle joint, 18, *19*, 67
annular pad of lens, 290
anomalies of bill, 26
anterior chamber of eye, 297
antitrochanter, 65
antitrochanteric facet of femur, 66
aorta, 218
aortic bodies, 263
aortic plexus, 278
aortic valve, 215
apical interscalar canal, 305
apteria, 36
apterial muscles, 23
APUD system, 208, 209
aqueous humour, 288, *297*
arachnoid granulations, 243
arachnoid mater, 237, 243
Archaeopteryx, *1*, 19, 26, 31, 55, 62, 84
archistriatum, 241, 254, *255*
arterial pressure, 215
arteriolae rectae, 182
arteriovenous anastomoses in skin, 27
arteriovenous shunt in lung, 141
artery or arteries
  axial, of feather, 30
  brachial, 218
  brachiocephalic, 218
  carotid body, 208
  caudal mesenteric, 220
  cerebral carotid, 218
  coeliac, 219
  common carotid, 218
  coronary, 218
  cranial mesenteric, *103*, 220
  cranial tibial, 220
  digital, 218, 220
  external carotid, 219
  external iliac, 220
  external ophthalmic, 218
  femoral, 220

319

INDEX

artery or arteries *contd.*
  internal carotid, 52, 202, *218*
  internal iliac, 221
  intralobular, of kidney, 176, *181*
  ischiadic, 220
  lateral caudal, 221
  metacarpal, 218
  metatarsal, 220
  oesophagotracheobronchial, 208
  ovarian, 220
  ovario-oviductal, 147, 220
  oviductal, 155, *220*
  pectoral, 218
  popliteal, 220
  pudendal, 190, *221*
  radial, 218
  renal, 181
  subclavian, 218
  testicular, 167, *220*
  ulnar, 218
  vaginal, 220
  vertebral, 53
arytenoid cartilages, 114
ascending pathways in spinal cord, 240
aspect ratio of wing, 18
asymmetry of
  external ear, 303
  eyeball, 285
  female gonads, 145, 166
  male phallus, 197
  testes, 166
atheromatous plaques in turkeys, 219
atlanto-occipital articulation, 52
atresia of ovarian follicle, 151
atria of parabronchi, 127
atrial muscles of parabronchi, *126*, 127, 138 141
atrioventricular bundle, 216
atrioventricular node, *216*, 278
auditory ossicles, 47
auditory pathways, *251*, 252, 308
auditory performance of birds, 307
autonomic nervous system, 275
avian pancreatic polypeptide, 210
axis, 52

back, 14
back of head, 10
bacterial fermentation in caeca, 105
balance organ, 308
banding, 17
basal interscalar canal, 305
basilar membrane, 306
basophils, 226
bastard wing, *17*, 60
bats, *5*, 31, 62, 58, 308
beak, 9, *12*, 24, 26
beak trimming, 25
bellows action of air sacs, *137*, 138, 142
belly, 14
bile ducts, 102, 105, *107*
bill, 9, *12*, 24, 26
  anomalies of, 26

bill tip organ, *24*, 259, 260
binocular vision, *284*, 293
bipolar cell, 292
bird's nest soup, 89
blastoderm, 157
blastodisc, 157
bleeding poultry at slaughter, 222
blind spot, 292
blinking, 297
blood cells, 224
blood coagulation, 225
blood samples, 222, 223
blood–gas barrier, 127
B lymphocytes, *225*, 232
body size restriction, 4
bone or bones
  angular, 47
  articular, *47*, 304
  basioccipital, 46
  basisphenoid, 46
  carpal, 60
  carpometacarpal, 60
  clavicle, 58
  coracoid, 58
  dental, 47
  dentary, 47
  epiotic, 46
  exoccipital, 46
  femur, 66
  fibula, 68
  frontal, 46
  furcula, 58
  humerus, 59
  ilium, 65
  ischium, 65
  jugal, 45
  maxillary, 45
  mesethmoid, 46
  metacarpal, 60
  metatarsal, 68
  nasal, 43
  opisthotic, 46
  orbitosphenoid, 46
  palatine, 45
  parasphenoid, 46
  parietal, 46
  patella, 67
  phalanges, 60, 68
  prearticular, 47
  prefrontal, 46
  premaxillary, *43*, 45
  prootic, 46
  pterygoid, 45
  pubis, 65
  pygostyle, 55
  quadrate, *45*, 47, 304
  quadratojugal, 45
  radius, 59
  ribs, 55
  scapula, 58
  sesamoids, 59, 68
  splenial, 47
  squamosal, 46

INDEX

sternum, 56
supra-angular, 47
supraoccipital, 46
synsacrum, 54
tarsal, 67, 68
tarsometatarsus, 18, *68*
tibiotarsus, 67
ulna, 59
vertebrae, *51*, 52, 53
vomer, 45
wishbone, 58
bony nasal openings, 43
bony orbit, 43, *46*
booted foot, 27
Bowman's capsule, 179
brachial glycogen body, 238
brachial plexus, 268
brachium conjunctivum, 248
brachium pontis, 247, 248
brailing, 60
brain, evolution of, 243
braincase, 46
breast, 14
breast blisters, 24
brood patch, 28
broodiness, 203
Brücke's muscle, 288
Brunner's glands, 102
bulbar axis of eye, 284
bumblefoot, 29
bursa of Fabricius, 232

caeca, 103
caecal tonsil, 104, *234*
cage layer fatigue, 55
cage layer osteoporosis, 72
calcification of shell, 155, 156, *157*
calcitonin, 207
canal of Schlemm, 287
cannibalism, 25
caponization, 173
cardiac muscle cells, 216
cardiac output, 215
cardioaccelerator fibres, 278
carina, 8, *56*, 62
carinates, 8, 56
carotid body, 208
carpal remix, 17
carpometacarpal spur, 26
caruncles, of skin of head, 27
casque, 24, 46
castration of cockerels, 173
caudal colliculus, *250*, 251
caudal group of air sacs, *139*, 140
caudal plexus, 274
caudal thoracic sac, 132
caudal vena cava, 222
caudal vertebrae, 54
caudodorsal secondary bronchi, 124
caudolateral secondary bronchi, 125
caudoputamen, 255
caudoventral secondary bronchi, 124
C-cells, 207

central area of retina, 293
centre of gravity, 67, *71*
cere, 25
cerebellospinal tract, 242
cerebellum, 248
  feedback pathways, 247, *249*, 250
  flocculonodular node, 248
  flocculus, 248
  folia, 248
  histology, 249
  lesions, 250
  lobules, 248
  peduncles, 247, *248*, 250
  primary fissure, 248
  projections, 249
  uvulonodular fissure, 248
  vermis, 248
cerebral aqueduct, 250
cerebral cortex, 254
cerebral hemisphere, 253
cerebrospinal fluid, 243
cerebrospinal system, 241
cervical enlargement, cord, 238
cervical ribs, *52*, 56
cervical sac, 131
cervical spinal nerves, 267
cervical vertebrae, 13, *52*
cervicothoracic lymph nodes, 233
chalaza, 156, *159*
chalaziferous region of oviduct, 152
cheek, 12
Chelonia, 4
chin, 12
choana, 84
cholinergic endings, 275
chondrodystrophy, 71
chorda tympani, 261
choroid layer of eyeball, 287
choroid plexus of lateral ventricle, 254
ciliary body, *287*, 297
ciliary ganglion, 258
ciliary muscles, 288
circadian rhythms and pineal gland, 211
cistern of scala vestibuli, 305
Clarke's column, 239, 240
classification of birds, 6
clavicular sac, 132
claws, 26
cleidoic egg, 184
climbing adaptations, 66, 71
climbing fibres of cerebellum, 249
clitoris, 192
cloaca, 187
cloacal bursa, 187, 191, *232*
cloacal drinking, 192
cloacal plexus, *274*, 277, 278
cloacal promontory, *171*, 192
cloacal reabsorption of water, 189
clubbed down, 35
coagulation of blood, 225
cochlea, 305
cochlear duct, *305*
cochlear ganglion, 262

cochlear window, 305
cochleocerebellar fibres, 249
coeliac ganglia, 278
coelurosaur, 3
coitus, 191, 193
collecting duct, 179, *180*
collecting tubules, 177, *180*
coloration of egg shells, 161
colour of iris, 290
colour of testis, 166
colour vision, 295
columella, 304
columellar muscle, 303
column of Terni, 239
comb, 27
commitment of forelimb to flight, 5, 14, 52
common hepatoenteric duct, 102, *107*
conducting system of heart, 216
cones of retina, *291*, 294
cones and basal caps of shell, 160
conjunctival sac, 301
connections of lungs to air sacs, 134
contamination of poultry carcase, 92, 107
contraction sequence of stomach, 99
control of airway calibre, *138*, 141
coprodeum, 188
coprourodeal fold, 189
coracoclavicular membrane, 62
cornea, 286
corpus luteum, 150
corpus striatum, 255
cortex of kidney, 179
cortex of ovary, 146
cortical region of renal lobule, 177
cortical tissue of adrenal, 209
cortical type of nephron, 179
costal cartilage, 56
costal grooves of lung, 121
costal process of vertebrae, 52
costoseptal muscle, *136*, 267
counter-current system of lung, *130*, 141, 142
Crampton's muscle, 288
cranial cervical ganglion, 277
cranial cnemial crest, 67, 68
cranial meninges, 243
cranial mesenteric ganglia, 278
cranial nerves, 256
    roots, 243
cranial thoracic sac, 132
cranial vena cava, 222
craniofacial hinge, 44
craniomedial secondary bronchi, 123
craniosacral nervous system, 275
creeper abnormality, 68
cricoid cartilage, 114
crissum, 14
crista basilaris, 306
crista neglecta, 310
cristae of semicircular ducts, 309
crocodiles, 4, 193, 195
crop, *90*, 93, 203
crop milk, *93*, 203
cropping of wattles, 27

cross-current system of lung, *130*, 140, 141, *142*
crowing, 115
crown, 10
crown stripe, 10
crural muscles, 68
crus, 18
crypts of Lieberkühn, 104
culmen, *9*, 24
culmination phase
    of epididymis, 169
    of ovary, 147
    of testis, 166, *170*
cupula, 310
curled tongue, 88
cuticle of gizzard, 97
cuticle of shell, 160
cysticoenteric duct, 102

D cells, 210
dead space, respiration, 116
decapitation and locomotion, 242
deep pectoral myopathy, 62
defaecation, *189*, 193, 195
denticulate ligaments, 237
dermal papillae of nail, 24
dermis, *23*, 24
descending pathways in spinal cord, 241
descending reticular formation, 253
desmognathous birds, 45
determinate layers, 149
detumescence, 190, *194*
dewlap, 27
diaphragm, 136
diastataxy, 17
diencephalon, 251
digestion in intestines, 105
digital pads, 29
digits of manus, 60
digits of pes, 19, *68*
dinosaurs, 2
direct connections of air sacs, 134
directional analysis by ear, 307
disc of latebra, 157
distal part of nephron, 179
distal vagal ganglion, 263
diving adaptations, 66, 68, 69, 71, 75, 110, 289, 300, 303
divisions of kidney, 175
dorsal cochlear decussation, 246
dorsal column of spinal cord, *238*, 240
dorsal fascicle of brachial plexus, *268*, 270
dorsal hepatic peritoneal cavities, 82
dorsal mesentery, 79
dorsal root ganglion, 265
dorsal white commissure, 239
dorsobronchi, 124
dorsolateral ascending bundle, 240
dorsolateral fasciculus, 241
dorsolateral secondary bronchi, 125
downstroke of wing, 58, 60, *62*, 76
downy barbs, 31, 35
drag, wing, 75, 76
drainage of cerebrospinal fluid, 243

driving force for diffusion of oxygen, 138
dubbing, 27
ductus brevis, 305
ductus deferens, 170
ductus reuniens, 309
duodenal glands, 102
duodenum, 101
dura mater, 237, 243
dural venous sinuses, 243
dust bathing, 31

ear coverts, 29, *302*
ear lobes, 27
ear region, 10
echo location, 308
eclipse plumage, 41
ectethmoid plate, 47
ectostriatum, 255
ectothermy, 4
efferent glomerular arteriole, 182
egg callosity, 25
egg formation, 156
egg laying, 151, *160*, 204
egg peritonitis, 82, *150*
egg production and thyroid gland, 204
egg size, 161
egg tooth, 25
ejaculation, *171*, 194, 197
elastic zones of facial bones, 44
elastic zones of mandible, 47
end-parabronchial gas, 141
endochondral bone, 73
endocrine cells of
  gut, 96, 97, 98, 105, *209*, 210
  ovary, 151
  stomach, 96, 97, 98
  testis, 169
endolymph, 305
endolymphatic duct, 309
endothermy, 4
energetics of birds, *5*, 114, 142, 215
enlarged hock disorder, 71
eosinophils, 226
epaxial muscles, *56*, 266
epidermis, *23*, 24
  of beak, 24
epididymal appendix, 169
epididymal duct, 169
epididymis, 169
epidural space, 237
epiphyseal disk, 73
epithalamus, 253
erythrocytes, 224
ethmoidal ganglion, 259, *261*
Eustachian tube, 305
evisceration of poultry carcasses, 92, 107, 175
evolution
  of avian brain, 243
  of birds, 1
  of flight, 4
evolutionary relationships to reptiles, 4
exchange tissue, 127, *140*, 141
excretion by kidney, 183

exercise capacity of birds, *5*, 114, 142, 215
expansor secundariorum muscle, 64
expiration, *137*, 140
extensors
  of elbow joint, 63
  of hip joint, 69
  of stifle, 69
external acoustic meatus, 10, 28, *302*
external features of birds, 9
external respiration, 137
extracolumellar cartilage, 303, 304
extravitelline lamina, 158
eye, 284
  movement pathways, *242*, 246, *251*
  movements in birds, 301
  of Ostrich, 284
  size, 284
  stripe, 12
eyelashes, 298
eyelids, 10, *297*

fasciculus lateralis prosencephali, 255
fasciculus proprius, 241
fat bodies, 23
female papilla of ductus deferens, 190
femoral diverticula, abdominal sac, 133
fertilization, 150
feathers
  afterfeathers, 31, 35, *36*
  barb ridges of, 33
  barbs of, 30
  barbules of, 31
  body, 29
  bristle, *38*, 298
  calamus of, 29
  carpal, 17
  colour of, 38
  contour, 29
  coverts, 15, *17*, 29
  dermal papilla of, 29, *32*
  diastataxy, 17
  down, 36
  downy barbs of, *31*, 35
  epidermal collar, 33
  filoplume, *36*
  flight, 15, *29*
  fluffy part of, 31
  follicle of, 32
  growth of, 33
  hypopennae, 36
  metacarpal, 16
  moulting of, 35, *39*
  muscles of, *23*, 33
  natal downs, 36
  notch of, *31*, 76
  overlap of, 16, 17
  pennaceous part of, 31
  pinfeather, 33
  powder down, 36
  primaries, *15*, 16, 29
  pulp caps of, 30
  rachis of, 30
  rectrices, *14*, 29, 55

feathers *contd.*
  remicle, 16
  remiges, *15*, 29
  riboflavin deficiency, 35
  replacement of, 35
  secondaries, *15*, 16, 29
  semiplumes, 36
  shaft of, 30
  sheath, 33
  spreading of, *57*, 63, *64*
  tail, *14*
  tectrices, 15
  umbilicus of, 29, 31
  vanes of, 31
  waterproofing of, 36
  wing, 15
fibrous tunic of eyeball, 286
flank, 14
flapping flight, 76
flat eyeball, 284
flexible junction of upper jaw, 43
flexors of hip joint, 69
flight, 75
flightless birds, 62
floating rib, *55*, 56
foam gland, 191
follicle of ovary, 148
follicle-stimulating hormone, 203
follicular atresia, ovary, 151
food pathway of Mallard oropharynx, 312
foot, 18
foot pads, 29
foot, so-called, 19
forehead, 10
foreneck, 14
forward-pointing toes, 19
fossa of scala vestibuli, 305
fourth ventricle, 243
foveae of retina, *293*, 294
fractures, *55*, 71, 72
frequency analysis by ear, 307
frontal process of turkey, 27
frontal shield of coot, 24
frostbite of feet, 220
frostbite of wattles, 27
funnel of infundibulum, 152

gait, 69, 71
gallbladder, *106*, 107
Gallornis, 3
ganglion cells of retina, *291*, 292
ganglion impar, 278
gape, 9
gaseous exchange, 140
gastric juice, 94
gastric proteolysis, 94, *98*
gastrin, 105
gastrocnemius muscle, 68, *69*
gastrointestinal endocrine cells, 209
gelatinous body, 238
general cortex of forebrain, 254
geniculate ganglion, 260
germinal disc, 157

germinal epithelium of ovary, 145
giblets, 107
gizzard, 93, 94, *97*, 99, 100
gland of nictitating membrane, 300
glandular part of stomach, 94
glenoid cavity, 58
globular eye, 285
globus pallidus, 255
glomera of paracloacal vascular body, 190
glomerular filtration, *183*, *184*
glomerulus, 179, *181*
glossopharyngeal ganglia, 262
glottis, 88, *115*
glycogen body, 238
gnathotheca, 9
goblet cells of cloaca, 189, 191
goblet cells of intestine, 105
gonys, 9
Grandry corpuscles, 35
granulosa cells of ovary, *148*, 150, 151
grasping foot, 21
grey matter, 239
grit in gizzard, 99
grooming, 14, 26, 52, 116
gular fluttering, 114
gular region, 9

Harderian gland, 300
Hassal's corpuscles, 232
hatching, 160
hatching muscle, 57
Haversian systems, 73
head, 9
heart, 214
  rate, 214
  size, 214
helicotrema, 305
hepatic ligaments, 80
hepatic lobule, 107
hepatocystic duct, 107
hepatoenteric duct, 107
Herbst corpuscles, *25*, 36, 192, 267
hernias into hepatic peritoneal cavity, 82
heterocoelous joint, 52, 53
heterodactyl foot, 19
heterophils, 225
high altitude flight, *6*, 86, 142
hilus of lung, 122
hilus of ovary, 147
hindbrain, 243
hindneck, 14
hippocampus, 253
hock joint, 18, *67*, 72
Hof, 225
holorhinal nostril, 43
holothecal foot, 27
homing, 279
hooks of bill, 25
hopping, 70
horizontal septum of coelom, 136
hormones of
  adrenal gland, 209
  gut endocrine cells, 210

INDEX 325

neurohypophysis, 204
ovary, 151
pancreatic islets, 210
parathyroid, 206
pars distalis, 203
testis, 169
thyroid, 204
ultimobranchial gland, 207
hovering flight, 77
humpback, 55
hydrochloric acid secretion, *96*, 98
hyobranchial apparatus, 49
hypaxial muscles, *56*, 267
hyperstriatum, 255
hypobaric chamber, 6
hypophysial blood vessels, 202
hypophysial functions, 203
hypophysis, 200
hypotarsus, 68, 69
hypothalamohypophysial tract, *202*, 203, 204
hypothalamus, 202, 204, *253*

ileorectal junction, 104
ileum, 102
ilioischiadic foramen, 66
ilium, 65
imbricated scales, 27
immunity, adaptive, *225*, 234, 235
impaction of crop, 92
impaction of gizzard, 100
incubation patches, 28
incus, 47, 304
indeterminate layers, 150
indirect connections of air sacs, 134
infraorbital sinus, 112
infundibula of lung, 127
infundibular cleft, *85*, 305
infundibulum of neurohypophysis, 202
infundibulum of oviduct, 150, *152*, 156
ingluvies, 90
inner ear, 305
innervation of
  abdominal muscles, 267
  alimentary tract, 275
  alular muscles, 270
  ampullary cristae, 262
  atrial muscles of lung, 275
  bill tip organ, 259, 260
  blood vessels of head, 277
  carotid body, 208, *263*
  carpal flexors, 268, 269
  ciliary body, *258*, 275, 277
  cloaca, 274
  crista basilaris of ear, 306
  crop, 263, 264
  digital flexors of manus, 269
  duodenum, 264
  expansor secundariorum, 270
  extensors of
    carpus, 270
    digits of manus, 270
    digits of pes, 273
    elbow, 270

hip joint, 273
intertarsal joint, 273
knee joint, 273
eyeball muscles, *258*, 260
eyelid muscles, 258, *298*
flexors of
  digits of pes, 273
  elbow joint, 268
  hip joint, 273
  intertarsal joint, 273
  knee joint, 273
  wing, 268
glands of nictitating membrane, 261, 275
glands of
  head, 277
  nasal cavity, 275
  nasal mucosa, 260, 261
heart, 217, 264, 275, 278
hypophysis, 202
intestines, 278
iris, 258, 275, 277
jaw muscles, 260, 261
lacrimal gland, *260*, 275
laryngeal muscles, 262
larynx, 263
liver, 264
lower beak, 25
lung, 264, 275
maculae of utricle and saccule, 262
muscle of rectrical bulb, 274
muscles of
  cloaca, 274
  nictitating membrane, 26, 29, *260*, 298
  respiration, 267
  shoulder joint, 268, 270
nasal gland, 261
oesophagus, 264
ovarian follicle, 148
oviduct, 156
phallus, 274
pharynx, 263
pineal gland, 211
pulmonary vessels, *128*, 141
renal portal valve, 182
salivary glands, *262*, 275
skin of head, 259, 260
stomachs, 264
syringeal muscles, 264
tail muscles, 274
taste buds, *262*, 312
tongue, 262, 263
trachea, 263
tracheal muscles, 264
upper beak, 25
urogenital ducts, 274
urogenital organs, 275, 277, 278
vent, 274
inspiration, 137, 140
insulation, 4, 36
insulin, 210
interaural spectrum, 307
intercalated discs, 218
intercarotid anastomosis, 218

intermediate segment of nephron, 179, 180
intermediate zone of stomach, 94, *98*
internal capsule of mammals, 255
internal laying, 82, *150*
internal nares, 45
interorbital septum, 46
interparabronchial septa, 127
interramal region, 9
interrectrical elastic ligament, 57
interremigial elastic ligament, 64
inter-renal tissue of adrenal, 209
interscapular region, 14
interstitial cells, ovary, 148, *151*
interstitial cells, testis, *169*, 170, 203
interstitiospinal pathway, 242
intertarsal joint, 18, 19, *67*
intestinal crypts, 104
intestinal folds, 104
intestinal peritoneal cavity, 81
intraocular pressure, 297
intraparabronchial vessels, 128
intravenous injection, 222, 223
involution of cloacal bursa, 232
iridiocytes, 290
iridociliary nerve, 258
iridocorneal angle, 297
iris, 290
islets of pancreas, 210
isthmus of oviduct, *153*, 156

jejunum, 102
jugal arch, 45
jugal process of maxilla, 45
juxtaglomerular complex, 180

keel, 8, *56*, 62
kidney, 175
kinesis, upper jaw, *43*, 44, 47, *48*, 49
kinky back, 55
knee, so-called, 18
knee joint, 18
knob of carpometacarpus, 60
koilin layer of gizzard, 97
kyphosis, 55

lacrimal apparatus, 299, 300
lacteals, 105
lagena, 305
lagenar ganglion, 262
large intestine, 103
laryngeal mound, *88*, 115
laryngeal skeleton, 114
larynx, 114
latebra, 157
lateral column of spinal cord, 238
lateral forebrain bundle, 255
lateral geniculate body, 252
lateral lemniscus, 247, *251*
lateral stripe, 12
lateral ventricle, 251, 254
laterobronchi, 124
laterodorsal secondary bronchi, 124
lateroventral secondary bronchi, 124

laying and medullary bone, 74
laying cycle, 146
lead poisoning, 101,
left atrioventricular valve, 215
leg proper, 18
lens, 290
Leydig cells, 169
Lepidosauria, 4
lice of eyelid, 297
lift, *75*, 76, 77
ligaments of jaw apparatus, 47
light adaptation, 290
limbic cortex, 254
limbus, 287
lipoid sebaceous secretions, 28
lips of vent, 192
Lissauer's tract, 241
liver, 106
lobate foot, *22*, 69
locomotion adaptations, 66, 70
loop of Henle, 177, 179, *180*
loral bristles, 12
loral line, 12
loral stripe, 12
lore, 10
lower jaw, 47
lower leg, 18
lumbar lymph nodes, 233
lumbar plexus, *270*
lumbosacral enlargement, *238*, 239
lumbosacral plexus, 270
lung
 elastic fibres, 127
 external appearance, 121
 position, 121
 volume, 123
 weight, 122
lymphatic folds, *171*, 191
lymphatic hearts, 229
lymphatic vessels, 229
lymph nodes, 233
lymphocytes, 225
lymphoid tissue, 231

macula
 of lagena, 305, 310
 of saccule, 310
 of utricule, 310
macula densa, 180
magnum, *152*, 156
malar stripe, 12
malleus, 47, 304
mamillary layer of shell, 159
mammalian type of nephron, 179
mandibular ganglion, 262
manipulation of foodstuffs by beak, 116
mantle, 14
manus, 60
Marek's disease, 279
marrow cavity, 73
matching perfusion and ventilation, 138, *141*
maturation division in female, *149*, 150
maturation division in male, 168

maturation of spermatozoa, 171
maturation phase of oogenesis, 146
maturation phase of spermatogenesis, 168
maxilla, 43, 45
mechanoreceptors of nail, 24
Meckel's diverticulum, 103
medial geniculate body, 252
medial lemniscal system, 248, 252
medial longitudinal bundle, 242, 246, 251
median coverts, 17
median eminence, 202, 253
mediastinum testis, 166
mediodorsal secondary bronchi, 124, 139, 140
medioventral secondary bronchi, 123, 139
medulla, of kidney, 179
medulla, of ovary, 146
medulla oblongata, 243
medullary bone, 73, 74, 206
medullary collecting tubules, 177, 178, 180
medullary cone, 177
medullary pyramid, 179
medullary region of renal lobe, 179
medullary reticular formation, 247
medullary tissue of adrenal, 209
medullary types of nephron, 177, 179
melanocytes of testis, 169
meningeal ramus of spinal nerve, 266
meninges, cranial, 243
meninges, spinal, 237
menisci of intertarsal joint, 68
Merkel cell, 25
mesencephalic colliculus, 250
mesencephalic tectum, 250
mesorchium, 166
mesovarium, 146
metabolic water, 114
metapatagium, 17, 64
metarsal pad, 29
metatarsal spur, 26
metatarsus, 18
micturition, 195
midbrain, 250
middle ear, 303
migration, 5, 114, 279
molybdenum deficiency, 72
monocular vision, 284
monocytes, 225
mossy fibres of cerebellum, 249
moulting and thyroid gland, 204
moulting of bill, 24
moustachial stripe, 12
movements of
  carpal joint, 63
  crus, 69
  digits of foot, 70
  digits of manus, 63
  elbow joint, 63
  eyeball, 301
  femur, 69
  humerus, 59
  intertarsal joint, 69
  knee joint, 69
  remiges, 64

  tail, 57
  tibia, 72
  wing, 60
mucin fibres of albumen, 159
mucosal folds of oviduct, 155
mucous secretion of duodenum, 102
Muller's muscle, 289
mural lymphoid nodules, 233
muscles
  abdominal wall, 57
  accommodation, 288
  alula, 63
  carpal joint, 63
  cloaca, 192
  crus, 69
  digits of foot, 70
  digits of manus, 63
  elbow joint, 63
  eyeball, 49, 301
  eyelids, 49, 298
  feather, 33
  hip joint, 68
  hock joint, 69
  hyobranchial apparatus, 49
  intertarsal joint, 69
  jaw, 47, 48
  knee joint, 69
  larynx, 51
  leg, 69
  metapatagium, 64
  neck, 57
  nictitating membrane, 49, 298
  pelvic limb, 69
  propatagium, 64
  rectrical bulb, 57
  remigial movement, 64
  respiration, 57, 137, 138
  shoulder joint, 60, 62, 63
  syrinx, 51, 120
  tail, 57
  thigh, 68
  third eyelid, 49
  tongue, 49, 88
  trachea, 51, 116
  trunk, 56
  wing, 60
  wing folding, 65
  wing spreading, 64, 65
muscular part of stomach, 94
myocardial cell, 216
myoglobin, 74

nail, 24
nape, 14
nasal bones, 43
nasal conchae, 111
nasal fossa, 10
nasal gland, 112
nasal septum, 110
nasal valve, 110
nasal-interorbital septum, 44, 45
nasolacrimal duct, 301
navigation, 279

INDEX

neck, 13
neck of latebra, 157
neocortex, 243, 253, 255, *256*
neopulmo, *126*, 139
neostriatum, 252, *255*
nephronal loop of Henle, 177, 179, *180*
nerve or nerves
  abducent, *260*, 298
  accessory, 264
  aortic, 263
  axillary, 270
  bicipital, 268
  bigeminal, 274
  cardiac, 278
  cardiac sympathetic, 278
  carotid body, 263
  caudal cardiac, of vagus, 264
  caudal coxal, 273
  caudal cutaneous femoral, 273
  cervical carotid, 277
  cervical cutaneous, 267
  ciliary, 289
  cochlear, 262
  cranial cardiac, 263
  cranial coxal, 273
  cranial crural cutaneous, 273
  cranial femoral cutaneous, 273
  deep fibular, 273
  descending cervical, 264
  descending oesophageal, 262
  dorsal cutaneous antebrachial, 270
  dorsal metacarpal, 270
  dorsal propatagial, 270
  facial, *260*, 312
  femoral, 273
  fibular, 273
  furcal, 272
  glossopharyngeal, *262*, 312
  greater superficial petrosal, 260
  hyomandibular, 261
  hypoglossal, 264
  hypoglossocervical, 264
  infraorbital, 260
  intercostal, 267
  intermediate caudal, 274
  intestinal, 274, 277, *278*
  intramandibular, 260
  iridociliary, 258
  ischiadic, 273
  lagenar, 262
  laryngopharyngeal, 262
  lateral caudal, 274
  lateral femoral cutaneous, 272
  lateral plantar, 273
  later sural, 273
  lingual, 262
  long ciliary, 258
  mandibular, *260*, 298
  maxillary, 259
  medial plantar, 273
  medial sural, 273
  median, 269
  medianoulnar, 268
  nasopalatine, 260
  obturator, 273
  oculomotor, *258*, 298
  olfactory, *256*, 311
  ophthalmic, 258
  optic, *258*, 292
  palatine, 260
  parafibular, 273
  pectoral, 268
  pelvic, 275
  pelvic limb, of, 270
  precarotid, 263
  pudendal, *274*, 275, 278
  pulmoesophageal, 264
  radial, 270
  recurrent, 263
  Remak's, 278
  spinal, *265*, 267
  splanchnic, 278
  sublingual, 260
  superficial fibular, 273
  supracoracoid, 268
  supraorbital, 260
  tail, of, 274
  tibial, 273
  trigeminal, 258
  trochlear, 258
  ulnar, 268
  vagus, 263
  vestibular, 262
  vestibulocochlear, *262*, 307, 310
  wing, of, 268
nervous layer of retina, 291
nesting behaviour, 14, 52, 116, 151
neural lobe, 202
neurohypophysis, 201
neurotensin, 105, 210
nictitating membrane, 298
nodose ganglion, 263
noradrenergic endings, 275
nostrils, 10, *110*
notarium, 53
nucleus or nuclei
  abducent nerve, of, 245
  accessory abducent, 245
  ambiguus, 246
  angular, 246
  auditory, 246
  basal, 253
  branchial efferent, 245
  caudal olivary, 249
  caudal salivatory, 246
  cerebellar, 249
  cochlear, 243
  cranial nerves, of, 245
  cuneate, *240*, 248
  dorsal motor vagal, 246
  Edinger–Westphal, 250
  facial nerve, of, 247
  glossopharyngeal nerve, of, 246
  gracile, *240*, 248
  hypoglossal nerve, of, 245
  infundibular, *202*, 253

intercollicular, 121, *251*
intermedius of spinal cord, *239*, 241
intermediate of medulla oblongata, *245*, 246
isthmo-optic, 251
lateral mesencephalic, 121, *250*, 251
magnocellular cochlear, 246
marginal, 238
mesencephalic trigeminal, 247
motor of accessory nerve, 246
motor trigeminal, 247
oculomotor, 250
olivary, 247
ovoidal, *251*, 252
paraventricular, *202*, 253
pontine, 247
preoptic, 253
principal optic, *252*, 255
principal trigeminal, 247
red, *241*, 251
rostral olivary, 247
solitary tract, of, *246*, 247
spinal tract of trigeminal, of, 247
substantia gelatinosa, of, 241
supraoptic, *202*, 253
trochlear, 250
nuptial phase of testis, 166
nutritional deficiencies, skeletal, 71

oblique septum, 136
obturator foramen, 66
occipital condyle, 52
ocular leucosis, 290
oesophageal sac, 92
oesophagus, *90*, 92, 93, 94, 264
oestrogens, *151*, 203
oil droplet, 295
oil gland, 13, *28*
olfaction, 112, *310*
olfactory bulb, 253, *311*
olfactory cell, 311
olfactory cortex, 254
olfactory epithelium, 310
olfactory region of nasal cavity, *112*, 310
olivocerebellar fibres, 249
oogenesis, 146
oogonium, 146
opening of left oviduct, 191
opening of ureter, 190
openings of urogenital ducts, 190
operculum of external ear, 10, *303*
operculum of nostril, 110
opisthocoelous joint, 54
optic chiasma, 251, 258
optic disc, 292
optic lobe, *250*, 251
optic tectum, *250*, 251
optic tract, 252, 258
oral cavity, 84
oral opening, 9
oral sacs, 89
orbital region, 10
orders of birds, 7
Oregon muscle disease, 62

organ of Corti, 306
orientation, 279
oropharynx, *84*, 89
osmiophilic bodies, atria, 127
os nervi optici, 287
ossification of cartilage bones, 73
ossification of tendons, 72
osteoblasts, *73*, 74
osteoclasts, *73*, 74, 207
osteodystrophy, 71
osteomalacia, 55
osteomyelitis, 55
ostium of air sac, 135
ovarian follicle, 146, *148*, 203
ovarian hilus, 147
ovarian hormones, 151
ovarian pocket, 150, 152
ovary, left, 145
oviduct, left, 152
  ligaments, 155
  glands, *154*, 155, 156
  obstruction, 157
  occluding membrane, *154*, 191
oviposition, 151, *160*, 204
ovomucin, 158
ovulation, 149, *150*, 203
ovum, 146, 150
owl's eye, 295
oxygen consumption in flight, 5
oxygen transport, 6, *215*
oxynticopeptic cell, 96
oxytocin, 204

pain transmission, 241
palate, 45, *84*
palatine bar, 45
palatoquadrate bridge, *45*, 48
palatovomeral complex, 46
paleopulmo, *126*, 139
paleostriatum augmentatum, 255
paleostriatum primitivum, 255
palisade layer of shell, 160
palmate foot, 21
palpebrae, 10
pamprodactyl foot, 19
pancreas, 105
pancreatic ducts, *102*, 105
pancreatic islets, 210
pancreatic juice, 105
pancreatic polypeptide cells, 210
panting, 114
papilla of ductus deferens, 170, *190*
papilla of ureter, 190
papillae of oropharynx, 88
parabronchi, 125
paracloacal vascular body, *190*, 193, 196
parahippocampal area, 254
paraplegia, 55
parathyroid glands, 204
parathyroid hormone, 74, *206*
parathyroid nodules, 207
paravertebral ganglia, 277
paravertebral trunk, 277

INDEX

parenchymatous zones of ovary, 146
pars distalis, 200
pars tuberalis, 200
partial pressure of carbon dioxide, 141
partial pressure of oxygen, 141
patellar sulcus of femur, 67
pecten, 296
pectinate ligament, 287
pectoral crest, *59*, 62
pectoral muscles, *62*, 74, 76
pellets from stomach, 100
pelvic girdle, 65
pelvic limb, 18
pelvic symphysis, 65
pelvis, 14, *65*
penetration of ovum by spermatozoa, 150
pepsin secretion, 96, 98
perching, 21, 70
perching foot, 21
perching muscle, 70
pericardium, 83
perilobular collecting tubules, 176, *180*
perilymph, *304*, 305
peripheral arterial resistance, 215
perirenal diverticula of abdominal sac, 133
peritoneal partitions, 79
peritonitis, 82
peritubular capillary plexus, *182*, 183
perivitelline lamina, *148*, 158
perosis, 71
pes, 18
pessulus, 117
petrosal ganglion, 262
phallus
  crocodile, 193, 195
  domestic fowl, 193
  drake, 196
  elastic ligament, 197
  elastic vascular body, 195
  erection, 190, *196*, 197, 198
  lymphatic cavities, 190
  female, 192, 194
  fibrocartilaginous body, 197
  fibrolymphatic bodies, *195*, 196
  intromittent, 195
  lymphatic folds, *171*, 191
  non-intromittent, 193
  Ostrich, 195
  phallic bodies, 193
  ratites, *193*, 195
  seminal groove, 197
  sulcus, *195*, 196
  turkey, 194
pharyngeal arches of embryo, 246, 247, 304
pharyngeal pouches of embryo, 204, 207
pharyngeal tonsil, 86
pharyngotympanic tubes, *85*, 305
pharynx, 84
photosensitive outer segment, 295
physical digestion of food, 93
pia mater, 237, 243
pigment epithelium of retina, 290, 291
pigmentation of shell, 160

pileum, 10
pineal gland, *210*, 211, 251, *253*
pinioning, 60
pipping, *25*, 57, 160
piriform cortex, 254
pituicytes, 202
pituitary gland, 200
plasma cells, *225*, 234
pleura, 136
plumping, 157
pneumatic foramen of humerus, 59
pneumatization of skeleton, 72, *135*
pneumatization of skull bones, 43, *46*
podotheca, 27
polar bodies, 149, 150
Polkissen cells, 180
polyspermy, 150
pons, 243
pontine reticular formation, 242
pontocerebellar pathway, 243, 247
pores of shell, 160
portal system of hypophysis, *202*, 203
posterior chamber, 297
postganglionic endings, 275
posthepatic septum, 80
postovulatory follicle, 150
postpatagium, 17, 64
preen gland, 13, *28*
preganglionic endings, 275
preganglionic sympathetic outflow, 239
premaxillonasal elastic zone, 45
prenuptial acceleration of ovary, 147
prepiriform cortex, 254
primary afferent neurons, 240, 251
primary branches of ureter, 183
primary bronchus, 123
primary oocyte, *146*, 149
primary sex cords, 145
primordial germ cells of female, 145
primordial germ cells of male, 166
Proavis, 4
procricoid cartilage, 114
proctodeal glands, 191
proctodeum, 191
progestagens, 151
prokinesis, 43, 48
prolactin, *93*, 203
  and crop milk, 93
propatagium, *17*, 64
propriospinal system, 241
proteinases in pancreatic juice, 105
proventriculus, 94, *95*, 98
proximal part of nephron, 179
psilopaedic species, 40
pterosaurs, 31, 58, 310
pterylae, 36
pterylosis, 36
ptilopaedic species, 40
pudendal plexus, 274
pulmonary aponeurosis, 136
pulmonary blood capillaries, 127
pulmonary circulation, 128
pulmonary fold, 136

INDEX

pulmonary rami of vagus, 264
pulmonary valve, 215
pupil shape, 290
Purkinje cells, *217*, 249
pyloric part of stomach, 94, *98*
pyramidal tract, 241
pyramidalis muscle, 49, *298*

quadratus muscle, 49, *298*
quill knobs, 16, 59
quintofrontal tract, *247*, 253, 255

radiation of land and water birds, 3
rami communicantes, 277
raptorial foot, 21
ratites, *8*, 56
receptacle of ductus deferens, 170
receptor cell of balance, 310
receptor cell of hearing, 306
rectocoprodeal fold, 188
rectrical bulb, *55*, 57
rectum, 104
recurrent bronchi, 135
red and white muscles, 74
reduction division, 146
refractory period of ovary, 147
regeneration phase of testis, 170
regression of ovarian follicle, 150
regulation of parabronchial ventilation, *138*, 141
regulation of perfusion of lung, 141
regurgitation in feeding, 93, 94
regurgitation of gastric juice in poultry, 96
releasing factors, 203
renal corpuscle, 179
renal cortex, 179
renal lobe, 178
renal lobule, 175, *176*
renal portal system, 222
renin-like substance, 181
reptilian features of skull, 43
reptilian type of nephron, 179
resistance to tracheal airflow, 116
resolving power of eye, 293, 294
respiratory mechanics, 137
respiratory region of nasal cavity, 111
restiform bodies, 248
rete mirabile ophthalmicum, 218
rete testis, 168
reticular formation of hindbrain, 247
reticulospinal pathways, 241
retina, 291
  efferent fibres, 251
retinal grain, 294
retinal vessels, absence of, 291, 294
Rexed's laminae, *239*, 241
rhamphotheca, 9, *24*
rhinencephalon, 254
rhinotheca, 9
rhomboidal sinus, 238
rhynchokinesis, 43, *45*, *49*
rib cage, 24
rickets, 55, *71*

rictus, 9, *27*
right atrioventricular ring, 216
right atrioventricular valve, 215
right gonad of female, 162
right ovary, fully functional, *145*, 150, *162*, 163
right oviduct, *162*, 191
ringing, 17
Roc of mythology, 146
rods, 291
roots, brachial plexus, 268
roots, cranial nerves, 243
rostral colliculus, 250
rostrum, 9
rotation of egg, 157
round window, 305
rubrospinal tract, 241
rump, 14
running adaptations, 66, 71
rupture of posthepatic septum, 82

saccoperitoneal membrane, 136
saccopleural membrane, 136
saccule, 308, 309
sacculocochlear duct, 309
sacral plexus, 270
sagittal eminence, 252, 253, 255
salivary glands, 88
salt gland, 112
sarcoplasmic reticulum, 216
scala media, 305
scala tympani, 305
scala vestibuli, 305
scales, 27
scaly leg, 27
schizorhinal nostril, 43
scleral ossicles, 286
scleral venous sinus, *287*, 297
sclerocorneal muscles, 288
scoliosis, 55
seasonal reconstruction of testis, 170
secondary branches of ureter, 179, *180*, 183
secondary bronchi, 123
secondary oocyte, 146
secondary sex cords, 146
secondary sexual characteristics of male, 169
secondary spermatocytes, 168
secretin cells, 210
semen, 171
semicircular ducts, 308
semicirular ducts, 308
seminal glomus, 171
seminal plasma, 171
seminiferous tubules, 166
septal concha, 112
septal sinus, 110
sex determination, 149
sexing
  by laparoscopy, 151
  day-old chicks, 193
  passerines, 171
shank, 19
shape of birds' eggs, 161

shape of pupil, 290
shell, 159
shell gland, 154
shell membranes, 154, *159*
shunts in lung, 141
shunts in renal portal system, 182
side, 14
side of neck, 14
sinuatrial node, *216*, 278
sinuatrial valvules, 215
sinus venosus, 215
sinusoids of liver, 107
skin, 23
skull, 43
slipped tendon, 71
slots of wing, 32
small intestine, 101
smooth muscle of airways, 140
smooth muscle of oviduct, 155
snakes, 4
snood of turkey, 27
soaring adaptations, 5, *18*, 57, 59, *62*
soft palate, 84
solitary lymphoid nodules, 233
somatic motor pathways, 242
somatosensory system, 240
somatostatin, 105, 210
sour crop in poultry, 94
spaces of Fontana, 287
spaces of irido-corneal angle, 287
species variations of air sacs, 133
speed of flight, 6
speed under water, 6
sperm-host glands, 154, *156*
spermatic fossulae, 154, *156*
spermatids, 168
spermatogenesis, 167, 168
spermatogonia, 168
spermatozoal maturation, 171
spermatozoon, 171
sphenopalatine ganglion, 260, *261*
sphincter of pupil, 290
spinal cord, 237
spinal meninges, 237
spinal nerves, 265
spinocerebellar tracts, 239, *240*, 249
spinoreticular pathways, 241
spinospinal pathways, 241
spinothalamic fibres, *241*, 252
splanchnic nerves, 278
spleen, 234
spondylolisthesis, 55
spongy layer of shell, 159
spreading of mandible, 47
spur-core, 68
spurs, 26
stalling, 60, 63, *76*
stapes, 304
statoconial membrane, 310
sternal bursa, 24
sternum, 56
stigma, 149
stomach, 94

stomach oil of petrels, 96
stomodeum, 200
storage
  of food by stomach, 94
  of spermatozoa in male, 171
  of spermatozoa in oviduct, 154, *156*
straight tubules of testis, 168
straining mechanism in ducks, 86
strata of yolk, 157
stratum granulosum of ovary, *148*, 150, 151
streamlining of body, 5
striatal complex, 255
subarachnoid space, 237, 243
submoustachial stripe, 12
substantia gelatinosa, 241
substantia propria, 286
subvertebral ganglia, 277
sucking movements of vent, 192
superciliary line, 12
superciliary stripe, 10
supernumerary male pronuclei, 150
supracoracoid muscle, 56, 58, *60*, 62, 63, 76
supracoracoid nerve, 268
surface area of blood–gas barrier, *138*, 142
surface tension in air capillaries, 138
surfactant, 127
sustentacular cells of testis, 168
sutures of skull, 43
swallowing, 89
swallowing action of larynx, 88
sweat glands, 28
swimming adaptations, 21, 62, 71
sympathetic chain, 277
sympathetic system, 277
syndactyl foot, 19
syringeal bulla, 119
syringeal cartilages, 117
syringeal function, 121
syringeal labia, *119*, 121
syrinx, 117

tactile transmission, 240, 241
tail, function in flight, 77
tapetum lucidum of choroid, 287
tapetum lucidum of iris, 290
tarsometatarsus, *18*, 19, *68*
tarsus, *18, 67*, 68
taste, 311
taste buds, 312
tectocerebellar pathways, 249
tectoral membrane, 306
tectospinal fibres, *242*, 251
tectothalamic fibres, 252
teeth, 84
tegmentum vasculosum, 305
telencephalic complex of birds, 255, *256*
temporal ciliary muscle, 289
temporal fossa of skull, 46
temporal resolution by ear, 307
tensor periorbitae muscle, 49, *298*
tertiary bronchi, 125
testa, 159
testis, 166

INDEX

testosterone, 169
thalamus, 252
thecae of ovarian follicle, 148
thecal interstitial cells, 151
thecal luteal cells, 150
thecodonts, 3
Therapsida, 4
thermoregulation, 28, *113*, 114, 220, 253
thigh, 18
third eyelid, 298
third ventricle, 202, 210
thoracic cage, 56, 58
thoracic ducts, 229
thoracic girdle, 58
thoracic vertebrae, 51, *53*
thoracolumbar nervous system, 277
thorax, 14
throat, 12
thrombocytes, 224
thymic corpuscles, 232
thymus, 231
thyroid gland, 204
tibial cartilage, *68*, 70
tibial dyschondroplasia, 71
tibiotarsal arterial rete, 220
tip vortex, 76
T lymphocytes, *225*, 232, 234
toes, 19
toilet claw, 26
tomium, *9*, 84
tongue, 86
tongue skeleton, 49
tortoises, 4
total O_2 diffusing capacity, 141
totipalmate foot, 21
trabecular reticulum, 287
trachea, 115
    bulb, 116
    cartilages, 115
    coils, 56, *115*
    dead space, 116
    membrane, 119
    sac, 116
tracking movement pathways, 242, 251
transparent fluid of semen, 171
transport of spermatozoa, 155
trapezoid body, 243, 246
tridactylism, 19
trigeminal ganglion, 258
trigeminocerebellar tracts, 249
trimming of beak, 25
triosseal canal, 58
truncobulbar node, 216
trunk, 14
trunks of brachial plexus, 268
T-tubules of myocardial cell, 216
tuber cinereum, 202, *253*
tubular eye, 285
tubular region of infundibulum, 152
tubular secretion of urates, 183, 184
tumescence, 190, 193, 195, *196*, 197, 198
tunica albuginea of ovary, 145, 146
tunica albuginea of testis, 166

"turkey syndrome '65", 71
turtles, 4
twisted leg deformity, 72
tympanic membrane, 303
tympaniform membranes, 118
tympanum, 117

ulceration of gizzard, 98
ulnocarporemigial aponeurosis, 64
ultimobranchial glands, 207
ultrasonic vibrations, 308
ultraviolet perception, *290*, 295
uncinate process, 56
undertail, 14
unicellular glands of oviduct, *154*, 156
unidirectional gas flow, *139*, 142
uniformity of structure of birds, 5
upper jaw, 43
upstroke of wing, 56, *58*, 60, *76*
urate clearance, 184
ureter, 183
ureteric papilla, 190
uric acid, 183
uricotelic excretion, 183
urinary bladder, 183, 233
urodeum, 189
urogenital ducts, 190
uroproctodeal fold, 189
uropygial gland, 13, *28*
uropygial wick, 28
uterus, *154*, 157
utricle, 308, 309

vagal ganglia, 263
vagina, *154*, 157
vagosympathetic trunk, 278
vallecula, 253
valve of pulmonary vein, 216
valving in lung, 140
vasa recta, 182
vascular body, *190*, 193, 196
vascular tunic of eyeball, 287
vascular zones of ovary, 146
vasoactive intestinal polypeptide, 105, 210
vein or veins
    adrenal, 222
    atrial, 128
    basilic, 222
    cardiac, 222
    caudal mesenteric, *182*, 223
    caudal renal, 183
    caudal tibial, 223
    coccygeomesenteric, *182*, 223
    common iliac, 182, 183, 222
    common pectoral, 222
    cranial renal, 183
    cranial tibial, 223
    deep ulnar, 222
    digital, 223
    dorsal metatarsal, 223
    efferent renal, 183
    external iliac, 222
    femoral, 222

vein or veins *contd.*
  hepatic, 222
  hepatic portal, 223
  interlobular, of kidney, 176, *182*
  internal iliac, 223
  interparabronchial, 129
  intralobular, of kidney, 176, *182*, 183
  intraparabronchial, 129
  ischiadic, 222
  jugular, 222
  occipital, 224
  ovarian, 147, 222
  oviductal, 155
  popliteal, 223
  pulmonary, 216, 222
  renal, 222
  renal portal, 182
  subclavian, 222
  superficial plantar metatarsal, 223
  testicular, 166, 222
  Thebesian, 222
  vertebral, 224
veneral disease in geese, 198
venous sinuses, intracranial, 224
vent, 14, 28, *192*
ventral column of spinal cord, 238
ventral fascicle of brachial plexus, 268
ventral hepatic peritoneal cavities, 82
ventral mesentery, 79
ventral white commissure, 239
ventrobronchi, 123
ventrolateral ascending bundle, 240
venulae rectae, 182
vertebrae, 51
vertebral diverticula of air sacs, 133
vestibular ganglion, 262
vestibular pathways, 251
vestibular region of nasal cavity, 111
vestibular window, 304
vestibule of primary bronchus, 123
vestibulospinal tracts *242*, 246
vestigial right oviduct, 191
villi, cloaca, 189

villi, intestine, 104
vinculum, 70
visceral motor pathways, 241, 242, 246, *253*
visual acuity, 284, 285, 292, 293, *294*
visual pathways, 251, 252, 255
visual pigments, 295
vitelline diverticulum, 103
vitellogenesis, 156
vitreous body, 297
vocalization, *121*, 251
volume of pulmonary capillary blood, 142

wading foot, 21
walking adaptations, 21, 69, *71*
wall of eyeball, 285
wall of ovarian follicle, 148
water economy, 105, *113*, 179, 184, 189, 192
watery-white eggs, 153
wattles, 27
whistle-kick, 57
white matter of spinal cord, 238
white yolk, 157
wind tunnel experiments, 5
wing, 18, *75*
  clips, 17
  loading, 4, *18*
  membrane, 31
  muscles, 60
  skeleton, 58
  slots, 18, *76*
  types, *18*, 75
Wülst, 253

yellow yolk, 157
yoke-toed foot, 19
yolk, 156, *157*
yolk membranes, 158
yolk spheres, 156

zona radiata of ovarian follicle, 148
zonular fibres, 287
zygodactyl foot, 19